T0146253

Kinematics

Kinematics
The Lost Origins of Einstein's Relativity

ALBERTO A. MARTÍNEZ

The Johns Hopkins University Press

BALTIMORE

The Johns Hopkins University Press
2715 North Charles Street
Baltimore, Maryland 21218-4363
www.press.jhu.edu

Library of Congress Cataloging-in-Publication Data

Martinez, Alberto A.
Kinematics : the lost origins of Einstein's relativity / Alberto A. Martínez.
p. cm.
Includes bibliographical references and index.
ISBN-13: 978-0-8018-9135-9 (hardcover : acid-free paper)
ISBN-10: 0-8018-9135-3 (hardcover : acid-free paper)
1. Kinematics—History. 2. Relativity (Physics)—History. I. Title.
QA841.M37 2009
531'.11209—dc22   2008028737

A catalog record for this book is available from the British Library.

*Special discounts are available for bulk purchases of this book. For more information,
please contact Special Sales at 410-516-6936 or specialsales@press.jhu.edu.*

The Johns Hopkins University Press uses environmentally friendly book
materials, including recycled text paper that is composed of at least 30 percent
post-consumer waste, whenever possible. All of our book papers are acid-free, and
our jackets and covers are printed on paper with recycled content.

Contents

Preface

Every student of science, even in high school, studies kinematics. It is the common property of all physicists. Engineers use it, and philosophers write about it ceaselessly. Yet, until now, there existed not a single book on the history of kinematics proper. Kinematics is said to be the science of motion, the geometry of motion. It caused much ado thanks to Albert Einstein's theory of relativity. In 1905 he complained that major problems that plagued physics stemmed from "insufficient consideration of kinematics." Physicists had failed to account neatly for experiments on moving magnets, wires, and light because they had neglected basic kinematics: concepts of length, speed, and time.

Einstein was not the first to be annoyed by ambiguities in kinematics while working on electromagnetic theory. Decades earlier, André-Marie Ampère, Heinrich Hertz, and others complained that kinematics was neglected. Both electrodynamics and kinematics were started, arguably, by Ampère, who coined the terms *electrodynamique* and *cinématique*. Radically new electrodynamics and kinematics, however, crystallized jointly in what became known as Einstein's special relativity. The history of electrodynamics has been investigated at length, especially in Olivier Darrigol's comprehensive work, *Electrodynamics from Ampère to Einstein* (2000). The present book unveils a convergent history: the rise of kinematics, mainly from Ampère to Einstein. It focuses on a time frame bounded by 1834 and 1905.

Unlike the science of the dynamics of electrical particles, the general science of motion has ancient roots, so the present work connects to various periods. Yet despite rich traditions in the study of motion, modern theoretical kinematics never became a professional discipline. Even when most physicists and most students of science had to study kinematics, few physicists specialized in it, and no journals, congresses, or societies developed around the subject. The innovative pursuit of the science of motion remained a sideline, a haphazard affliction of brooding nonconformists. Strangely, the same is somewhat true even today. After all, have you ever met a *kinematicist?*

There are two main traditions called kinematics, one in physics and the other in engineering, and this book concerns mainly the former. Because theoretical kinematics was scarcely advanced on the basis of experiments, this history does not dwell on particular technical developments, but neither does it celebrate pure theory; it does trace origins of the kinematics of machinery alongside its sibling, theoretical kinematics. A focus throughout is how perceptions and mechanical notions stimulated a field that too often was portrayed as purely abstract. The aspects of kinematics here discussed lay in a province of theoretical physics located between engineering and philosophy, and a considerable distance away from each.

By the way, this is not a work on the seemingly esoteric field known as the philosophy of space and time. Rather than seeking to ascertain logical subtleties in theoretical frameworks, the present work charts the growth of basic concepts and approaches employed in everyday physics. Still, students of various disciplines, including philosophy and engineering, often benefit from history of physics, just as they should enjoy this history of kinematics as well. It is hardly about the so-called foundations of the science of motion; rather, it is about the elements. It outlines the evolution of basic concepts that are common property of all physicists, introduced in elementary textbooks, concepts such as kinematics, reference frame, position, coordinate, speed, velocity, relativity, distance, displacement, duration, simultaneity, vector, and scalar.

This book traces how kinematics became the fundamental branch of physics, which arose in the 1830s from the now-disreputable practice of ranking the sciences. Overcoming an ancient prejudice against mechanical labor, physicists increasingly came to value engineers' analyses of motion as opposed to metaphysical abstractions. By the 1880s, though, some physicists misconstrued kinematics as the pure geometry of motion, independent of empirical knowledge. Engineering, mathematics, and even psychology pulled kinematics in various directions. From such tensions, physicists realized that the science of motion, presumed to be the most mature branch of physics, had been grossly neglected and barely developed. Scientists found vicious circularities in the usual definitions of motion, force, straightness, length, direction, rigidity, duration, and simultaneity. Some physicists even denounced the basic language of kinematics, coordinate algebra, as being intensely artificial, serving to destroy intuition, obstruct invention, and thus retard progress in physics.

Yet Einstein's theory of relativity emerged from algebraic kinematics. Using an unparalleled collection of sources, *Kinematics* traces its conceptual origins. Ironically, Einstein's theory led to the critical rejection of basic concepts of classical mechanics, which scientists had valued as logically perfect, while using algebra, which physicists for centuries had criticized and viewed with suspicion. Thus, modern kinematics inherited old and unresolved symbolical ambiguities. The neglected science of motion here reemerges as an intriguing and provocative field.

This book addresses general readers and historians who are already interested in learning more about the sporadic growth of the science of motion and the roots of Einstein's special relativity in particular. At the same time, it holds plenty of material that has scarcely been made available before; historians of science will find worthwhile "new" material here. A history of physics aimed at both interested laypersons and specialists is unusual, yet some books, such as Arthur I. Miller's *Albert Einstein's Special Theory of Relativity* (1981), have shown that relatively scholarly monographs can attract many readers. The present book is more accessible, but without committing mistakes that are unfortunately common in many popular and even scholarly histories of relativity. Specialists should please tolerate the pedagogical portions of this book, which include things that "everyone knows"—that is, only anyone who already knows them. And laypersons who approach this book with particular interests will find that it opens a variety of topics and insights that make the subject rewarding.

Not everything about the history of kinematics is in this book. Instead, it highlights topics with broad appeal. One chapter deals with the origins of Einstein's kinematics, a subject that draws the curiosity of many. Another chapter discusses the changing roles of the science of motion in the classifications of the disciplines throughout the ages, giving a large-scale picture of how the study of motion has been disjoined or interrelated to other branches of knowledge. Other chapters trace the emergence of essential concepts and principles. This book does not trace the formulation of advanced theorems in the kinematics of geometric figures or material systems. Nor does it discuss "kinematic" parts of either the theory of elasticity or moments and products of inertia. Instead, it traces the development and clarification of basic concepts, leading to the kind of thoughtful analysis commonly associated with the young Einstein.

This history of kinematics is also designed to illustrate several distinct historical perspectives. Many books on history of science use either a single historiographic approach or an admixture of approaches. In the best cases, they create comprehensive accounts. There remains, however, a need for accessible texts that illustrate, within a single book, a variety of distinct approaches. General readers interested in studying history of science can benefit from relatively clear-cut expositions of distinct approaches that serve to investigate and convey historical inquiries. Thus, readers can better sense what aspects of a narrative are artifacts of its particular approach and surmise what sides of a question the narrative does not address at all. The matter is analogous to a question dating back to the geometry of Greek antiquity: the distinction between analysis and synthesis. By most accounts, analysis was a method of investigation, whereas synthesis was a method of presenting results deductively. The great lack of the former in ancient texts led many students for centuries to wonder about what hidden methods of analysis the ancients used to discover propositions and proofs. Likewise, readers who are not yet aware of the diversity of historical approaches available can be misled by a narrative into thinking that a historical account is complete or definitive—that is, unless the account insinuates how the luster of its results depends on the angle taken, acknowledging alternative perspectives. For example, the focus on Einstein's life and work in the last three chapters achieves a degree of clarity at the expense of not analyzing deeply the important works of Henri Poincaré, especially his paper of 1906, which (as Scott Walter rightly points out) were arguably more influential in shaping the mathematical form of space-time physics later in the 1900s.

Each chapter takes a distinct approach. Kinematics involves some developments common to other sciences, but it can hardly be described as typical. So this work does not take a representative sample of major historical approaches but rather various approaches suitable for each particular topic. These approaches range from the large-scale perspective of conceptual schemes evolving over centuries to the narrow internal analysis of exact operations in a single scientific work. The narrative moves from the general to the particular; it could well have gone otherwise, but it begins with general matters so that readers with diverse backgrounds may readily place the subject matter in relation to broad frameworks and concerns.

To start, how did the study of motion, traditionally disdained and neglected, rise to become the purportedly fundamental branch of physics? Chapter 1 shows how kinematics rose to prominence owing partly to theorists' efforts to organize and interconnect disparate fields of knowledge. In 1834 Ampère defined kinematics as a key neglected science, but how was his ranking influenced by older and ancient hierarchies of the sciences? Philosophers such as Aristotle traditionally prized metaphysics over the study of mechanical motions, so how did the latter gain prominence? How did the science of motion subsume the various branches of physics? Among various findings, we will see that even mathematics became subsumed under mechanics, at least according to Isaac Newton, and we will see how even botany affected Ampère's views on physics. Following Francis Bacon and Jean d'Alembert, physicists overturned ancient conceptual hierarchies that mirrored social hierarchies that for centuries had vilified laborers and artisans. Hence this chapter takes a wide view of the history of ideas, progressing from the distant to the more recent; the narrative moves by intentional design, to reach its goal of tracing how Ampère's unlikely classification of the sciences ever became plausible. To use the proper word, we may call this a *teleological approach,* this mainly chronological narrative that gropes toward a preestablished destination.

Adding to the encyclopedic history of ideas, what more proximate factors influenced Ampère? Previously, leading engineers such as Agustín de Betancourt and Lazare Carnot had sought a "new" science of motion. Chapter 2 shows how engineers' study of machines, in Spain and in postrevolutionary France, suggested that a practical science of motion, independent of old-fashioned metaphysics, could ground physics in engineering. How did engineers justify the independence of mechanics from metaphysics? How did a new science of motion emerge despite the tension between scientists' drive to assert its independence and their drive to connect the disciplines? Contrary to chapter 1, the narrative moves from the proximate to the more distant, as if excavating layers of material cultural remnants and sedimentation. This *archaeological approach* goes mainly against the direction of chronology. The mention of a reference within a text leads the reader, just as it helped to lead the researcher. Historians often begin to approach the distant past by way of the more recent, and only afterward, when composing a narrative, do they order findings chronologically. But that process has unfortunate consequences; an account that begins from a subject's remote roots can lose the interest of some readers, and it obscures the direction of inquiry by conveying *first* what the historian found only *after* much detective work.

Here lies a paradox: Until the 1830s, most physicists believed that the science of motion was the one field of physics that had been most extensively developed, so why did many then come to think that, actually, the science of motion remained grossly neglected? Following Ampère's influence, how did the science of kinematics grow and propagate throughout Europe? Chapter 3 traces its sporadic growth from the 1830s until the early 1900s, as engineers, scientists, mathematicians, and philosophers pulled it in different directions. Kinematics originated as the study of motions as they appear in observations, yet various theorists soon misconstrued it as the pure geometry of motion, independent of experience. How did the kinematic outlook develop despite the diverse interests that splintered it into various disciplines? How did scientists come to redefine concepts that previously had seemed clear and unproblematic—concepts of motion, force, rest, space, time, reference frames, and relativity—some of which, they had said, should *not* be analyzed philosophically? Surprisingly, their analyses benefited from the novel fields of genetic psychology and critical history. For the most part, the narrative in chapter 3 is organized not by chronology, but by subject matter; it constitutes a sort of *genealogy,* tracing family relations among fields and ideas. Among the many individuals whose works are discussed are Isaac Newton, Immanuel Kant, Robert Willis, Henri Resal, Gustav Kirchhoff, Heinrich Hertz, Eugen Dühring, Carl Neumann, Johann Bernhard Stallo, Ernst Mach, Ludwig Lange, Walter Raleigh Browne, James Thomson, Auguste Calinon, James Clerk Maxwell, and Henri Poincaré.

Now, how did developments in mathematics affect kinematics? Just as physicists criticized old explanations of certain concepts, they also criticized traditional methods of coordinate algebra. They sought to improve algebra, as if it were not an ideally suited language for kinematics. Could physics change mathematics? According to the old hierarchies of knowledge, mathematics was independent of physics, yet the history of kinematics shows instances in which mathematics developed owing to physical notions. Thus, chapter 4 looks into the evolution of basic mathematical methods for representing motion during the 1800s. Why did physicists dislike coordinate algebra? How did vector theory arise as an alternative in the 1890s? Rather than analyze many subtleties of the formalisms themselves, the narrative consists of *a history of dissent.* It highlights objections by philosophers, mathematicians, and physicists against geometry, algebra, coordinates, quaternions, and vectors. It reveals controversies that transpired in their attempts to assert the "naturalness" of this or that way of expressing physical quantities or kinematic relations. Among the individuals whose arguments feature in this account are René Descartes, William Rowan Hamilton, Hermann Grassmann, Peter Guthrie Tait, James Clerk Maxwell, Josiah Willard Gibbs, and Oliver Heaviside.

And as physics triggered changes in mathematical concepts, how did such changes affect kinematics? Chapter 5 traces how physicists variedly used algebra, calculus, coordinates, and vectors to define and refine concepts of speed, velocity, distance, and displacement. Also, physicists struggled to clearly define the concepts of uniform motion, duration, and simultaneity. Hence, this chapter traces the conflicts between theorists who sought to base the science of motion on pure mathematics and abstract principles and those who sought to ground it on measurements and plausible experience. Here the narrative proceeds not chronologically but by topic; like many histories, it is *impressionistic*, it brings together various facets of related and converging developments. Among the many individuals whose works appear in this chapter are d'Alembert, Resal, Hamilton, Heaviside, Maxwell, Gibbs, Edwin Bidwell Wilson, Hertz, August Föppl, Mach, James Thomson, Calinon, Poincaré, and Ludwig Boltzmann.

Chapter 6 is a *biographical account* of how Albert Einstein came to formulate his relativistic kinematics. It consists of a comprehensive effort to bring together a wealth of extant bits of traces pertaining to how he struggled to solve problems that resulted in a new kinematics. Although we have no manuscripts of Einstein's researches before his work was published in 1905, this chapter draws from the previous years and the subsequent decades, systematically connecting documentary evidence with many retrospective accounts that were reported in papers, books, lectures, letters, recollections, biographies, interviews, and more. Among the individuals who shared in this part of Einstein's life were Michele Besso, Marcel Grossmann, Mileva Marić, Conrad Habicht, and Moritz Solovine, and hence they feature in the story. Among the individuals whose works most influenced Einstein were Kant, Mach, Maxwell, Armand H. Fizeau, Hendrik A. Lorentz, Poincaré, and David Hume. Among the many others who also appear significantly in this history are Newton, Arthur Schopenhauer, Hertz, Föppl, Albert A. Michelson, Ampère, and Emil Cohn. The narrative proceeds chronologically to a degree that has not been attempted previously in historical accounts of

Einstein's relativity. The result is both a partial biography and a case study of scientific creativity. To distinguish it from other accounts, it also notes errors that plague even recent scholarly histories of special relativity.

Despite Einstein's later fame, his original work generated immense interpretive difficulties and disagreements. Why? To understand how such questions emerged, chapter 7 consists of a detailed analysis of Einstein's expressions. There exist many accounts of special relativity, but very few try to explain how it appeared in its original form. Hence, this chapter consists of a line-by-line *textual analysis* of a work that many readers found difficult to understand for decades: the Kinematic Part of Einstein's "On the Electrodynamics of Moving Bodies" of 1905. It places the text in the context of other works, such as those of Kirchhoff, Lorentz, Poincaré, and Föppl; incorporates critical discussion of various earlier readings and interpretations that were made of this text; and thus includes references to the works of later commentators, such as Arthur I. Miller and Roberto Torretti. The mathematical parts of this analysis are carried out at a level of clarity that makes it accessible to anyone familiar with high school algebra and the rudiments of the calculus.

Finally, we must ask: How did the disjointed history of kinematics affect Einstein's special theory? Chapter 8 brings together several threads developed throughout the preceding, threads that did not converge. It elucidates issues that remained unresolved despite the emergence of relativistic kinematics. Einstein did not use the innovative kinematic concepts of vector algebra, so how did his preference for old-fashioned coordinates affect his kinematics? Also, like other physicists, Einstein hoped to ground physics on actual procedures of measurement, yet his definitions involved such highly idealized conditions that one should ask: To what extent did his work qualify as kinematics, as conceived by Ampère? Furthermore, Einstein's attention to physical symmetries was essential to his creativity, yet how was his theory affected by his use of old algebraic methods that did not capture all kinematic symmetries? This final chapter proceeds in the style of *critical history*, the kind of analysis pursued by writers such as Johann Bernhard Stallo and Ernst Mach. As history of science became a professional academic discipline, specialists have increasingly abstained from prominently intermingling personal judgments and concerns in the historical narratives. So too have I abstained from doing so, for the most part, in the preceding seven chapters. Fair though this is to the historical project, it unfortunately obscures motivations that lead historians to particular inquiries and, likewise, the relevance that the subject may have to readers today, even for nonspecialists. Accordingly, in this last chapter I discuss aspects of modern theoretical kinematics that remained unresolved, disjointed, and ambiguous. In particular, what were the effects of the fact that Einstein formulated modern kinematics in old algebraic methods that were not originally designed for the analysis of motion? The aim is to encourage continued creative analysis of the elements.

One aspect of old-fashioned history of science that has unfortunately been neglected over the past decades is the willingness to place particular scientific episodes in relation to distant periods and traditions. Often professional historians compose articles or books that focus on developments that transpired in a restricted geographical region, or even a single institutional setting, within a narrow window of time. By contrast, the present work freely involves references to individuals, works, and traditions in many countries and periods. The price for this luxury is that the degree of detail with which some of the subjects are treated is not as finely textured as the intricate accounts that characterize some historical works. But the present book is intended for many readers, especially those who are more immediately interested in the outstanding aspects of past developments rather than the minutiae that can be assembled about individuals, ideas, and institutions. Again, the guiding principle has been selectivity. Moreover, some sections do include more documentary material than has ever been assembled previously in historical accounts of the subjects in question.

Because historians deal mainly with texts, at least in fields where any archaeology of material remnants is minimal, the present book includes a wealth of quotations. Often I prefer to help the dead speak for themselves rather than paraphrase their expressions. Of course, such excerpts must be handled with care, as it is easy to distort an individual's ideas by looking only at a few words. Yet original passages convey at least definite glimpses of the past, rather than place all trust on the narrator's voice. And many words of long-dead writers are livelier and sharper than the stale monotonous droning of contemporary academic speech.

Given that there exists no other history of the kinematic roots of special relativity, I have seen no need to decorate this account with an overarching thesis or story. History need not be driven by the schemes of novels or the agendas of politicians. When a subject has been covered many times, new writers try to reinvent it, put a spin on it, or a new varnish, thus to claim that only they have rightly understood it. The special theory of relativity is that kind of subject. It used to be that many people found it extremely difficult to understand. Nowadays, very many believe themselves to understand it and its origins clearly. Accordingly, modern kinematics is relatively dead, inasmuch as many specialists behave as though it is a finished subject that holds no important ambiguities or defects. Yet there continue to exist important disagreements among specialists on fundamental questions of interpretation and matters of principle. What aspects of kinematics have been confirmed by experiment? What aspects are conventions? What aspects are hypotheses? What aspects are artifacts of algebraic representations? Specialists with patience and inclination to address such questions give different answers. Others hold divergent answers yet feel conviction that their own silent answer is evident and secure. The present work contributes to such discussions, especially on their historical side, not by providing fire, but by providing additional material fuel and light.

Finally, I abstained from framing general conclusions from some of the episodes to let readers pursue them further. I would much prefer that this book be read as a reliable sourcebook rather than as the apparently definitive last word on anything. Too often already kinematics has been treated as though there is nothing more to say or solve there. The readiness with which specialists often treat its elements as self-evident and unproblematic is grossly at odds with its history. As in the 1800s, many teachers and professors tend to rush impatiently over the elements of kinematics and view with disdain most attempts to analyze them in detail. And at advanced levels it seems pedantic to insist on distinctions between sibling concepts such as distance and displacement. Many physicists behave as if nothing of importance can come from analysis of the elements. That was the situation in the 1800s, and that is again the situation today.

In many ways this book began many years ago, so I have many people to thank. Perhaps I should do so chronologically. Maybe twenty years ago, my father Ronald Martínez Cuevas gifted me Bertrand Russell's *The ABC of Relativity*. I found it unintelligible. Like others who have walked similar paths, I proceeded to other books and writings. Among the professors who encouraged my early pursuit of the history of relativity were Mydiah Marianni at the University of Puerto Rico, and James Challey at Vassar College. Also, at New York University I owe thanks to Scott McPartland for valuable discussions and to Engelbert Schücking, for kindly teaching me much about Einstein's relativity theories as well as their history. There were also worthwhile conversations with Arnold Koslow at the City University of New York. Furthermore, I warmly thank Martin Tamny for kindly serving as my adviser for my master's thesis work at NYU on the concept of simultaneity in special relativity. Likewise, I thank Scott McPartland and Daniel Greenberger for their helpful and supportive evaluation of that work. Their kindness and assistance is not forgotten.

Then in 1995 Roger H. Stuewer welcomed me to the Program in History of Science and Technology at the University of Minnesota. As my adviser, Roger gave me wholehearted assistance and support. He was an exceptional teacher, while always granting me guidance, freedom, and encouragement to pursue my own research interests. In revising my dissertation on the history of kinematics, Roger was a diligent reader, an impeccable editor, and a straightforward, constructive critic. Thank you Roger, for your enthusiasm for history of science, which is both contagious and inspiring. I hope that my present work reflects some measure of your good labor as a teacher. I also thank Alan E. Shapiro, especially, for greatly helping me in my historical studies of mechanics, optics, and mathematics, in relation to the science of motion. He also taught me to better distinguish conjecture from evidence. I thank also the other members of my dissertation defense committee: Robert W. Seidel, Geoffrey Hellman, Ronald N. Giere, and Michel Janssen. The last, especially, was friendly and immensely helpful in reshaping my arguments into a more intelligible form, and kindly shared his expertise on the history of relativity. Along the way, my early research into the history of kinematics benefited also from conversations with Roberto Torretti and C. W. Francis Everitt. It was the latter who rightly turned my attention to Ampère. I also wish to thank Goran Prstic, Talia Fernós, Serge Rudaz, and John Norton for discussing with me several mathematical points. Likewise, I want to thank Justin Yonker, a fellow student and friend at the Program in History of Science and Technology, for the many times he listened to my ramblings and offered encouraging words and advice.

In January 2001 I moved to Washington, D.C., thanks to a fellowship to research the history of kinematics at the Dibner Library for the History of Science and Technology at the National Museum of American History, Smithsonian Institution. There, I thank the librarians and staff who kindly helped me locate and use various and rare materials, especially Ronald L. Brashear. I also appreciated helpful conversations with David Lindsay Roberts and Paul Forman. Subsequently, a postdoctoral fellowship enabled me to continue research at the Dibner Institute for the History of Science and Technology, at the Massachusetts Institute of Technology. There, George E. Smith, as acting director, helped me in many ways: providing perspectives on the science of mechanics since the time of Galileo and Newton, commenting on my interests, and, especially, welcoming me to his rich classes on the history of mechanics at Tufts University, as well as organizing a discussion group on the history of the laws of motion. Among the Dibner Fellows who were helpful in sharing their knowledge, insights, and opinions were Domenico Bertoloni Meli, Abigail Lustig, Elizabeth Cavicchi, Snait Gissis, Elaheh Keirandish, Gordon McOuat, Leonard Rosenband, Orna Harari Eshel, Andrew Janiak, Robert DiSalle, François Charette, Ronald Anderson, Sam Schweber, and Christopher Smeenk. Also, I warmly thank the several conversations and long emails with which Olivier Darrigol shared with me his expertise and insights on the history of special relativity. His exchanges were especially helpful in the preparation of chapters 6 and 7 of this book. Most recently, I have also received helpful comments on several chapters from Scott Walter.

I also acknowledge and give thanks for a grant-in-aid from the American Institute of Physics, which enabled me to repeatedly visit the Center for History of Physics at College Park, Maryland, in 2004, to carry out further research at the Niels Bohr Library.

Next, I continued research as a Fellow at the Center for Philosophy and History of Science of Boston University and, specifically, at the Center for Einstein Studies. I am thankful to the staff of the Howard Gottlieb Archive Center for granting me use and guidance with the duplicate copy of the Einstein Archive. Above all, I am thankful for the friendly and generous help that John Stachel gave me at the Center for Einstein Studies. I wrote most of chapter 6 during my tenure, and while there I also polished chapters 7 and 8, which benefited from John's helpful comments. It was a pleasure to use the wealth of materials collected by John over the years, and I enjoyed the time he shared with me, not least for his good humor and high spirits.

Then, during 2004 and 2005, I had the valuable opportunity to conclude my research while teaching at the California Institute of Technology. My project benefited there from resources in the Einstein Archive, which were kindly made available to me, at the Einstein Papers Project headed by Diana Kormos Buchwald. I also thank the Department of Humanities of Caltech for sponsoring research trips to Switzerland. Hence, I also thank the librarians of the Swiss Federal Polytechnic in Zurich, especially Christian John

Huber, Corina Tresch De Luca, and Astrid Forster, as well as the librarians of the Landesbibliothek in Bern, especially Andreas Berz, for their helpful and efficient assistance.

Permission to cite from selected documents from the Archives and Special Collections of the Bibliothek of the Eidgenössische Technische Hochschule, Zürich, has been granted and such documents remain property of the ETH, all rights reserved. Permission to cite from selected documents in The Albert Einstein Archives was granted by Princeton University Press and such documents remain property of the Hebrew University of Jerusalem, all rights reserved.

I am also glad to express my thanks to friends and colleagues who helped me in various generous ways: Eric Blair, Susan Boettcher, Erika Bsumek, Gail Davis, Carolyn Eastman, Randy Diehl, Roger Hart, Judy Hogan, Bruce Hunt, Neil Kamil, Brian Levack, Mary Long, Tracie Matysik, Jorge Pardo, James Sidbury, Alan Tully, Gregory Vincent, and James A. Wilson, Jr.

Finally—during the preparation for publication, I benefited from the libraries at the University of Texas at Austin. I also thank several anonymous readers for their thoughtful and useful comments. And, at Johns Hopkins University Press, I thank Robert J. Brugger, Josh L. Tong, Juliana McCarthy, Claire McCabe Tamberino, Robin Rennison, and Deborah Bors, as well as my careful copyeditor, Brian R. MacDonald. For helping me with the artwork for the cover I thank Corinne Roffler, Roberto Ortíz, Frank Benn, and Martha Sewall. I also thank Fumihide Kanaya for assistance with translations from Japanese, and I thank Hannah Siemens-Luthy for help in preparing the illustrations. And if my mother, Lillian Montalvo Conde, by this point, thinks that I won't include her, she is mistaken! But now she'd better read the book. I do thank her, my father, my family, and friends for their kindness.

Kinematics

CHAPTER ONE

Big Picture
Rise of a Rejected Science

By the early 1800s, physics was so complicated that to some people it looked like a tangled mess. To avoid confusion, a few theorists tried to reorganize physics. The science of motion, then, became portrayed as the one part underlying all others. That ranking was made by André-Marie Ampère in his *Essay on the Philosophy of the Sciences* of 1834. He was a self-educated eccentric, a physicist and mathematician who taught in Paris at the prestigious École Polytechnique and the Collège de France. He is best known for discovering magnetic effects between electrified wires, without magnets. But he also had far-reaching views: he sought to restructure all of physics.

In particular, Ampère felt that kinematics had been grossly neglected. He complained that writers of physics books generally "omitted discussion" of the fundamental concepts of the science of motion.[1] But how could he say that? For centuries motion had drawn the attention and analysis of philosophers. It had been a major subject of inquiry during the Scientific Revolution. The sprawling, flourishing science of mechanics itself was often defined as the science of motion. Isaac Newton even began his masterpiece of 1687 by noting that the ancients "esteemed the science of mechanics of greatest importance in the investigation of natural things."[2] Mechanics

---

[1] André-Marie Ampère, *Essai sur la Philosophie des Sciences, ou exposition analytique d'une classification naturelle de toutes les connaissances humaines* (Paris: Bachelier, Imprimeur-Libraire pour les Sciences, 1834; repr., Brussels: Culture et Civilisation, 1966), p. 50; translations by A. Martínez.

[2] Isaac Newton, *Philosophia Naturalis Principia Mathematica* (1687), *Mathematical Principles of Natural Philosophy*, trans. Andrew Motte in 1729 and rev. Florian Cajori (Berkeley: University of California Press, 1946), p. xvii.

became the one branch of physics that was *most* extensively developed. So why did Ampère complain that the science of motion was neglected?

By *kinematics* Ampère meant something different. "The science where movements are considered in themselves as we observe them in bodies in our environment, and especially in devices called machines," is his definition.[3] Although this sounds like mechanics, there were key differences: first, he defined kinematics to be the science of motion independent of the concept of force; second, he proposed that kinematics should study motion as it appears, as we observe it; and, third, he construed this science of motion as the fundamental branch of physical science.

Consider this ranking of kinematics as fundamental. One might ask, So what—isn't that what physicists always thought in the first place? No, for a long time writers regarded statics, the study of bodies at rest and forces in equilibrium, as fundamental and underlying the dependent science of motion, dynamics.

One might still ask, Why should we care whether Ampère thought that the study of motion is more fundamental than statics? After all, he wasn't even known as a writer on mechanics. What did *he* know about the proper way to subdivide mechanics or physics for that matter? Well, his opinions matter, because, although he is most remembered as a contributor to electrodynamics, he was one of the physicists who invested great efforts in all of history to devise a systematic ordering of all the branches of physics, and all sciences, and all arts, and actually of all fields of human knowledge.

The *Essay on the Philosophy of the Sciences* consisted precisely of an extensive classification of the sciences and other disciplines. Ampère described it as an "analytical exposition of a natural classification of all human knowledge." This scheme had grown from his attempts to categorize "general physics" and to distinguish it from other sciences.[4] He had pursued that project while preparing to teach a course on general and experimental physics at the Collège de France in 1829. There, student attendance was free, professors designed courses on anything, and there were no exams. At the time, Ampère was fifty-four years old, brooding, and sometimes melancholic, owing to the early tragic deaths of his sister, his father, and first wife, plus an awful second marriage. He had acquired a reputation as a stereotypically absent-minded professor: "He was reported to write equations on

---

[3] Ampère, *Essai*, p. 52.
[4] Ibid., p. v.

the back of carriages only to see them carried off into traffic. He was said to have thrown his watch in the Seine while he placed a stone in his pocket."[5] Yet he became increasingly focused on a wide view of things.

But "long before" Ampère busied himself trying to devise his classification scheme, he had become aware that a fundamental science of the study of motion was "generally omitted" at the beginning of "all the books" on the physical sciences.[6] To remedy that, he proposed to remodel the pure science of motion as the geometric study of all sorts of visible motions, independently of the study of the forces involved in their production. For this neglected science he proposed the name *cinématique* from the Greek κίνημα, for movement.[7]

So Ampère came to construe the science of motion as fundamental. But still, to our ears, talk of one science as being more fundamental than another may sound pointless, misguided, and uninteresting. Is physics more fundamental than biology? Is mathematics more fundamental than physics? Or is psychology more fundamental than physics? Or is it biopsychology, or psychobiology, or biosociology, or sociobiology? What difference does it all make? Really. What sort of problems, if any, can such questions possibly solve?

One historian warns, "Apart from librarians, few people today give much consideration to the classification of knowledge. . . . the topic is no longer academically fashionable."[8] Another historian comments, "Ampère's ostensive subject matter of classification is thus of little concern to many modern philosophers."[9] The oldfangled practice of constructing hierarchies of knowledge, ranking the sciences to show which claim primacy and autonomy over others and, in turn, how other sciences are derivative or dependent on such basic sciences, is steeped in disrepute. Nowadays such talk is mostly heard only in the rantings of zealous advocates of younger sciences seeking attention. In sciences more than a hundred years old, few scientists bother to indulge such pretentious proclamations openly. In the 1800s,

---

[5] James R. Hofmann, *André-Marie Ampère* (Oxford: Blackwell, 1995; repr., Cambridge: University Press, 1996), p. 142.

[6] Ampère, *Essai*, p. 50.

[7] Ibid., p. 52.

[8] Nicholas Fisher, "The Classification of the Sciences," in *Companion to the History of Modern Science*, ed. R. C. Olby, G. N. Cantor, J. R. Christie, and M. J. Hodge, (London: Routledge, 1989), p. 853.

[9] Hofmann, *Ampère*, p. 357.

however, the classification of the sciences was a subject pursued and debated by leading scientists and theorists in various fields. For example, Ampère, William Whewell, Auguste Comte, Herbert Spencer, Wilhelm Wundt, and Karl Pearson all proposed systematic classifications of the sciences. In the old days long since gone, classifying the various branches of knowledge was seen as a very important concern of science and philosophy.

In his *Essay*, Ampère repeatedly acknowledged influence from famous classifications advanced earlier by Jean d'Alembert and Denis Diderot in the mid-1700s and by Francis Bacon more than a century before them. To trace the origins of Ampère's big picture of the order of knowledge, then, we turn our attention to earlier "big pictures." Further still, to trace the rise of kinematics in an even wider perspective, in a long view of the history of ideas we may go back in time more than just two centuries before 1834. In fact, to properly convey just how big a transformation had finally taken place, once Ampère classified kinematics as fundamental, we must turn our attention all the way back to antiquity and specifically to Aristotle.

As Rest Precedes Motion

Now, of course, the prospect of considering Greek antiquity in general or Aristotle in particular might seem a hackneyed way to begin a discussion of more modern subjects. But the role of motion can best be introduced by paying attention to some old views. Aristotle lived some 340 years before the common era, yet many centuries later his works were still read with fascination by philosophically minded readers, including the young André-Marie Ampère. Moreover, in Aristotle's scheme we can readily see distinctions inherited by later classifications to various degrees. We focus here on the different degrees of truth ascribed to various branches of knowledge, their relations of independence or subordination, and the relation of these subjects to notions of motion.

Figure 1 outlines Aristotle's classification of knowledge. It incorporates both the steps in the origin and development of knowledge and Aristotle's ranking of the fields of knowledge. From the bottom, we have sense perception, as the initial step in the formation of knowledge. Then follow, successively, memory; experience, that is, the knowledge of individual things; practical actions, meaning willed behaviors responding to immediate necessity; and productive arts, those practices dealing with utility, profit, or enjoyment. Then we reach theoretical Knowledge, or *episteme*, free science,

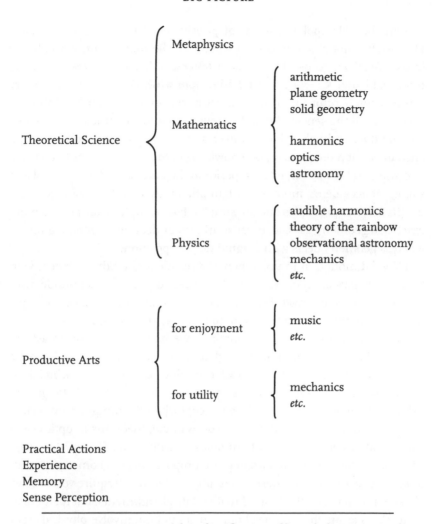

Figure 1. Aristotle's classification of knowledge

knowledge pursued for its own sake. Some branches deal with specific aspects of being, that which exists, and treat it as separable from the rest. For example, geometry abstracts the idea of figure, or magnitude, and deals with it as separate from matter. Meanwhile, highest of all, we have metaphysics, which treats being as being, without separating specific aspects of it to study.

In what sense does this system constitute a hierarchy? Well, Aristotle employed several criteria to express how the sciences stand in relation to one another. He spoke in terms of priority, exactitude, and universality. He used two notions of priority: some things, he said, are prior and better known relatively to us, things *nearer to sense;* other things are absolutely prior and better known and are *farther from sense.* For example, geometry was absolutely prior to mechanics, because geometry is farther from the experience of the senses. Aristotle strengthened this distinction by characterizing the absolutely prior as universal, whereas in sense perception we encounter but mere particulars. Knowledge of the universal was knowledge of causes, and it was valued as superior to the particular. Also, he spoke of some sciences being more exact than others. By this, he meant sciences that have fewer principles and are simpler. For example, arithmetic is more exact than geometry, he said, because arithmetic deals only with magnitude, whereas geometry deals with magnitude plus position.

How did motion fit in this scheme? Aristotle argued that sciences with fewer principles are more exact; and sciences concerning magnitude, with no motion, are *most* exact. He said that mathematical objects can be separated in thought from motion, and no falsity arises. Thus, he spoke of the "immovable principles of mathematics." We find, still, motion in astronomy, but that is uniform motion. And notice its ranking. Further down, he placed physics, concerned with substance, which contained the principle of motion and rest. It is not for one science to prove something belonging to a different science, Aristotle claimed, except when the things are so related that one is subordinate to the other. For example, theorems in optics were subordinate to geometry, and harmonics to arithmetic. Meanwhile, allegedly, you *cannot* use mechanics to prove propositions in geometry. This attitude was carried into modern times: physics teachers require you to solve physical problems on the basis of mathematical arguments, but mathematics teachers argue that you need not and should not involve physical arguments in the demonstration or proof of mathematical propositions.

Aristotle argued that "those sciences which are based on fewer principles arc more exact. . . . And if the things dealt with have magnitude, the science is most exact when they have no movement."[10] So why is greater priority, exactitude, or truth associated with the absence or lack of motion? Simply

[10] Aristotle, *Metaphysis,* book 1, 982a, lines 25–28; book 3, 1078a, lines 12–13; translation from *Mathematics in Aristotle,* ed. Thomas L. Heath (1949; repr., New York: Garland Publishing, 1980), p. 5. See also Joe Sachs, *Aristotle's Metaphysics, a New Translation* (Santa Fe: Green Lion Press, 1999), pp. 4 (982a), 258 (1078a).

put, some of the leading ancient philosophers, having once latched onto a notion of truth as that which is eternal, perfect, and unchanging, accordingly associated motion with change and thus ascribed greater truth to that which does not involve motion. Thus it was for Aristotle, and more so for Plato, his teacher, who seems to have denied the very possibility of a science of motion. For Plato, geometry was closer to metaphysics than to physics, as it concerned true knowledge, rather than that which comes to be or ceases to be in time. Similar arguments were voiced by others later.

A trend developed: there was no change, no motion in mathematics, or at least not in the most exact parts of mathematics, and motion in a science implied some degree of inexactitude. Mechanics involved the motion of material bodies, and therefore it was sometimes characterized as a sordid business. For example, the Greek philosopher and biographer Plutarch, writing in the first century AD, recounted that Plato denounced Eudoxus and Archytas for using mechanical arguments to prove some mathematical propositions:

> Eudoxus and Archytas had been the first originators of this far-fetched and highly prized art of mechanics, which they employed as an elegant illustration of geometrical truths, and as a means of sustaining experimentally, to the satisfaction of the senses, conclusions too intricate for proofs by words and diagrams. . . . But what of Plato's indignation against it as the mere corruption and annihilation of the one good geometry,—which was shamefully turning its back upon the unembodied objects of pure intelligence to recur to sensation, and to ask help (not to be obtained without base subservience and depravation) from matter; so it was that mechanics came to be separated from geometry, and repudiated and neglected by philosophers, took its place as a military art.[11]

Plutarch commented also on the attitude of Archimedes, one of the most famous geometers of antiquity, known also for his mechanical inventions and studies. He claimed that Archimedes "would not deign to leave behind any commentary or writing on such subjects [mechanics]; but, repudiat-

---

[11] Plutarch, "Life of Marcellus" (ca. AD 100), in *Plutarch's Lives of Illustrious Men*, translated by John Dryden and others, and corrected from the Greek and revised by Arthur H. Clough (1876; repr., Boston: Little, Brown, 1928), p. 221.

ing as sordid and ignoble the whole trade of engineering, and every sort of art that lends itself to mere use or profit, he placed his whole affection and ambition in those purer speculations where there can be no reference to the vulgar needs of life; studies, the superiority of which to all others is unquestioned . . . geometry."

Thus, many theorists deemed geometry to be prior and superior to astronomy and mechanics, as though its allegedly static character was more fundamental than the branches of knowledge that involved motion and change. There were a few exceptions, however. When Newton wrote his *Principia*, as mentioned previously, he appealed to an ancient view that allotted a more favorable position to mechanics: "The ancients (as we are told by Pappus) esteemed the science of mechanics of greatest importance in the investigation of natural things."[12] The greatest importance? What is there in the extant writings of Pappus of Alexandria, around AD 300, concerning the position of mechanics relative to the branches of knowledge? Not much. But he did claim that mechanical science "takes almost first place in dealing with the nature of material elements of the universe."[13]

Pappus followed the tradition of the school of mechanics established by Heron of Alexandria, whose works were not available in the Latin West and hence were unknown to Newton. Still, Heron's conception is worth mentioning because it illustrates a significant exception to the way in which the most influential theorists construed the relationships between mechanics and mathematics. In Heron's account of mechanics, he listed the mathematical disciplines as branches of mechanics (figure 2).[14] Here geometry appeared as but a kind of mechanics. But heedless of this exceptional conception, most theorists classified mechanics as inferior and subordinate to mathematics. There prevailed a bias against concepts of motion in the classifications of knowledge in general and against the discipline of mechanics in particular.

Proclus, in the fifth century AD, argued that geometry is superior to mechanics because the latter deals with objects in the sense world, whereas

---

[12] Newton, *Principia* (1687), ed. Cajori, p. xvii.
[13] Pappus of Alexandria, *Book 7 of the Collection*, edited with translation and commentary by Alexander Jones (New York: Springer-Verlag, 1986).
[14] Hero of Alexandria, *Les mécaniques, ou, L'élévateur des corps lourds*, Arabic text of Qusta Ibn Luqa, established and translated by B. Carra de Vaux, introduction by D. R. Hill and commentaries by A. G. Drachman (Paris: Belles Lettres, 1998).

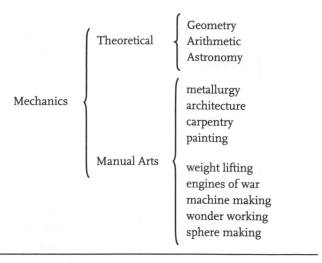

Figure 2. A classification scheme by Heron

mathematical being is "unchangeable, superior to things that move about in nature," and he claimed that "geometry comes into being before spherics [astronomy], as rest precedes motion."[15] Following a classification of the branches of mathematics formulated by Geminus in the first century BC, Proclus gave a higher standing to the branches of mathematics that do not involve motion, as depicted in figure 3. Notice the lower position of mechanics, by comparison to arithmetic and geometry. Proclus recounted also the scheme of some ancient Pythagoreans, mystic theorists and mathematicians whom he seemed to admire (figure 4). We thus hear that they too divided knowledge into that which is stationary and that which is in motion.

Like Aristotle, Themistus, another Greek writer, reiterated that solid geometry is more exact than astronomy because there is no motion in the former. Later, in the sixth century, the Christian philosopher John Philoponus characterized Aristotle with the view that "mathematical things are unmoved," that there is no generation, no change in them: "Mathematical

---

[15] Proclus, *Eis proton Eukleidou stoicheion biblon* (ca. 450); *A Commentary on the First Book of Euclid's Elements,* translated and with an introduction and notes by Glenn R. Morrow (Princeton: Princeton University Press, 1970), p. 30. See also Plato, *Republic* (ca. 380 BC), trans. G. M. A. Grube and rev. C. D. C. Reeve (Indianapolis: Hackett, 1992), pp. 199–201.

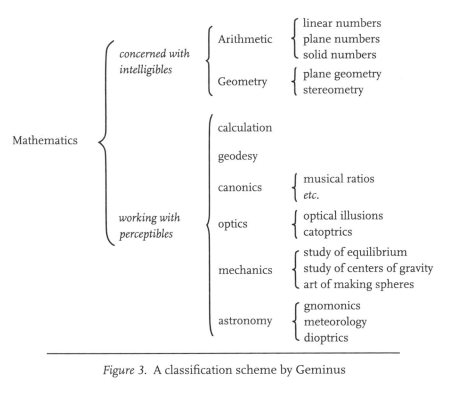

Figure 3. A classification scheme by Geminus

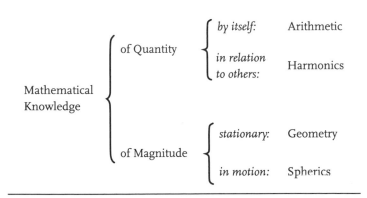

Figure 4. A classification scheme ascribed to the Pythagoreans

objects neither come to move nor are for the sake of something."[16] Philoponus too argued that intellectual and immaterial things are more exact than material knowledge.

Multiple variations of Aristotle's classification scheme were propagated during the Middle Ages by Christian writers in the Latin West and by Muslim writers in the Middle East. To illustrate the Islamic intellectual tradition, consider the classification scheme developed by al-Farabi, a famous commentator on Aristotle's works and founder and major exponent of the Peripatetic school of Islamic philosophy.[17]

Al-Farabi wrote and taught mainly in Baghdad and Damascus in the first half of the tenth century, helping to initiate a period of intense philosophical and mathematical studies among Muslims. Al-Farabi formulated a classification and enumeration of knowledge that emphasized the ordering of the disciplines as a hierarchy (figure 5). He proposed a threefold division of the philosophical sciences, as in the scheme of Aristotle. Al-Farabi claimed that among the "intermediate" or "mathematical" disciplines, the principles of arithmetic are the easiest to comprehend. Following it, in order of increasing difficulty, he placed geometry, optics, astronomy, music, and mechanics. Thus, in some regards, his classification concerned the sequence in which subjects ought to be studied in order to be properly understood.

More important, though, his classification also reflects the hierarchy he ascribed to all that exists. For example, he construed celestial bodies as having perfect and eternal forms and motions and described them as superior to terrestrial bodies. Thus, astronomy had a higher ranking over the studies of mechanical subjects and practical arts. Likewise, theoretical arithmetic and geometry were deemed superior to all the former because their subject matter, quantities and magnitudes, was supposed to be absolutely independent of material things. In the disciplines that one might designate as mechanics, the sciences of weights and ingenious devices were not each divided into theoretical and practical parts, as they were not considered to have a purely theoretical side. Following al-Farabi, several other Muslim

---

[16] Philoponus, *On Aristotle's Physics 2*, trans. A. R. Lacey (Ithaca: Cornell University Press, 1993), pp. 108, 140.

[17] The following account of al-Farabi's classification is based mainly on the work of Osman Bakar, *Classification of Knowledge in Islam* (1992; Cambridge: Islamic Texts Society, 1998).

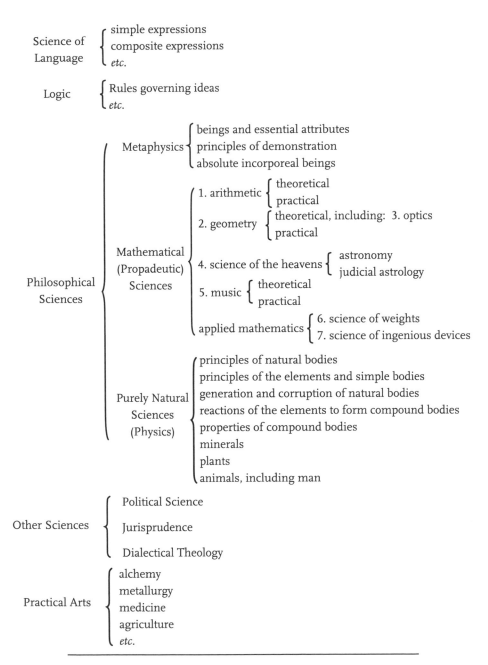

Figure 5. Hierarchy of knowledge according to al-Farabi

scholars formulated various classifications of knowledge, primarily on the basis of his scheme and the seminal works of Aristotle.

Meanwhile, in the Christian tradition, the general classification of Aristotle was disseminated by Boethius in his *De Trinitate*. It was later adopted and refined by Thomas Aquinas around the mid-thirteenth century. Not only did Thomas reiterate Aristotle's threefold division of the theoretical or "speculative" sciences, but he also reified the low standing of the mechanical or manual branches of knowledge (see figure 6). Moreover, even in his day, he still found it necessary to discuss the question of whether science can even treat motion, or whether, following Plato, true knowledge excludes that which is in motion: "Every science has to do with what is necessary. But whatever is moved, as such is contingent, as is proved in the *Metaphysics*. Therefore, no science can treat of what is subject to motion. . . . every science concerns that which is universal. Therefore, natural science does not treat of what is in motion."[18]

Following arguments advanced by Aristotle, however, Thomas rejected such views and concluded that science properly concerns itself not only with what is necessary and unchanging but also with what is contingent and subject to movement. Nonetheless, his division of the sciences was starkly hierarchical. He followed Aristotle's notion that the more complicated and concrete sciences are subordinated to the simpler and more abstract sciences. Thus, he argued, for example, that "the principles of mathematics are applicable to natural things, but not vice versa, because physics presupposes mathematics; but the converse is not true."[19] He claimed that mathematical principles can be used to study motion, but "the mathematician as such does not treat of anything subject to motion."

This bias against mechanics was carried, to varying degrees, throughout the centuries. Meanwhile, mathematics ranked well above all "gross and base" material things.[20] Eventually, some theorists came to reconsider the role and function of mechanics. Of those who did, we may highlight the

[18] Thomas Aquinas, *Super Boetium de Trinitate, Expositio* (ca. 1256); *The Division and Methods of the Sciences: Questions V and VI of His Commentary on the De Trinitate of Boethius*, trans. Armand Maurer, 3rd rev. ed. (Toronto: Pontifical Institute of Mediaeval Studies, 1963), article 2, p. 19.

[19] Ibid., article 3, p. 37.

[20] John Dee, "Mathematicall Præface specifying the chiefe Mathematicall Sciences," in *The Elements of Geometrie of the most Auncient Philosopher Euclide of Megara*, trans. H. Billingsley (London: John Daye, 1570), preface.

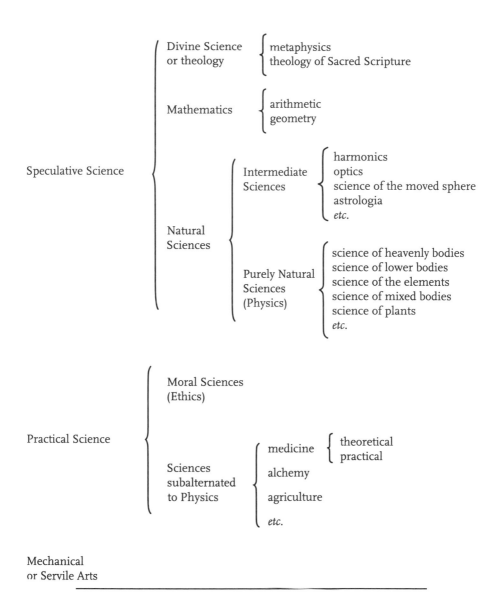

Figure 6. Division of the sciences by St. Thomas Aquinas

contribution of Francis Bacon, who formulated one of the main classifications that Ampère, centuries later, sought to supersede.

Francis Bacon first published his classification of knowledge in 1605, under the title *Of the Proficience and Advancement of Learning*. He expressed various dissatisfactions with traditional classification schemes. He did not discuss anything quite like what we would call "a science of motion," and even discussion of mechanics as a science is absent. Bacon did write, however, about what he called "Historia Mechanica":

> For History of Nature wrought, or MECHANICAL, I find some collections Made of Agriculture, and likewise of Manual Arts, but commonly with a rejection of experiments familiar and vulgar. For it is esteemed a kind of dishonour unto Learning, to descend to enquirie or Meditation upon Matters Mechanicall; except they bee such as may be thought secrets, rarities and special subtilties: which humour of vaine and supercilious arrogance is justly derided. . . . But if my judgement bee of any weight, the use of HISTORIEMECHANICAL, is of all others the most radicall, and fundamental towardes Natural Philosophie, such Natural Philosophie, as shall not vanish in the fume of subtile, sublime or delectable speculation, but such as shall bee operative to the endowment, and benefit of Man's life. . . . But furder, it will give a more true and reall illumination concerning Causes and Axioms, then is hitherto attained. . . . So the passages of Nature cannot appear so fully in the liberties of Nature, as in the trialls and vexations of Art.[21]

In 1623 Bacon's book was reissued in Latin, as a much expanded and revised text.[22] There Bacon included extensive discussion of the roles of motion and mechanics in the advancement and classification of knowledge. He complained that the philosophy of the Greeks, especially as formulated by Aristotle, had neglected the importance of useful knowledge and thus had been fruitful of controversies and barren of effects. By contrast, he

---

[21] Francis Bacon, *The tvvoo bookes of Francis Bacon: of the Proficience and Aduancement of Learning, Divine and Humane* (London: Henrie Tomes, 1605), Second Booke, pp. 9–10.

[22] Francisci Baronis de Verulamio, Vice-Comitis Sancti Albani, *De Dignitate et Augmentis Scientiarum Libri IX*, ed. William Rawley (London, 1623; Paris: P. Mettayer, 1624).

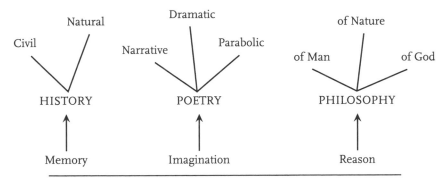

Figure 7. The branches of knowledge according to Bacon

argued, the mechanical arts had advanced constantly toward perfection. Accordingly, he formulated a plan for the advancement of all kinds of knowledge. To do this, he proposed a general division of learning according to what he deemed to be the three faculties of the mind: memory, imagination, and reason. From these, he claimed, stemmed the fundamental fields of knowledge: from memory came history, from imagination came poesy, and from reason came philosophy (figure 7).

Bacon construed certain dependence and interconnections among the branches of knowledge, just as he deemed the faculties of the mind to work in unison. Hence, figure 8 illustrates the relations that Bacon emphasized among the subfields of natural philosophy and natural history. However, Bacon's classification is difficult to convey in a single diagram. He used various metaphors to describe the organization of knowledge: he wrote about knowledge as a tree, a ladder, a pyramid, and more. Accordingly, the editions of his works that included a diagram did not convey all relations in question. Bacon's scheme is not a strict hierarchy, though it involves some indications of how one field of knowledge benefits from advances in another. For example, he explained that the knowledge obtained in the natural history of the heavens served to enrich the knowledge employed in the investigation of concrete physics.

In his book of 1623, we find both motion and mechanics in several forms. Bacon divided natural philosophy into physics and metaphysics. To each of these so-called speculative sciences, he assigned an operative part, mechanics and magics, respectively. Thus "mechanics," as part of physics, assists experimentally in the investigation of nature. The term appears, also, as a subdivision of natural history and of the arts, a mechanics that

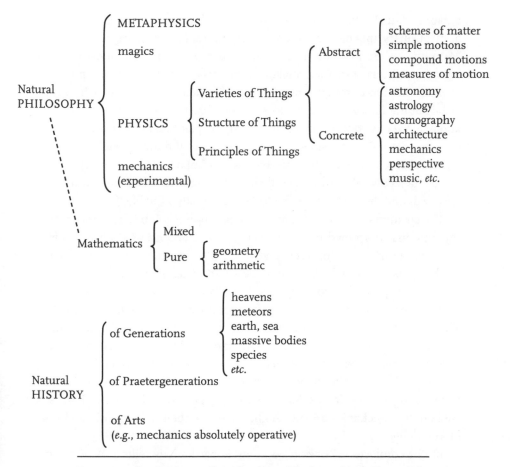

Figure 8. Part of Bacon's classification of the branches of knowledge

Bacon characterized as "absolutely operative." It was the mechanics of the artisan, craftsman, or the construction worker, a manual practice employed for utility rather than for physical inquiry, and itself independent and prior to the mechanics of philosophy.

In Bacon's scheme, motion is not associated with the inexactitude of knowledge. For one, it played prominent roles in abstract physics. The concepts of neither motion nor mechanics were used here in a pejorative sense. Here the high ranking of mathematics is gone. Mathematics appeared as a "great appendix" to natural philosophy, "its proper position, not among the substantial sciences." Bacon explained that because he valued not only truth and order but also the utility of knowledge, he had to construe mathematics as an auxiliary appendage to physics, mechanics, metaphysics, and magics: "And this we are in some measure obliged to do, from the fondness and towering notions of mathematicians, who would have their science preside over physics. It is a strange fatality, that mathematics and logic, which ought to be but the handmaids to physics, should boast their certainty before it, and even exercise dominion against it. But the place and dignity of this science is a secondary consideration with regard to the thing itself."[23] Other writers soon complained that mechanics was unfairly "most neglected" by mathematicians, objecting that "those common arts, which are so much despised" and disdained as "ignoble," were actually valuable.[24]

The position of astronomy shows Bacon's appreciation of it, rather than the rank that it seemed to already occupy: "Astronomy, as it now stands, loses its dignity by being reckoned among the mathematical arts, for it ought in justice to make the most noble part of physics." Bacon reported the physical part of astronomy, celestial physics, as "wanting," as "absolutely lame and incomplete, though its dignity claims the highest regard."[25] The desire for a more physical astronomy was soon satisfied thanks to the works and mechanical accounts of Johannes Kepler, René Descartes, and Newton. Philosophers became increasingly convinced that celestial bodies are made of matter, that they have weight, and that they behave as if subjected to mechanical forces. So, as astronomy became celestial mechanics, we find that mechanics ascends higher than ever before in the hierarchies of knowledge.

In this connection, Newton's account is particularly telling. We find in his view that mechanics played a role of priority relative to mathematics. In fact, for Newton mathematics appeared as a branch of mechanics. Newton

---

[23] Francis Bacon, *Physical and Metaphysical Works of Bacon including his Dignity and Advancement of Learning, in nine books, and his Novum Organum, or, Precepts for the interpretation of nature*, ed. Joseph Devey (London: Henry G. Bohn, 1853), book III, chap. VI, p. 149. An early English translation, of 1639, reads "for the wantonesse and arrogancy of *Mathematicians*, who could be content that *this science*, might even command and over rule *Physique*." In *Of the Advancement and Proficience of Learning, or the Partitions of Sciences, IX Bookes*, trans. Gilbert Wats (Oxford: Lichfield, for Rob. Young and Ed. Forrest, 1640), p. 173.

[24] John Wilkins, *Mathemagicall Magick or the Wonders that may be performed by Mechanicall Geometry* (London: M. F. for Sa:Gellibrand, 1648), p. A4.

[25] Bacon, *Advancement of Learning*, ed. Devey, book III, chap. IV, pp. 128, 126.

followed not Aristotle or Plato but the less influential tradition stemming from Alexandria, from the works of Heron and Pappus. Following them, he divided mechanics into a rational and a practical part and construed the branches of mathematics as branches of rational mechanics.[26] For example, in an unpublished treatise on geometry dating to 1693, Newton elucidates his position by claiming that geometry "has no power" over mechanical construction "but merely speculates upon it." He rejects the old claim that geometry is more exact than mechanics and even ascribes to mechanics the origins of mathematics: "I say that nearly all the mathematical sciences originated from mechanics. It is mechanical to count different things placed before one and to count continued parts of the same thing, and to carry out arithmetical operations on paper. It is mechanical to observe the motions of the stars. It is mechanical to treat of musical instruments, and so on. These are all practical mechanics, and speculations grounded in them are rational mechanics."[27] Thus, mechanics in particular and motion in general were gaining importance in the system of knowledge. Sometimes statics and motion became intermingled, as in the *Cyclopaedia* of 1728 and 1738, which briefly characterized statics as "the doctrine of Motion," as if the two were the same, and it noted that the usual partitioning of the sciences was "wholly arbitrary; and might be altered, perhaps not without advantage."[28]

In the Age of Enlightenment, its most famous product concerning the systematization of knowledge was the *Encyclopédie* of Denis Diderot and Jean d'Alembert, its first volume published in 1751. Again, this was one of the main classification schemes that Ampère sought to supersede. In designing this comprehensive and so-called *Rational Dictionary of Sciences and Arts*, its editors were most impressed by Bacon's work. Thus, Bacon's scheme was immensely influential and admired for more than a century by knowledgeable individuals of the period. The editors of the *Encyclopédie* spoke of Bacon as "an extraordinary genius."[29] They formulated a similar

---

[26] Newton, *Principia*, preface.

[27] Newton, *Geometry* (ca. 1693), published and translated in *The Mathematical Papers of Isaac Newton*, vol. 7, *1691–1695*, ed. Derek T. Whiteside (Cambridge: Cambridge University Press, 1976), p. 341.

[28] Ephraim Chambers, *Cyclopaedia, or, An Universal Dictionary of Arts and Sciences*, 2nd ed., vol. 1 (London: D. Midwinter, A Bettesworth and C. Hitch, et al., 1738), pp. iv, ix.

[29] Denis Diderot and Jean d'Alembert, eds., *Encyclopédie, ou Dictionnaire raisonné des sciences, des arts et des métiers, par une société de gens de lettres* (Paris, 1751), vol. 1, p. 2.

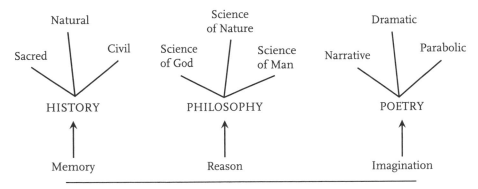

Figure 9. The division of knowledge according to Diderot and d'Alembert

account of the branches of human knowledge, dividing all knowledge into three great partitions of sciences and arts corresponding to three human faculties (figure 9). This scheme is essentially Bacon's with just a few modifications. Searching for the science of motion, we turn again to natural philosophy, now called synonymously the science of nature, as depicted in figure 10.

Here the science of nature was divided into metaphysics, mathematics, physics, and chemistry. Botany, anatomy, medicine, and more appear here as branches of physics. Judicial astrology and natural magic are still present. In mathematics we again have some of the old favorites: harmonics, optics, and astronomy appear here, but the order seems inverted from that of antiquity: acoustics, optics, and astronomy. And we now find mechanics as occupying a position above the others; it now sits just beneath the pure studies of arithmetic and geometry.

D'Alembert's diagram by itself does not suffice to express the exalted position that mechanics had gained in relation to other branches of physical inquiry. To illustrate this point, we may remark on the extent to which many philosophers of nature regarded mechanics as a discipline that included many others. For example, William Emerson, in his *The Principles of Mechanics* of 1754, sketched the history of mechanics by stating that ever since antiquity mechanics had been neglected, that is, until mechanical clocks were devised in about 1650: "At this time several of the most eminent mathematicians began to consider Mechanics. And by their study and industry, have prodigiously enlarged its bounds; and made it a most comprehensive science. It extends through heaven and earth; the whole universe and every

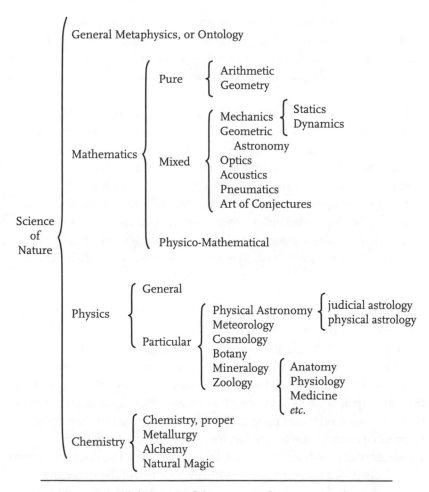

Figure 10. Subdivisions of the science of nature according to
d'Alembert and Diderot

part of it is its subject. Not one particle of matter but what comes under its
laws. For what else is there in the visible world but matter and motion; and
the properties and affections of both these are the subject of Mechanics."[30]

---

[30] William Emerson, *The Principles of Mechanics; Explaining and Demonstrating The General Laws of Motion, The Laws of Gravity, Motion of Descending Bodies, Projectiles, Mechanic Powers, Pendulums, Centers of Gravity, &c. Strength and Stress of Timber, Hydrostatics, and, Construction of Machines* (London: W. Innys and J. Richardson, 1754), pp. v–vi.

Emerson subsumed under mechanics the works of architecture, clocks, vehicles, navigation, husbandry, military arts, wind works, waterworks, musical instruments, and more. "This is a science of such importance, that without it we could hardly eat bread, or lye dry in our beds":

> By Mechanics we come to understand the motions of the parts of an animal body; the use of the nerves, muscles, bones, joints, and vessels. All which have been made so plain, as proves an animal body to be nothing but a mechanical engine. But this part of mechanics, called anatomy, is a subject of itself. Upon mechanics is also founded the motions of all the celestial bodies, their periods, times, and revolutions. Without mechanics a general cannot go to war, nor besiege a town, or fortify a place. And the meanest artificer must work mechanically or not work at all. So that all persons whatever are indebted to this art; from the king down to the cobbler.

Thus we may surmise the extent to which some philosophers construed the wide-reaching domain of mechanics.

By this point it should also be clear that many hierarchies of knowledge, to various extents, reflected traditional social hierarchies or biases. Politicians and aristocrats worked more with words and less with manual labor than did servants and slaves. Religions often encouraged people to value more the immaterial over material constructions and innovations. Philosophers often celebrated works of metaphysics and abstract mathematics over the works of engineering, manufacture, and industry. The disciplines were often divided into those that are essentially activities of the mind and those that involve physical operations and labor, and often the latter were viewed with disdain. Thus, the rise of mechanics to prominent positions in the purported order of nature and knowledge reflected a positive reappraisal of the role of laborers, artisans, and craftspersons in society and in the advancement of knowledge. This is but a brief simplification, of course, yet it nonetheless points to the realm of social history that was undeniably entwined with the evolution of the classifications of knowledge. At various points, some philosophers attributed the lower status widely ascribed to some fields of knowledge to conventional social divisions. Thus, for example, d'Alembert commented on the traditional "division of Arts into liberal and mechanical, & the superiority that one accords to the first over the second. This superiority is without a doubt unjust in many respects," as he claimed that it was based on "prejudices, all ridiculous." D'Alembert even attributed such prejudices to an ancient struggle on the part of the organized but weaker members of society to undermine the primitive rule of the physically strong:

> Bodily force, having been the first principle that rendered useless the right that all men have of being equal, the weakest, having always been greater in number, joined together to repress it. They hence established, by recourse to laws & various sorts of governments, an inequality of convention in which force has ceased to be the principle. This latter inequality being well affirmed, the men, in coming together with reason to conserve it, have not ceased to reclaim secretly against it by that desire of superiority that nothing has been able to destroy in them. Therefore they have sought a kind of compensation in an inequality less arbitrary; and bodily force, chained by the laws, can no longer offer any means of superiority, they have been reduced to seek in the difference of spirits a principle of inequality also natural, more serene, & more useful to society. Thus the most noble part of our being has in some manner avenged the first advantages that the most vile part had usurped; & the talents of the spirit have generally been recognized as superior to those of the body. The mechanical Arts depend on manual operation . . . a species of routine, having been left to those among men who prejudices have placed in the most inferior class. The indigence that has forced such men to apply themselves to labor, more often than inclination or genius have dragged them there, has become subsequently a reason to despise them, as much as it darkens all that accompanies it. . . . But society, in respecting with justice the great geniuses who enlighten it, should not at all vilify the hands of those who serve it.[31]

Thus, d'Alembert rejected the traditional and ancient prejudice against mechanical labor.

In his classification, the division of mechanics into statics and dynamics explicitly shows what by then and generally thereafter was the way of dividing mechanics. Statics is the study of physical magnitudes at rest, weights or forces in equilibrium; there is no motion there. Dynamics is the science

---

[31] *Encyclopédie*, vol. 1, p. xiii; translation by A. Martínez.

of motion caused by forces. But statics comes first. This order recapitulates the genealogy of the sciences, because statics had been developed since antiquity by Archimedes and others, whereas dynamics arose much later. Hence, this ordering still echoed the ancient classifications that viewed motion as playing a secondary role.

D'Alembert and Diderot acknowledged that one could find other ways, involving various advantages, of classifying sciences and arts. On the one hand, they complained that, unfortunately, there were many theorists "who willfully place in the center of all sciences the one with which they themselves are occupied, just as the first men placed themselves in the center of the world, persuaded that the universe had been made for them."[32] On the other hand, d'Alembert and Diderot admitted that their own arrangement involved a significant degree of arbitrariness, and it was this arbitrariness that bothered Ampère.

Ampère's Chimera

Before Ampère's work, the science of statics was often regarded as the most fundamental physical science. This traditional ranking appeared again, for example, in Siméon Denis Poisson's *Treatise on Mechanics*, published in a second edition in 1833, just a year before the publication of Ampère's *Essay*. Poisson ordered the subjects of mechanics in accordance with the sequence followed in the courses of mechanics as then taught at the Polytechnique in Paris. The subject of statics appeared as more fundamental than that of dynamics because statics studied only the conditions of equilibrium of material bodies, whereas dynamics studied the motions taken by any body "when the forces applied to it are not in equilibrium."[33] According to Poisson, geometers were accustomed to "reducing all questions of movement to simple problems of equilibrium." Again, this approach echoed the history of mechanics. Nonetheless, Poisson realized that at least for pedagogical purposes it was preferable to introduce basic concepts of dynamics before the more advanced aspects of statics. Thus, Poisson introduced dynamics between the first and second parts of statics. In his introduction, he skipped over the question of motions that do not involve forces by stating

[32] Ibid., p. xv.
[33] Siméon Denis Poisson, *Traité de Mécanique;* 2nd ed., vol. 1, Imprimeur-Libraire pour les mathématiques, no. 55 (Paris: Bachelier, 1833), p. 3.

in passing that "matter never moves spontaneously."[34] Of course, Poisson accepted that matter in motion continues to move uniformly so long as nothing interferes. But given this idea of natural uniform motion, it might seem reasonable to define a study of motion independently of the notions of force or equilibrium.

Ampère complained that most classifications, especially Bacon's and that of the *Encyclopédie,* organized knowledge by arbitrary principles. The issue concerned not only the true relations of the sciences, as displayed in any diagram, but also their arrangement in text form. The alphabetical arrangement of subjects in the *Encyclopédie* is a case in point. Even today, if you open the first volume, say, of a children's encyclopedia, expecting an entrée to the universe of human knowledge, and turn to the first page and read "aardvark," you know immediately that you require some sort of guide to provide an orientation in the corridors of knowledge. Otherwise, you are lost as if in a labyrinth. D'Alembert and Diderot were aware of this danger, they discussed it. But they trusted the encyclopedic tree to guide you along, like a map. Ampère wanted more; he wanted specific knowledge to be organized according to its interrelations with other branches. It bothered him that in the *Encyclopédie* sciences that are interrelated are treated separately. For example, in figure 10, geometric astronomy and physical astronomy appear separately.

When young, Ampère intensively studied the *Encyclopédie* and memorized many of its articles. By 1815, when he was forty, he had devised a diagram reordering the disciplines.[35] In 1819 and 1820 he gave a course at the Faculté de Lettres of Paris to publicly present his ideas on classification. And by 1829 he was thoroughly engrossed with classification schemes, while preparing a course on general and experimental physics at the Collège de France. He then wanted to solve two main questions: How is general physics distinguished from other sciences, and what are the different branches of general physics? More generally, he wondered, what do we understand precisely by a *science?*

He sought to elucidate the relations of physics to neighboring sciences—in a broader sense, the relation of physical truths to the ensemble of human knowledge. To warrant this general classification he gave several reasons.

---

[34] Ibid., p. 2.
[35] Ampère to J. Roux, 28 March 1817, A. Ampère and Jean-Jacques Ampère, *Correspondance et Souvenirs de 1805 à 1864,* ed. H. Cheuvreux (Paris: J. Hetzel, 1875), vol. 1, p. 116.

Above all, he argued, "*Classification should serve the progress of the sciences.*"
He also claimed that classification would serve:

> *To define the sciences; to know their object and relative importance.*
> *To better communicate notions and truths.*
> *To interconnect the sciences, to understand how they can help one another.*
> *To identify gaps in our scientific understanding.*
> *To establish divisions in classes and sections of a scientific society.*
> *To form the plan of a good library, or a good bibliography.*
> *To aid instruction in schools; for a coherent sequence of subjects.*
> *To construct a methodical encyclopedia.*[36]

These goals, he argued, would best be served by a natural ordering than by
arbitrary partitions. Already, Ampère had experience in trying to ascertain
the natural order of things. For one, in 1816 he had published a classification of the known chemical elements, in a cyclical arrangement, which, to
his disappointment, was promptly rejected by chemists.[37] More importantly,
seeking a model of natural classification for the fields of knowledge, he
found it in natural history.

Since his youth, Ampère had been interested in botany. He particularly
enjoyed Jean-Jacques Rousseau's *Letters on Botany*, which gave detailed instructions on how to examine and classify plant specimens, while criticizing
the artificial classifications promulgated by Carl Linnaeus. Ampère also appreciated the approach of Georges Louis Leclerc comte de Buffon, who also
aspired to a "natural" classification of plants. Another prominent classification scheme was formulated by Bernard de Jussieu, botanist of the King's
Garden in Paris. Jussieu organized species of plants into natural families
according to the embryonic traits of their seeds. Jussieu's classification was
published in a refined and elaborate form by his nephew Antoine-Laurent de
Jussieu in his *Genera Plantarum* of 1789, the year of the French Revolution.[38]

---

[36] Ampère, *Essai*, pp. 18–24.

[37] Ampère, "Suite d'une Classification naturelle pour les Corps simples," *Annales de Chimie et de Physique* 2 (1816), 5–32, 105–125. For an overview, see Hofmann, *Ampère*, pp. 206–212.

[38] Antonii Laurentii de Jussieu, *Genera Plantarum secundum ordines naturales disposita juxta methodum in Horto regio Parisiensi exaratam anno MDCCLXXIV* (Paris: Herissant, 1789).

After the Jacobins came to power in 1792, they suppressed the opposition government in Lyon, jailing its members, including Ampère's father, who earlier had certified arrest warrants against Jacobins. In the Reign of Terror, Jean-Jacques Ampère was taken to the guillotine. When the seventeen-year-old André-Marie learned that his beloved father had been decapitated, he was devastated. He was overwhelmed by mental and physical disability that lasted eighteen months and from which his interest in botany helped him escape.[39] He was drawn by the pastime of seeking orderly patterns in nature. He was nearsighted but unaware of it for years, so walks in the countryside gave him relief to focus on the complexion of nature close at hand.

Later in life, when composing his classification of the branches of knowledge, he drew inspiration mainly from the classification of plants formulated by Bernard de Jussieu and his nephew.[40] One of the remarkable things about their scheme was the degree of continuity into which they managed to arrange the many diverse varieties of known plants.[41] Ampère was impressed by this continuity, and he saw that it was lacking in the classifications of knowledge.

Ampère was not the first to seek continuity among the sciences. Bacon, for example, stressed the importance of interconnections. In his fourth book of *The Advancement of Learning*, concerning man and civil philosophy, he commented: "And here we must admonish mankind, that all divisions of the sciences are to be understood and employed, so as only to mark out and distinguish, not tear, separate, or make any solution of continuity in their body; the contrary practice having rendered particular sciences barren, empty, and erroneous, whilst they are not fed, supported, and kept right by their common parent. Thus we find Cicero [in *De Oratore*] complaining of Socrates, that he first disjoined Philosophy from rhetoric, which is thence become a frothy, talkative art."[42] Bacon was concerned that philosophy itself was becoming increasingly disjoined from experience, so that it too would become a "frothy" and increasingly irrelevant discipline. (Incidentally, his fear was borne out as philosophers became ever concerned with

---

[39] Hofmann, *Ampère*, pp. 21–24.
[40] Ampère, *Essai*, pp. xij–xv, xxxv.
[41] Ibid., p. 14; Peter F. Stevens, *The Development of Biological Systematics: Antoine-Laurent de Jussieu, Nature, and the Natural System* (New York: Columbia University Press, 1994), pp. 25–26, 43, 59.
[42] Bacon, *Advancement of Learning*, ed. Devey (1853), book IV, p. 151.

problems of their own interest rather than assisting the development of other branches of knowledge.)

Now, continuity of the sciences is not the same as unity. The idea that the fields of knowledge should be united was very popular in Europe, in the Islamic tradition, and elsewhere. Uniting the sciences was a widespread desire; another was to show that there should be no sharp distinctions between neighboring sciences. Bacon's construal of the *divisions* of knowledge as connecting rather than separating the sciences is particularly appropriate when we pause to consider the word *knowledge*. In English there exists no plural for this word, which may foster the impression that those who classify the sciences seek to carve divisions in knowledge. Yet in French, for example, the plural term *connaissances* carries the realization that the many sciences are not necessarily connected and, hence, that one must labor to link them or to find interconnections. That labor, carried out to eliminate discontinuities, was pursued by Ampère. The continuity of knowledge was desired too by the editors of the *Encyclopédie*. They preferred a tree of knowledge with the greatest number of connections. In the most natural arrangement, they said, subjects should follow one another by *insensible nuances*.[43] But they admitted that their own ordering had still very many gaps, because so much remained unknown.

Ampère was disturbed by such gaps, so he searched for an organizing principle that would let him discern the fields of knowledge that remained unknown. He fancied himself a naturalist seeking some organizing principle that might allow him to anticipate the form of species that would fill gaps in the continuous order of living things. Jussieu's classification had offered the possibility of predicting species of plants having intermediate forms between species separated by certain gaps.[44] To group the sciences, Ampère claimed to begin not from preconceived general groups but by tracing similarities among specific sciences. Like a classifier of plant species, he sought interrelations, trying to find a natural order.

Finally, however, his classification was structured by symmetric divisions that give it a rather artificial look. Figure 11 translates diagrams from his book in the exact format. He divided the totality of human knowledge into two kingdoms, one pertaining to the cosmos, the other to the human mind. He divided each kingdom into two sub-kingdoms; each sub-kingdom into two branches; those into two sub-branches; and each of those divided suc-

[43] "Discours Préliminaire des Editeurs," pp. xv, xxxv.
[44] Stevens, *The Development of Biological Systematics,* p. 27.

DIVISION of ALL KNOWLEDGE into TWO KINGDOMS

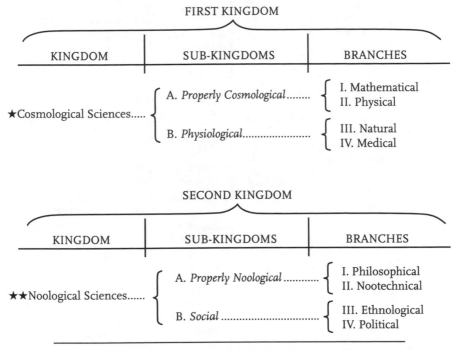

Figure 11. Ampère's classification of human knowledge

cessively into pairs of sciences of first, second, and third order. Thus, he listed 128 disciplines while admitting even further divisions: sciences of fourth order, and so on.

Now, because we are looking for the place of kinematics in this grand scheme, consider the "properly cosmological sciences" in their two major branches depicted in figure 12, particularly the upper part of his subdivision. The diagram illustrates the continuity he likened to the natural scheme of Jussieu. Ampère had a penchant for coining new words from Greek roots. Most of his neologisms did not enjoy the good luck of being adopted by other scientists, and even his biographers criticized them as "bizarre."[45] His term kinematics was an exception.

[45] C.-A. Valson, *La Vie et Travaux d'André-Marie Ampère,* 2nd ed. (Lyon: Imprimeur-Librarie de l'Archevêté et des Facultés Catholiques de Lyon, 1897), p. 301.

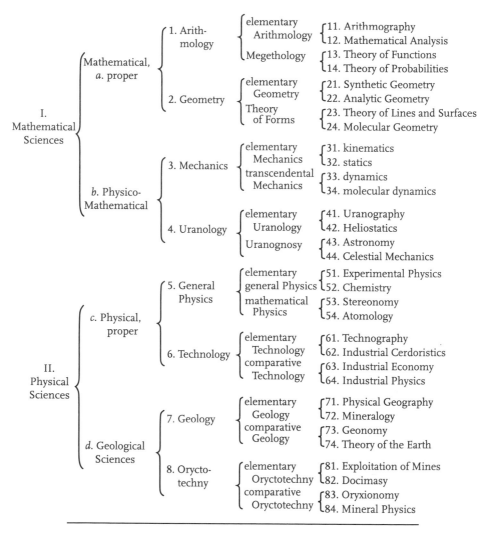

Figure 12. Ampère's classification of the mathematical and physical sciences

In the sciences of third order that compose mechanics, we find kinematics, statics, dynamics, molecular mechanics. Each studies parts of mechanics. About kinematics, Ampère claimed: "Such a science should contain all there is to say about different sorts of motions, independently of forces that might produce them." This brief excerpt suffices for now; Ampère's main concern and emphasis were on the description of motion as it is observed and on the description of motions irrespective of forces. Because this conception is described further in the next chapters, for now we need only contrast it to the other three branches of mechanics as he conceived them:

> *Statics.* It should follow kinematics, where we treat, on the contrary, forces independently of movements. . . . Statics should not come but after kinematics, because the idea of motion is that which is given to immediate observation, while we do not see the forces that produce the movements that we witness, and that we cannot equally conclude their existence but from that of the observed movements.

> *Dynamics.* After kinematics has studied movements independently of forces, and statics has treated the latter independently of the former, it remains to consider both simultaneously, to compare forces with the movements that they produce, and to deduce from that comparison the laws . . . , according to which, the motions being given, one calculates the forces capable of producing them, or on the contrary, one determines the movements when the forces are known.

> *Molecular Mechanics.* Finally, another science . . . of which there does not exist a general treatise, but of which different parts are found dispersed in various memoirs or some specialized works, due to the most distinguished mathematicians, who transport the same laws obtained in dynamics of isolated points or of a finite body, to the molecules that compose bodies, so that one finds in the equilibrium and movements of those molecules the causes of phenomena that bodies present.[46]

These four branches of mechanics conveyed the sequence in which Ampère said they ought to be studied and investigated, with kinematics first.

Although Ampère's classification of the order and interdependence of the subfields of physics offered novel perspectives, aspects of his approach were quite consonant with ancient classification schemes. Aristotle had organized the fields of knowledge by construing the more complex disciplines as dependent upon those based on fewer principles. Thus, for example, both

---

[46] Ampère, *Essai,* pp. 50–55; translation by A. Martínez.

Aristotle and Ampère considered arithmetic as more fundamental than geometry because the former involved only considerations of magnitude, whereas the latter dealt also with position in space.[47] And, like Aristotle, Ampère construed the study of motion as less fundamental than geometry, because geometry did not involve motion. For Aristotle, the mathematical science of astronomy, *spherics,* was more fundamental than mechanics because motions in the heavens were natural and uniform, whereas earthly mechanical motions were "enforced." Accordingly, for Ampère kinematics was more fundamental than dynamics. Still, by Ampère's day, progress in scientific knowledge ruled out many features of the Aristotelian scheme. For instance, the study of celestial motions could no longer be considered as intrinsically distinct from terrestrial mechanics. More importantly, Ampère diverged from ancient tradition by granting superior standing to kinematics over statics, for it had been common to regard rest as prior to motion.

The four divisions of mechanics are a fraction of the system Ampère proposed. So how did that relate to the whole? Because he was seeking some organizing principle, like Bacon he turned his attention to the "investigation of aspects founded on the very nature of our intelligence."[48] After some time, he succeeded in finding a key principle, and he was so impressed by it that he even recorded the date of his "discovery" in his *Essay:* "Only after having finished, at least for all the ensemble of cosmological sciences, my classification just as it is presented in this work; it's only after having made it known, in my classes at the Collège de France, and in the *Revue Encyclopédique,* that I discovered this principle on the month of August 1832."[49]

What Ampère found were four fundamental points of view that he argued should systematically be applied to every subject. He called these views:

1. *autoptic:* concerns that which is given to immediate observation
2. *cryptoristic:* seeks what is hidden from observation, by way of interpretation
3. *troponomic:* relates the object to others, by comparisons, to deduce laws

---

[47] Aristotle, *Analytica Posteriora.* A. 27, 87a, lines 31–37; see *Mathematics in Aristotle,* ed. Heath, p. 11. See also Plato, *Republic* (ca. 380 BC), trans. Grube and rev. Reeve, pp. 193–200.
[48] Ampère, *Essai,* p. 72.
[49] Ibid.

4. *cryptologic:* seeks what is hidden by using notions from the prior viewpoints[50]

These four viewpoints can be seen to underlie the four branches of mechanics described previously. Likewise, Ampère appealed to these four viewpoints to subdivide the other fields of knowledge. For example, zoology included (1) zoography, the study of mammals and insects, from the external point of view of their forms; (2) animal anatomy, the study that looks into the internal structure of such animals, by cutting them open; (3) comparative anatomy ("zoonomy"), the study of the general laws of the organization of the bodies and organs of animals; and (4) animal physiology, studying the causes of life, the formation and functions of organs.

We began by noting that Ampère was searching for a natural classification of the sciences, and to his satisfaction he found it. That is, he found the key that would allow him to order all knowledge systematically: the four points of view. He identified this realization as one of the greatest achievements of his life. So is that it? Was that the reason why kinematics earned first place among the physical sciences?

Because Ampère formulated his four points after he had developed his scheme, we would have to believe, with him, that these points had been operating subconsciously all along, guiding his formulations. But we need not accept the alleged naturalness of Ampère's four perspectives, just as scarcely any physicists seem to have accepted his theory of knowledge. As one biographer commented, "From one perspective, the entire enterprise was a misguided and quixotic pursuit of a chimera. Ampère's 'natural' classification may well strike the modern eye as a thoroughly artificial and contrived sequence of disciplines reflective only of the academic categories operative in Ampère's own milieu."[51]

Much more popular than Ampère's was an ordering proposed by Auguste Comte in 1830 that generated extensive attention and discussion. In his *Course on Positive Philosophy,* Comte outlined a simple scheme.[52] He designated six sciences as fundamental and organized them in a strictly hierarchical order: mathematics, astronomy, physics, chemistry, physiology, and social physics. Comte argued that each science necessarily depends on

---

[50] Ibid., pp. 41–45.
[51] Hofmann, *Ampère,* p. 366.
[52] Auguste Comte, *Cours de Philosophie Positive,* 6 vols. (Paris: Bachelier, 1830–1842).

those that precede it, in that order. Hence, he claimed that the sequence of learning should necessarily follow that order. Any thorough understanding of one science had to depend on understanding those which precede it. Comte believed that his arrangement of the sciences followed the order in which things in the universe had come into being. Thus, for example, one could not have physiology without chemistry. However, some philosophers and scientists objected to his arrangement, claiming, for example, that the universe described by astronomy could not exist before physics.[53] Still, Comte's outlook was very influential, especially his decision to exclude theology and metaphysics from the canon of valid knowledge. He argued that all knowledge develops in three stages: beginning in a "theological" stage in which explanations are supernatural, advancing to a "metaphysical" stage in which explanations are abstract, and finally progressing to a "positive" stage where explanations are scientific, such as the mechanics determined by the laws formulated mathematically by Newton. Many writers were impressed by Comte's approach,[54] though not Ampère, who deigned to not even mention Comte in his *Essay* or in its second volume, published after his death.[55] Ampère greatly valued metaphysics and allocated it in his classification of the noological sciences. Moreover, presumably, Comte's hierarchy may have seemed hopelessly simplistic to Ampère.

Yet Comte's works became popular, while most scientists disregarded Ampère's system of knowledge. Nor did they accept Ampère's four points of view. But if they disdained his ordering principle, why did so many adopt his notion of kinematics as the fundamental physical science?

Was the fundamental role of the science of motion just a philosophical fashion?

Various sciences were viewed as the "fundamental" branch of physical science, depending of what we mean by that. The first ranking has been occupied by harmonics, metaphysics, astronomy, mechanics, statics, kinematics, and so on. So what do we learn? Did theorists classify the sciences

---

[53] Herbert Spencer, *The Classification of the Sciences, to which are added Reasons for Dissenting from the Philosophy of M. Comte*, 3rd ed. (London and Edinburgh: Williams and Norgate, 1871); e.g., preface to the second ed., pp. ii–iii, 26.

[54] G. H. Lewes, *Comte's Philosophy of the Sciences: Being an Exposition of the Cours de Philosophie Positive of Auguste Comte* (London: Henry G. Bohn, 1853), pp. 40, 46, 121.

[55] André-Marie Ampère, *Essai sur la Philosophie des Sciences, Seconde Partie*, ed. Jean-Jacques Ampère (Paris: Bachelier, 1843).

arbitrarily? No, kinematics rose in importance mainly owing to physical reasons. That story is not covered here, but it is familiar enough, and we can appreciate its extent only if we take, again, a long view of the history of science. It is the story of how, throughout the ages, scientists failed to discover any sort of physical interaction that is transmitted in no time and involves no motion. It is the story of how they gradually realized that everything that seems to be at rest can, with better reason, be said to be moving, and that many qualities and sensible phenomena, such as sounds, color, heat, and magnetism, can be understood in terms of invisible motions. Such were the reasons physicists gave for their esteem of the science of motion. So it was *despite* the ancient philosophical biases *against* concepts of motion in science that the advancement of explanations of natural phenomena based precisely on motion gradually helped raise the science of motion to a prominent role.

In hindsight, the classifications of knowledge can serve as an index of the progress of scientific understanding. There is a question of truth in the organization of the fields of knowledge. There were, indeed, natural reasons why physicists gave a central position to concepts of motion in the explanation of phenomena, instead of, say, concepts of flavor or smell. This does not mean, of course, that the science of motion must necessarily be regarded as the fundamental branch of physics. It just means that part of the reasons why kinematics rose in importance was the wealth of conceptual and experimental ways in which scientists found that they could understand and control physical phenomena by carefully describing motions.

As stated earlier, the classification of the sciences is now rarely pursued by philosophers and scientists. The practice is out of fashion, even frowned upon. It seems that if someone now argues for the primacy of a science, it is usually, the primacy of *their* science. But the preceding account shows that this game of "my science is more fundamental than yours" is *not* what some theorists in ages past were doing. In particular, think of the late 1800s and the growing attempts to construct an electromagnetic worldview. And consider Ampère in the early nineteenth century, and notice not only how much his diagram includes but what it does not include. Some of the ancient favorites, harmonics and even optics, are nowhere to be seen. Nor is there, explicitly, a science of electricity and magnetism here. So when Ampère classified kinematics as a fundamental science, he was not classifying *his* science. Never did he even write a paper on kinematics. In a way, it was nobody's science; after all, no one had written a treatise on it.

Ampère stressed its importance because he thought it remained neglected. So did Hertz, decades later, and Einstein and others. In this way, Ampère's classification did help the progress of science. As we will see, it helped to identify an important gap in physics, in a way that encouraged physicists to correct it. Those were, after all, two of the key functions expected of the natural classifications of the sciences.

Along with kinematics, there was another neglected science, the importance of which Ampère highlighted, namely, the science of classifying sciences. Ampère called it mathesiology and classified it as a subfield of pedagogy.[56] It grew throughout the centuries, sometimes motivated by educational concerns, other times by philosophical concerns about knowledge or about the natural order of things. Sometimes it stalled, perhaps because existing classifications were sufficiently useful, or because broad conceptual frameworks occasionally became stagnant when people turned their attention to other things. Yet part of the present value of the old hierarchies and classifications of the fields of knowledge is that they allow us to grasp ways in which past theorists ordered their world.

Given the neglect of the study of the classification of the disciplines, it is hence fitting that presently most people routinely abstain from attempting to specify the way in which they use the word science. Many indulge a reticence to even attempt to define their own specific field of study. Many specialists are more interested in claiming the autonomy of their science than in finding how they stand to gain from other sciences. We also find that even among the subfields of a single science, specialists often fail to understand each other's works or their relevance.

[56] Ampère, *Essai* (1834), pp. xxxij–xxxiij, 31–32.

Nowadays, despite the ease with which information is obtained, there is a lack of order that easily allows any study to seem to be of a central importance. Nearly anything can seem linked to whatever one studies. Thus, researchers and students of many sciences can become engulfed in the impression that their own field lies at the center of things. It is a deceptively pleasant condition. One old metaphor well describes this perspective, where everyone seems to be at the center, far from the periphery, and yet interconnected to all. It is not that of the tree of knowledge, nor that of the pyramid, the ladder, the hierarchy, or the map. It is the labyrinth.

CHAPTER TWO

Where to Begin?
Invisible Causes or Visible Motions

What were the origins of Ampère's notion of kinematics? The first chapter said little toward answering this question. There we traced the rise of the study of motion in the classifications of knowledge throughout the centuries but not the proximate influences that encouraged Ampère to define kinematics as the fundamental branch of physics. The account of the classifications of the sciences was a first approximation, a bird's-eye view—introducing kinematics by mapping it in relation to other fields. By contrast, we now approach its origins to a closer degree, focusing mainly on a period of decades rather than centuries, to identify relatively recent precursors to Ampère's outlook.

What roots can we trace for Ampère's conception of the science of motion? To answer this question we start not with antiquity but with 1834; and, instead of presenting a chronological sequence, we invert the direction of history to move now from the past to the more distant past, as in field archaeology, digging up layers of ever earlier antecedents. We begin with Ampère's book of 1834 and then trace its connections to previous works. Whereas historians usually hide this process by inverting the sequence to follow the arrow of time, in this chapter we read history by moving against the direction of time.

When Ampère formulated his notion of kinematics, he admitted that he was not the first to conceive of such a science. In particular, the notion of studying motion irrespective of forces had a long history. Some relatively recent works had some direct effect on his conception, as well as some other relevant predecessors in that tradition. In his *Essay on the Philosophy of the Sciences,* Ampère admitted that he had not quite invented the science of kinematics. He briefly mentioned only two sources that served to illustrate, at least in part, his conception of kinematics—"what Carnot has written about movement considered geometrically, and the *Essay* on the composition of machines by Lanz and Betancourt."[1] Thus, we have but three names, traces disclosed by Ampère that serve as leads to pursue in order to uncover the proximate influences that helped shape his conception. The works in question illustrate the beginnings of what grew to become two distinct kinds of kinematics. The relevant writings of Lazare Carnot pertain to the study of the geometry of motion as an abstract part of physics, what physicists sometimes call theoretical kinematics. The work of José María Lanz and Agustín de Betancourt concerned the study of the motions of actual machines, what became known as industrial kinematics. Our history of kinematics treats physics rather than engineering, so it focuses on the former tradition, but first we consider the study of the motions of machines.

In addition to general statements about theoretical kinematics (discussed in chapter 1), Ampère also made certain specific claims about the kinematic study of mechanical devices:

It should then study the different instruments with the help of which one can change a movement into another; such that by comprehending, as is used, those instruments under the name of machines, one will need to define machine, not as one does ordinarily, *an instrument with which one can change the direction and intensity of a given force,* but instead as *an instrument with which one can change the direction and speed of a given motion.* One thus makes this definition independent of the consideration of the forces that act on the machine, a consideration that cannot serve but to distract the attention of whoever seeks to understand the mechanism. To formulate a clear idea, for example, of the gears with which the minutes-pointer of a clock marks twelve turns, while the hours-pointer only marks but one, does one need to busy oneself with the force that sets the clock in motion? The effect of the gears, inasmuch as it rules the interaction of the speeds of the two pointers, does it not remain the same, whether the motion is due to a force whichever other than that of the ordinary motor; when it is, for example, with a finger that one moves the pointer of minutes?[2]

---

[1] André-Marie Ampère, *Essai sur la Philosophie des Sciences, ou exposition analytique d'une classification naturelle de toutes les connaissances humaines* (Paris: Bachelier, Imprimeur-Libraire pour les Sciences, 1834), p. 51; see also p. 21.
[2] Ibid., pp. 51–52.

This is an example of what Ampère meant by the study of motions irrespective of the concept of forces. He sought to encourage a kind of study of motion that would be free from the widespread concept of force that had proliferated in mechanics since the late 1600s. That old approach followed the ancient tradition of basing mechanics on questions of statics, the equilibrium of forces. "A treatise where one would consider thus all the movements independently of the forces that can produce them will be of an extreme utility in instruction, in presenting the difficulties that can offer the play of diverse machines, without the spirit of the student having to defeat at the same time those difficulties which may result from considerations relative to the equilibrium of forces." When he wrote these lines, there existed already a well-known treatise that, on the main, described the study of mechanical motions along such lines. The definitive edition of the work of Lanz and Betancourt was published in 1819, under the title (literally translated) *Essay on the Composition of Machines.* It was the second edition of an earlier work and was reissued in an English translation in 1820, under the title *Analytical Essay on the Construction of Machines.*

Machinery in Motion

Widely used in technical schools throughout Europe, the *Essay* of Lanz and Betancourt aimed to provide a systematic catalog and analysis of the known kinds of mechanical motions, demonstrating how their combinations serve to compose a great variety of contraptions. Many works on machines had existed before, but none had such a broad systematic scope. By contrast, already in antiquity, geometric knowledge, at least, had been synthesized into a major systematic work, Euclid's *Elements,* around 300 BC, in which all geometric figures were constructed on the basis of points, straight lines, and circles. Throughout the centuries, however, no treatise on mechanics had synthesized the workings of virtually all known machines in terms of elementary parts and motions. Lanz and Betancourt distinguished three kinds of elementary mechanical motions: rectilinear motions, circular motions, and motions constrained along given curves. They subdivided each of these three kinds of motion into two varieties: "continuous" motions (i.e., unidirectional motions) and "alternating" motions (i.e., motions that change directions). They then assembled a table listing systematically every known interconnection of mechanical contraptions that served to convert one kind of motion into another (see figure 13).

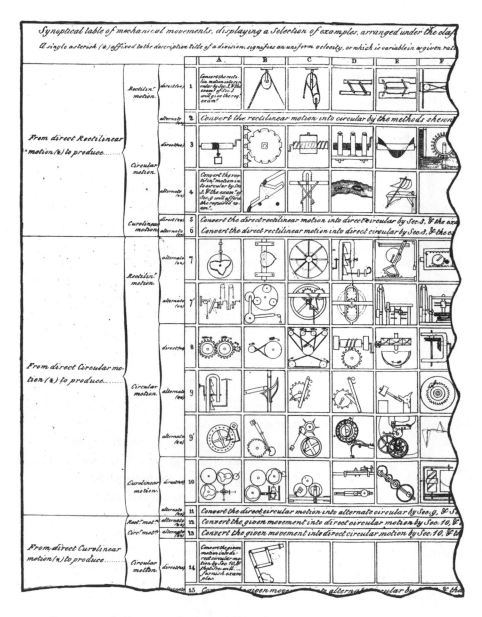

Figure 13. Small excerpt from the table of the composition of machines of Lanz and Betancourt, showing combinations of mechanical motions

They illustrated various ways of converting rectilinear continuous motions into other rectilinear continuous motions, rectilinear alternating motions, circular continuous motions, circular alternating motions, and so on. For example, a hanging pulley serves to convert the rectilinear motion of a string being pulled down through it in one direction, into a rectilinear motion in a different direction, upward along the string on the other side of the pulley. At the outset of their book, Lanz and Betancourt explained: "The movements that one employs in the arts are rectilinear, or circular, or determined by given curves, they can be continuous or alternating (back and forth), and one can consequently combine them two by two in fifteen different ways, and in twenty-one if one combines each of those movements with itself. Every machine has as its aim to modify or communicate one or several of those twenty-one motions."[3]

Although the work of Lanz and Betancourt concerned mainly the question of the transmission and transformation of motions, it did not fully meet the strictures that Ampère later placed on his conception of kinematics. In particular, it did discuss, however briefly, some considerations about forces and how to ascertain them. The authors explained that as long as a machine is in static equilibrium, one can analyze it merely on the basis of the intensity of the forces involved, whereas when it is in motion one can analyze it in terms of both the forces and the spaces traversed by the parts of the machine. Moreover, the second edition of the *Essay* included more dynamical considerations than its predecessor.

The earlier and first edition of the *Essay on the Composition of Machines* was published in 1808 together with a similar work by Jean-Nicolas Hachette, titled *Program of the Elementary Course on Machines for the Year 1808*. Independently, Hachette, professor of the imperial Polytechnique in Paris, had developed a similar though preliminary outline describing "elementary machines" and various mechanisms that serve to convert motions into one another.[4] Hachette divided mechanical motions into rectilinear and circular motions, and he subdivided each of those into continuous or alternating (see figure 14).

[3] José María Lanz and Agustín de Betancourt, *Essai sur la Composition des machines* (Paris: Bachelier, 1819), p. 1.
[4] Jean-Nicolas Hachette, *Programme du cours élémentaire des machines pour l'an 1808* (Paris: Imprimerie Impériale, 1808).

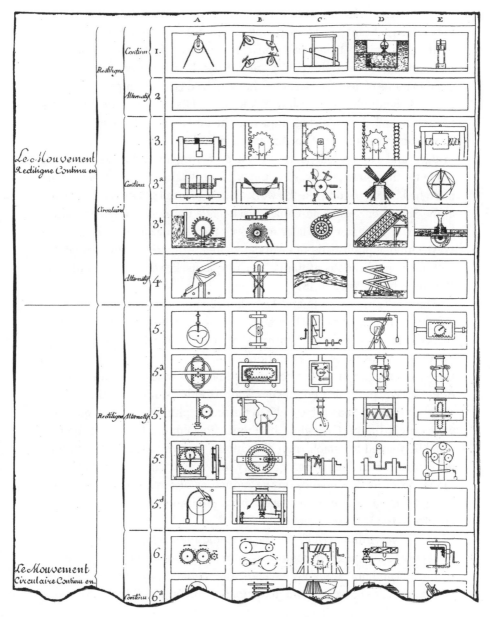

Figure 14. Excerpt from Hachette's table of basic mechanisms

His systematic enumeration of possible combinations of motions served to reveal significant gaps; for example, it pointed out the lack of knowledge of any mechanism that would convert rectilinear continuous motion directly into rectilinear alternating motion. Still, by employing the diverse combinations of devices, Hachette characterized gears, pumps, the hydraulic press, windmills, pulleys, levers, and more. His work responded directly to the needs that the Polytechnique had just established in 1806 for a new course on the elements of construction and machines employed in mines, bridges, and dams. While Hachette was preparing his outline as a teaching guide for the course, he heard of the work of Betancourt and Lanz. Betancourt had begun to compose the work already before 1805 in Madrid, Spain, at the School of Roads and Canals (Escuela de Caminos y Canales), where Lanz was a professor and later took over the writing project.[5] Hence, the Polytechnique published both works together. Ampère became acquainted with these publications at the Polytechnique itself, where he taught from 1804 until he resigned in 1828.

The works of Hachette and Lanz and Betancourt coincided in essentially the same plan because they stemmed from a common influence. Both works followed a basic plan formulated years earlier by Gaspard Monge for the study of mechanical motions. The famous Monge helped found the Polytechnique in 1794, when it was known first as the Central School of Public Works (École Centrale des Travaux Publics; it was renamed École Polytechnique in September 1795, to reflect the plurality of techniques that were taught there). Following the French Revolution, leaders of the new republic wanted to establish an educational system based on "revolutionary courses" in mathematics, chemistry, and more.[6] They sought to liberate France from its dependence on the industries of foreign nations. Among them, Monge was a leading proponent of the importance of teaching mechanics to train future engineers.

Monge was influenced by the encyclopedic approach to learning. Hence, he sought to design a pedagogical program that would educate engineers in theoretical and practical knowledge in an integrated way. He distinguished two major fields of instruction: mathematics and physics. Figure 15 conveys

[5] Antonio Rumeu de Armas, *El Real Gabinete de Máquinas del Buen Retiro* (Madrid: Editorial Castalia, 1990), pp. 61–66.

[6] A. L. Fourcy, *Histoire de l'École Polytechnique* (Paris: École Polytechnique, 1828); reprinted, with notes by J. G. Dhombres (Paris: Belin, 1987), pp. 389–390.

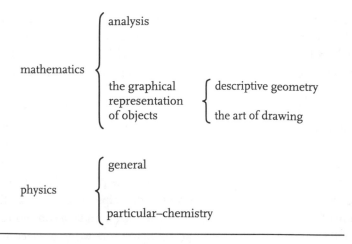

Figure 15. General pedagogic plan of Monge

his broad scheme for the main disciplines that engineers would study.[7] A noteworthy point is that he included drawing as a branch of mathematics.

The outstanding paragon of Monge's program was his innovative course on "descriptive geometry." He characterized the main objective of descriptive geometry as "representing with exactitude, within drawings that have but two dimensions, objects that have three."[8] A large part of his *Descriptive Geometry* (1799) involved the generation of surfaces by the motion of particular curves, to construct, for example, a tangent plane on a point on a given surface. This kind of geometry had national value, and especially because of its applications to war, it was partly a secretive or classified matter. For example, Monge used it to prepare the plates for his work of 1793, *Description of the Art of Constructing Cannons,* but he was forbidden from explaining there how he had prepared them.[9]

Monge had begun to develop descriptive geometry in the 1760s, while he taught at the royal school of engineering at Mézières, France. It was then

[7] J. Languins, "Sur la première Organisation de l'École polytechnique. Texte de l'arrêté du 6 frimaire An III," *Revue d'Histoire des Sciences* (1980), pp. 298–313.

[8] Gaspard Monge, "Lessons," ed. B. Belhoste, R. Laurent, J. Sakarovitch, and R. Taton, in *L'École Normale de l'An III, Leçons de Mathématiques, Laplace, Monge,* ed. J. Dhombres (Paris: Dunod, 1992), pp. 267–459; avant-propos.

[9] Kirsti Andersen and I. Grattan-Guinness, "Descriptive Geometry," *Companion Encyclopedia of the History and Philosophy of the Mathematical Sciences,* vol. 2, ed. I. Grattan-Guinness (London: Routledge, 1994), pp. 887–896, p. 889.

only a section of his course on stereotomy, the art of tracing the forms into which stones will be cut to be used for construction. For the new Central School, in 1794, Monge inverted the relationship between stereotomy and descriptive geometry by making the latter engulf the former, along with many other practices.[10]

Because Monge was keenly interested in the teaching of multifarious practical applications of mathematics, his innovative course on descriptive geometry at the Polytechnique involved an extensive range of subjects requiring the combination of intellectual and manual labors. Figure 16 displays the subfields that Monge attributed to the potentially enormous field of descriptive geometry. He conceived of descriptive geometry as the integration of the ensemble of arts of construction all under a common geometric method. Beginning students at the new École were expected to spend half of their time studying descriptive geometry. Stereotomy appeared now as the general introduction of methods and rules of descriptive geometry.

Among the many fields, the diagrammatic study of the "principal composite machines" is what later developed into the discipline of the kinematics of machinery. Monge's course on descriptive geometry allocated two months to the study of the description of the "elements" of machines, and especially of those used in public works. According to his pupil Hachette, Monge explained: "One understands, by *elements of machines,* the devices by which one changes the direction of movements, those by which one can generate some from others, the forward motion along a straight line, the motion of rotation, the alternating motion of *back and forth.* One senses that the most complicated machines are solely the results of combinations of such individual devices, of which one should try to make a complete enumeration."[11]

Thus, the study of machines acquired a character of research centered on classification: to catalog and organize systematically all kinds of machines. Monge dedicated the last lessons of his course on descriptive geometry to presenting various mechanisms by which one may convert different kinds of motion into one another, as well as mechanisms that facilitate various

[10] Joël Sakarovitch, "La géométrie descriptive, une reine déchue," in *La Formation Polytechnicienne, 1794–1994,* ed. B. Belhoste, A. Dahan Dalmedico, and A. Picon (Paris: Dunod, 1994), pp. 77–93.

[11] Hachette, *Programme du cours élémentaire,* p. v; Lanz and Betancourt, *Essai sur la Composition des Machines.*

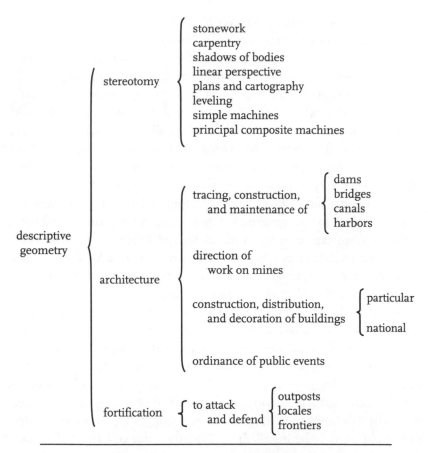

Figure 16. Subfields of descriptive geometry, according to Monge

kinds of motions, and to describing machines employed in the arts, or pow-
ered by animals, or by water, wind, and other "forces of nature." Again,
Monge's notion of the study of mechanical motions can be distinguished
from Ampère's in regards to the role of the concept of force. For Monge, the
essential property of machines was the transformation and transmission of
what he called "live force," which scientists later called energy.

In any case, Monge's course on machines was adjourned promptly in
1795 because the duration of course work at the Polytechnique was reduced
from three years to two by law, effectively limiting the study of machines
mainly to those employed in the exploitation of mines. The French had gone
to war and needed to train military engineers as quickly as possible. More-

over, Monge was enlisted in the army of his friend Napoléon Bonaparte, in campaigns that took him far away from Paris, to Italy and Egypt.

Hachette's close collaborations with Monge placed him in a privileged position from which to carry out Monge's plan. Hachette had begun as the assistant (*répétiteur*) of Monge at the École and was promoted to professor in 1799. But meanwhile, Agustín de Betancourt too was acquainted with Monge and his approach to mechanics, plus he had an advantage over Hachette in devising a more thorough account of the composition of mechanical motions—namely, that he was one of the few individuals in all of Europe who was personally familiar with very many kinds of machines.

In 1785 Betancourt, who was already a leading hydraulics engineer, had convinced the Spanish government to support his efforts to establish a school of engineering in Spain, the School of Roads and Canals (established only much later, in 1802), and to develop a great collection of models of machines used in public works and industry. Betancourt led a workshop of engineers and craftsmen who labored to construct, in much reduced but exactly scaled measurements, the parts of every machine and structure they procured, even down to the number and dimensions of nails. Various cranes, windmills, vapor machines, water pumps, bridges, and more were fashioned imitating particular machines and structures found throughout Europe. Accordingly, the *Essay on the Composition of Machines* included not merely idealized machines but abundant references noting the cities where particular machines were in use (and even suggesting other sites where they could usefully be installed). In 1792 the royal collection of models, the Real Gabinete de Maquinas, was opened to the public in Madrid at the palace of Buen Retiro, under the auspices of King Carlos IV, and later it was employed at the School of Roads and Canals, to aid in the instruction of engineers.

And what became of this material collection for the inventory and study of mechanical motions? In 1808 when Madrid was overtaken by Bonaparte's troops, the palace of Buen Retiro was bombarded by French artillery. The damaged collection of machines was then carelessly relocated to the Real Academia de Bellas Artes and soon afterwards to the basement of the palace of Buenavista. There the collection remained neglected until 1813 when Madrid was liberated, and in 1814 the collection was again relocated. Although many models were broken or lost, what remained was eventually transferred back to the reestablished School of Roads and Canals. By the 1850s the collection was in disarray and increasingly dispersed. Ultimately, not even a single model was known to survive. Strangely, all disappeared without a trace: more than 270 working models of machines, crafted with utmost detail, and, more than 350 drawings framed in mahogany and high security glass, and 99 illustrated memoirs about machines, construction procedures, and public works.[12] The search for any remnants or documentary traces of the extraordinary collection of machines, lost but to records and hearsay, became an object of historical research and antiquarian speculation.

The idea that it is preferable, wherever possible, to study machines as they perform their functions rather than in any static representation had precedents other than Betancourt's project for the educational use of models. Monge had advocated the idea that the fundamental nature of a machine is not the equilibrium of its parts but their movement. In this tradition, the prominent mathematician and physicist Leonhard Euler had played an influential role. Already in the mid-eighteenth century, he had argued that motion is the principal property of machines and that hence they ought best to be studied in motion rather than at rest in any state of equilibrium of forces. Furthermore, in the *Theatrum Machinarum*, published in 1724, Jacob Leopold wrote several sections on the conversion of motions—circular into rectilinear, back-and-forth motion into circular motion, and so on—closely related to the specific plan later advanced by Monge and developed by Hachette and Lanz and Betancourt.

Geometry in Motion

A similar investigation can reveal the roots of theoretical kinematics. As mentioned before, Ampère explicitly referred to the writings of Carnot as an illustration and source of his notion of kinematics. Lazare-Nicolas-Marguerite Carnot was a military engineer who, after the revolution of 1789, had taken a leading part in successful war campaigns throughout France. Hence, Carnot also served in the French Directory government, and even as minister of war for Bonaparte briefly in 1800, a position that he resigned as he boldly opposed the establishment of the empire under Bonaparte, who crowned himself emperor of France in 1804. Carnot was also a leading figure in founding the École Polytechnique with Monge, and alongside his involvement in military engineering and politics, he found time to contribute to the disciplines of geometry and mechanics.

[12] Rumeu de Armas, *El Real Gabinete de Máquinas del Buen Retiro*, pp. 109–220.

What was Carnot's conception of a science of motion? In his *Geometry of Position* of 1803, Carnot defined what he called "the geometry of position" as a study that described the disposition or orientation of geometric figures in relation to one another by simple mutations of their parts. He proposed a new way of applying algebra to the analysis of geometric problems, by a revision of the function of positive and negative signs (rejecting the usual notion of isolated negative quantities). He distinguished his conception of a geometry of position from another nascent study known already as the "geometry of situation."

The so-called geometry of situation, a subject that had not yet been developed systematically, was the name given by some writers to puzzles concerning trajectories over predetermined paths. For example, the name was associated with a well-known problem that Euler had analyzed and solved in 1735. It consisted of finding a method for determining whether certain trajectories are possible. The popular problem that drew Euler's attention was how to figure out which trajectory one must take in order to cross over bridges placed along a winding river in such a way that one does not cross any one bridge twice (see figure 17).

In a river in Königsberg (now Kalingrad) in Prussia, an island called the Kneipfhof lay near the point in which the river bifurcates in two, and the sections of land were connected by seven bridges. To determine if someone could take a path that would cross each bridge once and none more than once, Euler formulated a method for solving such problems whatever the shape of the river, the distribution of its branches, and the number of bridges.[13] (This kind of problem helped motivate the development of the branches of mathematics that later became known as graph theory and topology.) Euler first became interested in such kinds of problems owing to his acquaintance with ideas explored by Gottfried Wilhelm Leibniz. Euler shared the impression that, whereas ordinary traditional geometry, dealing with magnitudes or quantities, had been extensively developed, there existed the possibility of another kind of geometry, dealing not with magnitudes but with situation. Leibniz had argued that common algebraic and

---

[13] Leonhard Euler, "Solutio Problematis ad Geometriam Situs Pertinentis," *Commentarii Academiae Scientiarum Imperialis Petropolitanae* 8 (1741), 128–140; Euler, *Opera Omnia*, ser. 1, vol. 7 (1923), pp. 1–10; "Solution of a Problem Belonging to the 'Geometry of Position,'" reissued and translated by Peter Wolff, in *Breakthroughs in Mathematics* (New York: New American Library, 1970), pp. 197–206.

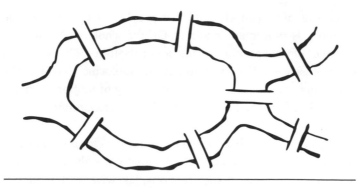

Figure 17. Euler's problem

geometric methods were insufficient for the "analysis of situation," that is, the analysis of the relative positions of parts of a given figure, or even the analysis of their changing positions. He believed that one should be able to "handle mechanics in nearly the same way as geometry." After sketching the basic elements for a new method of symbolic analysis, he hoped that, with its use, "one would be able to cast into symbols, that only would be letters of the alphabet, the description of a mechanism, however it might be composed, which would give the intellect a method of knowing it distinctly and easily with all its parts, and even with their use and movements."[14] This goal, as suggested roughly by Leibniz, led Euler and later Carnot to develop new symbolic methods of geometric analysis.

Regarding the geometry of situation, Carnot commented that, once developed,

> this branch of geometry would properly serve to describe in a simple and uniform fashion the diverse procedures of the arts, and in this regard, it could become infinitely useful. But it seems to me that the name geometry *of situation* is less convenient than, for example, that of geometry *of transposition;* since in effect the movement or transposition of the parts of the system enter as an essential element in

[14] Gottfried Wilhelm Leibniz, letter to Christiaan Huygens, 8 September 1679, in *Christiaani Hugenii aliorumque seculi XVII. virorum celebrium exercitationes mathematicae et philosophicae,* ed. Uylenbroek (1833), fasc. I, p. 10; translated in Hermann Grassmann, *A New Branch of Mathematics, The Ausdehnungslehre of 1844 and Other Works,* trans. Lloyd C. Kannenberg (Chicago: Open Court, 1995), p. 404.

all questions of its concern, and that properly it is to the geometry of position, as movement is to rest. On the whole, that geometry of situation or of transposition stands not by itself but is the least part of a very extensive science, a very important one, which has never been treated. This science is, in general, the theory of motion, considered in abstraction from the forces that produce or transmit it.[15]

Thus, we find that Carnot had indeed formulated a conception of a geometric study of motions independent of considerations about any forces that generate, transmit, or modify such motions. Likewise, such motions would be studied irrespective of the mass of the bodies involved. Carnot gave a couple of examples:

If I propose to move a knight in the game of chess over all the squares on the chessboard, without passing twice over any given one, it doesn't concern me at all what is the mass of the knight and the force that I employ to move it. Likewise, as a thread forms the nets of a knitting, this does not concern at all the laws of action and reaction, nor of the force with which the thread is stretched: in the end it's the same with all machines where the aim is not to economize forces but to establish such or such relations among the directions or the speeds of the different points of a system. I will return in another work to the notion of such movements of which I have already given the limits, *geometrical movements,* and of which the theory embodies the passage from geometry to mechanics.

Carnot never wrote the work he anticipated on the geometry of motion. His references to forces, and action and reaction, pertained of course mainly to the theory of mechanics of Isaac Newton, who posited three laws of motion. The immediate point is to note that at stake was the matter of defining a descriptive science of motion independent of the question of the forces or actions that bodies exert on each other.

Carnot also sketched his conception of a science of motion in his *Fundamental Principles of Equilibrium and Movement* of 1803. There, Carnot re-

[15] Lazare N. M. Carnot, *Géométrie de Position* (Paris: J. Duprat, Libraire pour les Mathématiques, 1803), p. xxxvij.

marked on two traditional but distinct ways of envisaging mechanics in regard to its principles:

> The first is to consider it [mechanics] *as the theory of forces,* that is to say, of causes that engender motions. The second is to consider it *as the theory of motions* themselves. In the first case, one establishes the reasoning about whichever causes that impress or tend to impress movement to bodies on which one supposes them to be applied. In the second, one regards the motion as already impressed, acquired and residing in the bodies: and one seeks only what are the laws according to which the motions acquired propagate, are modified or destroyed in each circumstance.[16]

Indeed, various writers defined mechanics by giving a central role to either the concept of forces or the concept of motion. Carnot argued that each approach has advantages and disadvantages. The approach focused on forces had the advantage of being widely espoused as mathematically simpler, though it had the disadvantage, said Carnot, "of being founded on a metaphysical and obscure notion, which is that of *force.*"[17] He argued that the notion of force could hardly be understood because it was based on the ambiguous notion of *cause.*

He claimed that in the exact language of mathematics, itself necessary for physics, operations involving forces or causes scarcely made sense. Of course, scientists spoke of the force of gravity, but this example remained mysterious, an invisible agency that was inferred only on the basis of actually observed motions. Rather, the basic notion of force seemed to be grounded on direct personal experience: the sensation of resistance experienced when pushing or pulling a heavy object, or the impacts felt from the collision of bodies against one's own body. But rather than address such sensations of force, Carnot focused his complaints on the notion of force as the cause of movement as willed by a person or animal. He asked, Are the causes of a bodily motion the result of the will or of the physical constitution of a human or animal? Lacking any answer to this kind of question,

---

[16] Lazare N. M. Carnot, *Principes Fondamentaux de l'Équilibre et du Mouvement* (Paris: Crapelet, 1803), p. xj.
[17] Ibid., pp. xj–xij.

it seemed meaningless to talk about mathematical relations among willed actions. A physical cause did not seem clearly quantifiable to him. For example, he made no sense of the notion of doubling or tripling an action of the will or a physical constitution. He claimed that so long as the notion of cause was posited as a principal idea underlying the notion of force, then all uses of the word force would convey an absolutely inevitable obscurity. So long as forces were seen as causes, and motions as effects, such that the former precede the latter, the notion of forces seemed metaphysical and mathematically inaccessible. In contradistinction, it seemed clearly meaningful to speak about perceptible quantities of motion and to measure them.

Another reason why the notion of force seemed mysterious was that it was often allowed to stand alone, as though forces could exist independently of material bodies. The idea that the study of motion is fundamentally independent of other branches of physics emerged against traditional interpretations where motion was explained in terms of other "more fundamental" agencies. Given the popularity and success of Newton's approach to mechanics, the concept of force had been employed often as just such a fundamental agency. Carnot commented that by admitting the notion of force as a metaphysical axiom, one allowed the idea that a force could allegedly act on a body from any arbitrary direction without any perceptible effect; many supposed forces could allegedly act on a body simultaneously, without affecting it, like fictions. By contrast, any one body could not be said to move but in *one* direction at a time, and none could be said to have several motions at once. There could exist no motion independent of a material body. For Carnot, the usual notion of force entailed metaphysics, whereas notions of motion necessarily required appeals to physical experience.

In assessing the hierarchical relations between physical experience and metaphysics in the classifications of knowledge, many philosophers ascribed a higher standing of truth to metaphysical knowledge over sense impressions. Within that scheme, the science of mechanics would seem to be more certain by having its fundamental concepts pertain to metaphysics rather than experience. The notion of force had been granted that role, loosely in accord with the ancient framework of Aristotle and others who deemed metaphysical causes to be prior to physical phenomena. But Carnot rejected the seemingly metaphysical foundation of mechanics, embodied in the notion of force, in favor of what for him was the empirical notion of movement. Thus, his work was a step in the direction of inverting the grounds on which the certainty of mechanics was justified.

Still, unlike Ampère's conception of kinematics, Carnot did admit a notion of force into his formulation of mechanics. Like other writers, he noted that one way to make sense of the concept of force was to identify its action by the quantity of motion that is observed in a body. This approach seemed to justify the appeal to forces by relying directly on the alternative conception of mechanics as the science of the communication of motions. To pursue this way of employing the notion of force, Carnot admitted that he would not distinguish between causes and their effects. In his account, forces did not precede motion but were instead a particular way of speaking about and analyzing motions. Nonetheless, he did distinguish the part of the science of motion that deals with geometric movements alone and not with forces or actions. Thus, in his *Fundamental Principles of Equilibrium and Movement,* he advocated what he called a new generalization—namely,

> to substitute *virtual velocities* that are infinitely small, with finite velocities that I call *geometric;* I've kept that basis in this present edition. There results a sort of new theory about a class of movements that is less the subject of mechanics than of geometry. Such geometric movements are those that the different parts of a system of bodies can take without counteracting one another, and which consequently do not depend on the action and reaction of the bodies, but only on the conditions of their connections, which can be determined by geometry alone and independently of the rules of dynamics.[18]

The previous edition implied in this passage refers to an earlier work by Carnot, a substantially different book titled *Essay on Machines in General,* first published twenty years earlier, in 1783. Around 1800, he intended merely to issue a revised edition of that work, but extensive alterations and additions led to the change in outlook and title. To explain what he meant by geometric motions, we can consider his earlier work, but first we should touch upon recent works of one of the authors, whose contributions and approach elicited both admiration and critical attention by writers such as Carnot.

Joseph-Louis Lagrange was the leading authority on mathematics and analytical mechanics. He served as founding professor of analysis at the Polytechnique from 1794 until 1799. Carnot's notion of a study of transposition, defined as *a branch of geometry,* has a kinship to some of Lagrange's

---

[18] Ibid., p. x.

earlier notions of analytical mechanics. In a treatise of 1799, on the theory of functions, Lagrange envisaged mechanics as a branch of geometric analysis. He proposed the notion of mechanics as a geometry of four dimensions: "Here the functions are referred essentially to time, which we will designate always by t; and as the position of a point in space depends on three rectangular coordinates x, y, z, those coordinates, in the problems of mechanics, will be taken as being functions of t. Thus, one can regard mechanics as a geometry in four dimensions, and analyze mechanics as an extension of geometrical analysis."[19] (In an article of 1754, Jean d'Alembert had already noted that time might be construed as a fourth dimension, and he attributed this idea to someone he did not name.[20]

Lagrange construed the notions of space, speed, and accelerative force as particular functions of time. He claimed to have deduced the principles and fundamental equations of motion from the pure theory of algebraic functions. He demonstrated indeed that the theory of mechanics could be formulated mathematically without employing the notions of infinitesimally small quantities used in the differential calculus. But more significant to the questions at hand, Lagrange also advocated the usual notion about machines as devices that transmit and modify "motive powers" or forces. This approach, as we have seen, was increasingly replaced by that of Carnot and others. Moreover, Lagrange's most prominent work on mechanics, his *Analytical Mechanics* of 1788, was based systematically upon the science of statics and specifically upon the notions of force and virtual velocities, which Carnot and others sought to supersede.

Rather than deal with particular machines or mechanisms like Lanz and Betancourt, Carnot's early *Essay on Machines in General* of 1783 concerned a higher level of abstraction, establishing properties common to all machines. Carnot claimed that the science of machines in general could be reduced to the question of how to determine, given the virtual movements of a system of interacting bodies (i.e., the motions each body would take if it was free), the real movements that will take place. This focus on virtual velocities was something he abandoned in his later work. Also, this early work

---

[19] Joseph Louis Lagrange, *Théorie des Fonctions Analytiques, contenant les Principes du Calcul Différentiel, dégagés de toute considération d'Infiniment Petits ou d'Évanouissans, de Limites ou de Fluxions, et réduits à l'Analyse Algébrique des Quantités Finies* (Paris: Imprimerie de la République, 1799–1800), p. 223.

[20] Jean d'Alembert, "Dimension," *Encyclopédie*, vol. 4 (Paris, 1754), p. 1010.

included abundant discussion on forces, weights, equilibrium, centers of gravity, collisions, pressure, intensity of action, "conservation of live forces," the impossibility of perpetual motion, and other subject matter proper to dynamics. Nonetheless, already in this early work he was concerned with the "communication of movements," and with what he called "geometrical movements." Already he claimed that there exists a class of motions that "can be determined by the lone principles of geometry, & are absolutely independent of the rules of Dynamics."[21]

By geometric movements, he meant motions that do not interfere with one another. For example, Carnot designated two bodies, separated by a constant distance and traveling in the same direction with a common speed, as having geometric motions because their motions do not interfere with one another; moreover, their motions can be reversed and still would not interfere. By contrast, two bodies moving away from each other would not constitute so-called geometric movements because if their motions are reversed they would be made to collide or stop by their mutual impenetrability. Accordingly, you can surmise that his approach served to describe some mechanical motions by referring to the parts of machines that move in unison and are able to take equivalent but opposite motions. Now, of course, many machine parts are in contact with one another, so Carnot had to employ a fiction—that is, he drew a further distinction between movements that are treated as geometric by supposition and others that are "absolutely" geometric, but we need not delve into such matters here.

The leads that Ampère explicitly gave us as precedents to his notion of kinematics do not end in themselves but refer us farther to additional sources. For example, the expression "the communication of movement" had been used by philosophers and geometers before Carnot, including Jean Bernoulli, among others.[22] Also, the idea that one can study motion irrespective of forces has a long history. If we pursue precursors to Carnot, we find Euler, for example, who entertained the idea of separating the theory of motion from mechanics in general. In a well-known memoir of 1775, Euler wrote:

---

[21] Lazare N. M. Carnot, *Essai sur les machines en général*, nouvelle édition (Dijon: Defay, 1786), p. iij.

[22] Jean Bernoulli, *Discours sur les loix de la communication du mouvement: qui a merité les Eloges de l'Academie Royale des Sciences aux années 1724. & 1726. & qui a concouru á l'occasion des Prix distribuez dans lesdites années* (Paris: Chez Claude Jombert, 1727).

The investigation of the motion of a rigid body may be conveniently separated into two parts, the one geometrical, the other mechanical. In the first part, the transference of the body from a given position to any other position must be investigated without respect to the causes of the motion, and must be represented by analytical formulæ, which will define the position of each point of the body after the transference with respect to its initial place. This investigation will therefore be referable solely to geometry, or rather stereotomy. . . .

It is clear that by the separation of this part of the question from the other, which belongs properly to Mechanics, the determination of the motion from dynamical principles will be made much easier than if the two parts were undertaken conjointly.[23]

Others even objected radically to the notion of force itself, although we do not review that history here.

In the 1680s, a century before Carnot's comparatively minor essay on machines, Isaac Newton had developed his conception of mechanics on the idea that physical motions can be analyzed by means of the mathematical concept of force.[24] This concept served to account systematically for changes in the motion of any observed objects or processes, without requiring any certain knowledge of their physical causes. Thus, for example, whereas Newton demonstrated that the force acting to keep the moon in orbit around the Earth is the same force that is responsible for the motion of falling bodies, namely, "gravitational attraction," his physics did not disclose the causal mechanism responsible for this force. In the opening pages of his *Mathematical Principles of Natural Philosophy* of 1687, Newton cautioned readers about the sense in which he used the term *force:*

I here design only to give a mathematical notion of these forces, without considering their physical causes and seats. . . . I likewise . . . use the words attraction, impulse, or propensity of any sort towards a centre, promiscuously, and indifferently, one for another; considering

---

[23] Leonhard Euler, "Formulae generales pro translatione quacunque corporum rigidorum" (1775), *Novi Commentarii Academiae Scientiarum Petropolitanae,* vol. 20 (1776), pp. 189–207; also in *Opera Omnia,* ser. II, vol. 9, pp. 84–98.

[24] Isaac Newton, *Philosophiae Naturalis Principia Mathematica* (1687), trans. A. Motte in 1729 and rev. Florian Cajori (Berkeley: University of California Press, 1934), preface.

these forces not physically, but mathematically; wherefore the reader is not to imagine that by those words I anywhere take upon me to define the kind, or the manner of any action, the causes or the physical reason thereof, or that I attribute forces, in a true and physical sense, to certain centres (which are only mathematical points); when I speak of centres as attracting, or as endued with attractive powers.[25]

Newton's approach postponed the investigation of causes in favor of securing an understanding of the mathematical principles involved in physical processes. This approach to mechanics became so popular that many physicists came to construe the idea of force as a fundamental concept of physics. Yet, as physicists realized that motion could be described mathematically without appealing even to the concept of force, the study of motion was gradually reconstrued as a more refined field. The motions of any given object could well be described geometrically without any consideration of the mass of the object or, say, whether the object is solid or hollow, heavy or light.

A common way of understanding motion was to conceive of it as an effect produced by a force. In this interpretation, force was construed as more fundamental than movement; force appeared as the requisite cause of motion.[26] But this conception diverged from Newton's original use of the concept of force. Newton had conceived the concept of force as *distinct* from the idea of physical cause. He relied on the concept of force to account systematically for physical phenomena, that is, to deduce observed motions from mathematical principles, facilitating subsequent investigations of the possible causes of such phenomena. Unfortunately, in the investigations of Newton and his followers, many questions of the identification of physical *causes* remained unresolved. The natural inclination to think in terms of causes often led physicists to employ the idea of force as if it were actually the cause of a given phenomenon. Any production of motion, as well as any change in a body's motion, was taken to signal the action of a force. As stated before, force was often conceived as independent of motion, as though forces could exist without motion. This increasingly metaphysical concept became increasingly repugnant to certain empirically minded theorists such as Carnot.

---

[25] Ibid., p. 8.

[26] See, e.g., Siméon Denis Poisson, *Traité de Mécanique*, 2nd ed., vol. 1, Imprimeur-Libraire pour les mathématiques, no. 55 (Paris: Bachelier, 1833), p. 2.

A further reason why the concept of force was marginalized from the elementary science of motion was that it always had suffered from ambiguity. Even Newton and his contemporaries had been unsatisfied with the concept of force insofar as it described dynamic relationships between objects separated in space without indicating the physical cause of such action. The idea of an object exerting a force on another *distant* object, without the mediation of any physical agencies, seemed inconceivable. Newton asked that this idea not be ascribed to him, arguing that the notion "that one body may act upon another at a distance through a *vacuum*, without the mediation of anything else, by and through which their action and force may be conveyed from one to another, is to me so great an absurdity, that I believe no man who has in philosophical matters a competent faculty of thinking, can ever fall into it."[27] Despite Newton's own criticisms, the idea of action-at-a-distance gradually acquired an important role in physics and was sometimes introduced as a fundamental idea alongside the concepts of space, time, and mass. Regardless, the important point to make for our present line of inquiry is that we find in Newton, and even in the works of Galileo Galilei and others long before Ampère, contributions to the study of motion irrespective of forces or causes. To various degrees, Ampère's conception of kinematics stemmed from such old traditions.

Furthermore, if we turn again to the early 1800s, we find also proximate influences that were not acknowledged in print by Ampère. So far, we have guided our archaeology of Ampère's kinematics by pursuing the traces that he himself made explicit. But we should also consider other possible sources. We may turn, for example, to the works of his peers, where we find significant coincidences. In particular, when Ampère began teaching at the Polytechnique in 1804, one of the professors of mechanics was the well-known Gaspard Clair François Marie Riche de Prony, the founding professor of mechanics there since 1795.

Prony was a dedicated engineer who had operated a school for geographers and who also actively served as director of the School of Bridges and Roads (École des Ponts et Chaussées) since 1798. Already in 1800 Prony had published a treatise concerning mechanics divided into various parts. He there advanced what he considered a new plan of exposition of the "philo-

---

[27] Isaac Newton, letter to Bentley, 28 February, 1692/3, in *The Correspondence of Isaac Newton*, vol. 3, *1688–1694*, ed. H. W. Turnbull (Cambridge: Cambridge University Press, 1961), p. 254.

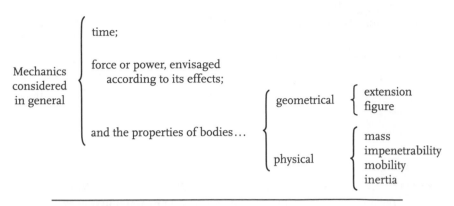

Figure 18. Prony's partitions of philosophical mechanics

sophical" part of mechanics, based on his previous experience of teaching the subject for five years to engineering students at the École. Prony formulated his division of the subject matter of mechanics in a diagram, as conveyed in figure 18. Then he began his treatment of the elementary concepts of mechanics not by considering the combination of forces in no time (as one would proceed in statics) or by the combination of forces in time (as in dynamics), but by considering motion irrespective of forces—specifically, he proffered, "let's consider first *mobility* and *time,* and let's occupy ourselves with questions concerning the isolated movement of a point, abstraction made of the causes that produce the movement, and also not considering in the line traversed nothing but the length of that line, without any regard to its curvature."[28] Thus he began with the spaces traversed by moving points in specific intervals of time. Subsequently, he added consideration of *force, mass, impenetrability,* and *inertia* of bodies; and only afterward he added consideration of the properties of *extension* and *figure* of bodies. At first sight, Prony's approach is reminiscent of Ampère's conception of kinematics, to an extent. As with other predecessors, though, we can identify differences—for example, that Prony first treated moving mathematical points and then soon attributed the properties of force and impenetrability to them, rather than first dealing extensively with the study

[28] Journal de l'École Polytechnique, Tome III, septième & 8me cahiers contenant les trois premières parties d'un ouvrage du C.en Prony, *Mécanique Philosophique, ou l'Analyse Raisonée des Diverses Parties de la Science de l'équilibre et du Mouvement* (Paris: Imprimerie de la République, 1800), p. 3.

of the observable motions of ordinary bodies, like machines, as Ampère would have it.

Nevertheless, if perhaps it seems that the key elements of Ampère's conception all derived from the contributions of others, then what was original in his conception, aside from bringing such notions together in a particular way?

For one, is it merely that he gave the name kinematics to the science of the geometry of motion? Even in this regard we find precedents. In particular, we have been looking at works in mechanics, but by considering again works in the classification of knowledge, we find a suggestive expression in one of the contemporary works that Ampère did not deign to mention in his *Essai*. Remember that Ampère said he derived the term *cinématique* from the Greek word for motion κίνημα. But already in 1823, in a work in French on the classification and nomenclature of the arts and sciences, George Bentham had used a similar term, *cinésioscopique,* based on the Greek words κίνησις, for "movement," and σκοπέω, for "I observe."[29] Bentham, an English botanist, had based his work on a classification scheme formulated by his uncle, the prominent jurist Jeremy Bentham (yes, the one who had his own dead body embalmed and displayed in a glass box, so that you may see him, even today, as a statue of himself, in a hallway in University College, London). The combination of concepts of motion and observation, in one Greek neologism, within a classification of the sciences, suggests a certain kinship with Ampère's conception. Indeed, Ampère used several of the same unusual terms as Bentham, such as *arithmology* and *noological sciences*. And both their schemes were based on bifurcating divisions. However, we again find key differences. For example, Bentham used his term to refer to the observation of the movements of humans and animals, as part of the study of the organs of living animated bodies, that is, a branch of contemporary zoology—natural history rather than mechanical philosophy of nature.

At any rate, the more sources we consider in the history of the study of motion, the less does it seem that Ampère was the originator of a new science. The same is true in many other sciences. If we choose one particular aspect of Ampère's conception as the key element that distinguished his

---

[29] George Bentham, *Essai sur la nomenclature et la classification des principales branches d'art-et-science. Ouvrage extrait du Chrestomathia de Jérémie Benthum* (Paris: Bossange Frères, Libraires, 1823), p. 113.

conception, we find that other individuals had advanced similar notions earlier. By contrast, if we decide that to designate anything as kinematics it *must* include strictly and exclusively all the elements that Ampère included in it, we will find that no earlier writers had that exact conception. Moreover, we will find that scarcely any *later* physicists adopted that exact conception either. The same is true in many other sciences; the reputed originators of a science actually defined it as something significantly distinct from what later writers, practitioners, and teachers took it to be. In the case of kinematics, we will see that after Ampère various theorists construed it in conflicting ways, sometimes giving the impression that they were not even dealing with the same subject. But it was this very plasticity in the definition of a field of study that underlay the origin of kinematics itself. Because, after all, Ampère was just one more scientist, in a long tradition, who took a preexisting science, the study of motion, and tried to redefine it in a way that would assist its continued development.

To conclude this chapter, we highlight three points from the preceding discussion. The first point concerns the creation of new sciences or fields of study. The works of Ampère, Carnot, and Lanz and Betancourt illustrate a juncture in the development of a science, when, despite its apparent maturity as judged by most specialists, some individuals reconsider it from a perspective that makes it seem as though it has hardly developed at all. From that change of perspective, scientists realize that they can develop a new systematic analysis of certain matters. In particular, by restricting the science of motion to the study of motions irrespective of forces, scientists found that they could extend knowledge in novel ways.

The impression that one subdiscipline of a science is more fundamental than another has served sometimes to justify its systematic development. Thus, we find historically that some scientists advanced a line of inquiry by claiming its independence from certain elements of science. But we can also set aside the question of which discipline is more fundamental, and simply attempt to extend a science by focusing on the interrelations of just certain elements. For example, consider again Prony's division of mechanics, as illustrated in figure 18. If we take every listed element or property as separable from the others, we can then proceed to formulate specialized lines of inquiry. To formulate a sort of kinematics, perhaps, we can develop systematically the possible relationships among *figures, mobility,* and *time.* To develop a statics, we can develop systematically the interrelations among *mass* and *forces*. To develop a kind of dynamics where we study moving points that do not collide, we can focus on the concepts of *mobility, time,* and *forces,* whereas if we want to formulate a theory that involves collisions, we will consider also *impenetrability.* We may also include *extension,* to study the dynamics of bodies rather than of mere mathematical points. The important thing to realize is that the exclusion of some concepts in theoretical analysis can lead not only to the solution of specific problems where some factors are unknown but also to the development of entire branches of science.

Of course, the inclusion of many concepts can also often help the development of science, but the point is that the divisions that were traced by theorists were not frivolous but served the goal of understanding the possible kinds of interrelations among specific factors or elements of study. If you try to isolate combinations of a science's basic elements, you may find that certain kinds of possible or actual interrelations among those selected elements remain to be investigated systematically.

A second point to highlight from the historical account is the question of the significance of the role of the concept of force in the foundations of physics. After all, you may wonder, Who cares and what difference does it make whether we regard force as more fundamental than motion or vice versa? Do physicists nowadays bother about such things? Were past physicists just wasting their time debating such questions? The historical dimension of such questions calls for a sympathetic effort to understand why they were granted great importance for centuries. My narrative focuses on questions about convenience in the organization and teaching of knowledge. For some writers it seemed easier to study motion irrespective of forces first, because at least motion could be measured by observations. But there is another matter, that is, the question of whether the concepts that are employed in physics refer to anything that actually exists in the world. To most physicists, if not all, it seemed clear that concepts of motion actually described real changes observed to transpire in the relations among objects. But the concept of force seemed more ambiguous. Of course, one could always experience pressure, violent impacts, and the like. But what if such sensations were all constituted merely by the invisible motions of very small particles in our bodies? What if force were just a fiction, a word used to name the unexplained bodily sensations we experience? Or, in scientific theories, what if the mathematical concept of force were just a fictitious construct devised to help facilitate a systematic accounting of motions?

Nowadays, in elementary textbooks, force is often defined as the product of mass and acceleration, $F = ma$. If we imagine, as did some physicists in the distant past, that mass is a concept that represents merely the quantity of matter in a body, for example, not necessarily its weight but the number of molecules in it, and we now arbitrarily multiply that number by the acceleration of the body, does the product represent anything more than just matter in motion? Or is force a property or agency that exists independently of matter? What if we multiply an acceleration by the *thickness* of the body in question? Does that constitute a new kind of thing that really exists in nature, or is it just an artificial concept? Or what if we divide the speed of a body by its mass and multiply that by its volume?

The question then is whether we believe that various concepts in physical science represent actual things or physical relations *to various degrees,* or whether we believe that all concepts are exactly and equally valid in that they all represent things and relations that are not merely imaginary. Thus, when *some* physicists entertained doubts about the concept of force, it was not merely that they thought it would be easier to begin the study of mechanics without appealing to forces, but also that perhaps mechanics would be based on more physical grounds if it were based on concepts that at least had a seemingly close relation to experience.

The way to engage past physicists' concerns is to seriously ask the questions that they asked. Does the concept of force represent anything other than matter in motion? Can there be force where there is no matter? Or, if you prefer to pursue historical questions, To what extent did the reservations voiced by past physicists regarding forces in the foundations of mechanics reflect disbelief in the independent existence of forces?

Finally, consider a third point to highlight from the historical account in this chapter. Doubtless, a noteworthy point is the recurring presence of a certain institution in the emergence of kinematics: the École Polytechnique. It was founded by Monge and Carnot, both of whom made seminal contributions in what their contemporaries came to construe as kinematics. It was the institution that commissioned and published the texts of Hachette, Lanz and Betancourt, and Prony. It was the institution where Ampère began his teaching career in 1804. Soon after Ampère arrived at the Polytechnique, Napoléon Bonaparte militarized the school in November 1804 and declared himself emperor a month later. The Polytechnique became highly bureaucratic, and Ampère taught students who were required to wear uniforms, march to class, and carry out daily drills. It was an institution that conjoined abstract theoretical frameworks and technical material practices. Its forceful blend of the abstract and the concrete was implanted from the outset, when the leading mathematician Lagrange was appointed as founding professor of analysis, while the engineer Prony was appointed as founding professor of mechanics. The Polytechnique was mainly a school for educating civilian and military engineers, which emphasized the importance of practical scientific knowledge despite an intensely mathematical curriculum. Is it not apparent that to some extent the mixture of conceptions that Ampère called kinematics stemmed from the peculiar nature of the Polytechnique?

CHAPTER THREE

Ambiguous Truths

The Allegedly Pure Science of Motion

Ampère died in 1836. Had he lived longer, he may have been pleased to see that from 1838 to 1840, the geometer Jean-Victor Poncelet, who had been his student at the Polytechnique, tried to carry out his conception of the science of kinematics in his lessons at the Faculté de Sciences of Paris. Poncelet applied the theory of the principal means of transmission of motion to the devices that continually indicate or trace motion, proper for experimental observations.[1] Meanwhile, Robert Willis, a professor of the University of Cambridge, wrote a book titled *Principles of Mechanism*, first published in 1841, where he analyzed a great variety of mechanisms exhibiting how they serve to transform motions, irrespective of the concepts of force and work. Willis employed the term *mechanism* to designate all considerations about machinery concerning only relations among motions, and he echoed Ampère's impression of the relative neglect of kinematics: "Machinery as a modifier of force, has in the science of Mechanics occupied the attention of nearly every mathematician of eminence who has arisen in the world; but, by some strange chance, very few have attempted to give a scientific form to the attractive and valuable results of mechanism."[2] Willis appreciated contributions such as the work of Lanz and Betancourt, but he noted that they did not go far beyond a descriptive inventory of the kinematic elements of machines. Following Ampère's prescription, Willis argued that the study of mechanical motions should be developed further, to make it into a science,

---

[1] See, e.g., Jean Victor Poncelet, *Mécanique Industrielle*, 2 parts (Brussels, 1839), and Poncelet, *Introduction à la mécanique industrielle, physique ou expérimentale*, 3rd ed. (Paris: Xavier Kretz, 1870).

[2] Robert Willis, *Principles of Mechanism*, 2nd ed. (London: Longmans, Green, 1870), p. v.

by formulating the mathematical laws that would enable the systematic investigation and calculation of the relations and combinations of motions. That was the plan he carried out in his *Principles.*

Divergence: Engineers, Mathematicians, and Physicists

Willis introduced several important refinements to the plan advanced by Monge, Ampère, and others. Rather than defining "machine" as an instrument that changes the direction and velocity of *a given motion,* Willis treated it as an instrument that produces *relations of motion* between two parts. Willis realized that, for the most part, the connection between the parts of a machine, and thus the mathematical "law" that relates them, are independent of the *actual* motions of such parts but do depend on the relative velocities between them. Remember, Ampère had illustrated the kinematic study of mechanical motion by means of an example concerning a clock: the conjoined minutes pointer and the hours pointer move in constant proportions regardless of whether they are moved by a motor mechanism or by one's finger. Willis took this example to illustrate further that this clock, ordinarily described as a device that converts one continuous circular motion into another continuous circular motion, could also double as a device that converts alternating circular motion into alternating circular motion, if only we move one of the pointers back and forth. Thus, the same device would occupy two places in the system of Lanz and Betancourt. Yet Willis showed that in his system this device would occupy a single place, namely, an example of a relation between two mechanical parts (the pointers) where their velocity ratio and their directional interrelation are constant. Both pointers move in the same direction at any given moment and the proportions of their displacements are constant regardless of whether the pointers move in one circular direction or back and forth. In this way, Willis eliminated redundancies that otherwise encumbered treatments such as that of Lanz and Betancourt.

Moreover, rather than espouse the classification of machines by whether motions are rectilinear, or circular, or alternating, Willis based it primarily on the velocity ratio (constant or varying) and the directional relation (constant or varying). However, in a second edition of his *Principles,* Willis modified his classification to be based primarily on the modes of communication of motion, that is, the kinds of contact, such as rolling or sliding, that connect the parts of a contrivance. At any rate, Willis's account involved a clear separation of the notions of force, time, and motion, such that motions could be analyzed without reference to measures of force and time.

Other books on kinematics followed. Such early works on kinematics concerned mainly the study of machinery, that is, the field that became known as industrial kinematics. Promptly in 1841, the same year as Willis's book was first published, William Whewell published his *Mechanics of Engineering,* where he intermingled Willis's system on the classification of the communication of motion with dynamical considerations (e.g., on the effects of forces and resistance in machines). Later, other books appeared, incorporating Willis's kinematic scheme to various extents. Among such works, published before the appearance of Willis's second edition in 1870, were the following:

Tom Richard, *Aide-Mémoire des Ingénieurs* (1848)

Charles Laboulaye, *Traité de Cinématique* (1849)

Thomas Tate, *The Elements of Mechanism* (1851)

Thomas C. E. Baker, *Elements of Mechanism* (1852)

William John Macquorn Rankine, *A Manual of Applied Mechanics* (1858)

Charles François Girault, *Éléments de géométrie appliquée à la transformation du mouvement dans les machines* (1858)

Thomas Minchin Goodeve, *Elements of Mechanism* (1860)

William Fairbairn, *Treatise on Mills and Mill-Work; Part 1: On the principles of mechanism and on prime movers* (1861)

Julien N. Haton de la Goupillière, *Traité des mécanismes, renfermant la théorie géométrique des organes et celle des résistances* (1864)

Jean Baptiste C. J. Bélanger, *Traité de Cinématique* (1864)

J. Edmond E. Bour, *Cours de Mécanique et Machines* (1865)

William John Macquorn Rankine, *A Manual of Machinery and Mill-work* (1869)

Several of these books were promptly reissued in subsequent editions, such as those of Baker, Fairbairn, Laboulaye, and Goodeve. Yet this tradition did not focus on the theoretical part of kinematics, the part that pertained to fundamental physics more than to its applications. Willis, for example, acknowledged that his treatise did not aim to carry out entirely "the able and comprehensive views of Ampère," because he confined it to deal exclusively with machinery and not "the more abstract generalities of motion."[3] Likewise, Laboulaye celebrated the works of Ampère, Willis, and others, and he noted, "Nowadays the notions of kinematics are professed widely," as scientists rushed to fill the visible "lacuna of kinematics in the structure of science," yet he too focused on the theory of mechanisms, clockwork, and practical machinery.[4]

Meanwhile, a few writers chose to begin their textbooks on mechanics with general discussions of kinematics, following Poncelet, as did, for example, Barré de Saint-Venant in 1851.[5] Also, some writers on practical mechanics wrote elementary presentations of theoretical kinematics. One such writer was Arthur Jules Morin, a pupil of Poncelet and graduate of the Polytechnique, where he too had been a student of Ampère.

In the 1830s, Morin worked and wrote about advances in mechanical engineering at Metz—mainly about friction, the transmission of motion by impact, and the use of projectiles to penetrate uneven media. He also wrote about hydraulic wheels, turbines, and rules describing the movements of water. In the late 1830s, Morin was also commissioned by the Ministries of War and of Public Works to study the destructive effects of dragged vehicles on roads. Such various experiences placed him in a position to produce general surveys of practical mechanics. In 1837 he published a manual on practical mechanics for use by artillery officers and by civil and military engineers, which became very popular.[6] He also wrote a brief study on chronometric devices useful for the description of motions.[7] As part of his series of *Lessons on Practical Mechanics*, Morin wrote a textbook titled *Geometrical Notions on Movements and their Transformations, or, Elements of Kinematics.*

---

[3] Ibid., p. xiii.

[4] Charles Laboulaye, *Traité de cinématique ou théorie des mécanismes* (1849); rev. ed. (Paris: E. Lacroix/H. Plon, 1861), p. xv.

[5] Adhémar Jean Claude Barré de Saint-Venant, *Principes de mécanique fondés sur la cinématique* (Paris: Bachelier, 1851).

[6] Arthur Morin, *Aide-mémoire de mécanique pratique à l'usage des officiers d'artillerie et des ingénieurs civils et militaires* (Metz: Mme. Theil, 1837). This work was promptly reissued: *Hilfsbuch für praktische Mechanik,* translated into German by C. Holtzmann (Karlsruhe: Groos, 1838). By 1871 six augmented French editions had been published.

[7] Arthur Morin, *Description des appareils chronométriques à style, propres à la présentation graphique et à la détermination des lois du mouvement, et des appareils dynamométriques propres à mesurer l'effort ou le travail développé par les moteurs animés ou inanimés et par les organes de transmission du mouvement dans les machines* (Metz: S. Lamort, 1838).

In this work, Morin noted that the study of motions from a geometric point of view was of great interest for science and industry and that it should comprise an immensity of mechanisms. It consisted of elements and practical methods that should be known before commencing properly the study of mechanics. Morin construed kinematics as "the mechanics of the laborer," and he deemed that the study of forces, work, and its effects formed the science of the engineer. Morin began his account of kinematics by discussing the measurement of time by old devices such as the clepsydra and the pendulum or by other chronometric apparatus.[8] Then he discussed motion in general before proceeding to discuss mechanical devices for guiding motions, such as tracks, pulleys, and levers.

In Germany, the kinematics of machines advanced only later, thanks finally to the labors of Franz Reuleaux. He had become interested in the principles of machine design while he was studying at the Polytechnic in Karlsruhe, where he learned of the urgency of systematizing the elements of mechanism from Professor Ferdinand Redtenbacher. Following further studies in Berlin and Bonn, Reuleaux became an engineer and professor of machine design at the Polytechnic in Zurich, Switzerland, in 1856. There he formulated innovative ideas on the kinematics of machinery. He began to lecture on kinematics at Zurich in 1864, but in 1865 he accepted a position at the Royal Industrial Academy in Berlin. He used the word *kinematics*, like Willis's term *mechanism*, to mean the science of the constrained motions of machines, without reference to measures of time or force. He published his systematization of mechanisms in his so-called *Theoretical Kinematics* of 1875.

Although Reuleaux wrote about mechanisms, he approached the topic from the theoretical side, seeking the geometric essentials and commonalities in machines. He argued that "in any department thoroughly elucidated by Science the truly practical coincides with the theoretical, if the theory be right."[9] Among his findings were the following. The elements of machines

---

[8] Arthur Morin, *Leçons de Mécanique Pratique. Notions Géométriques sur les Mouvements et leurs transformations, ou, Élements de Cinématique*, 2nd ed. (Paris: L. Hachette, 1857), pp. 1–6.

[9] Franz Reuleaux, *Theoretische Kinematik* (1875), reissued as *The Kinematics of Machinery: Outlines of a Theory of Machines*, translated by Alexander Blackie W. Kennedy (London: Macmillan, 1876); reissued, with a new introduction by Eugene S. Ferguson (New York: Dover Publications, 1963), p. 3.

always act in pairs, never singly (and so he claimed that machines should not be defined as consisting of elements but rather of pairs of elements). He effectively divided all kinds of mechanical components into six classes, a classification scheme that became widely followed. Reuleaux emphasized the inner simplicity of the elements of machinery and the analytical principles that apply in virtually any mechanical device. For example, he showed that pairs of elements, when joined in a closed sequence, constitute a "kinematic chain" that can be formed into various mechanisms in precisely as many ways as there are pairs of elements (he called each pair a link), by making any chosen pair stationary. He showed also that the fixed link of any mechanism, its pedestal, is kinematically identical to the moving links, from which it followed that linkages can be inverted, by successively fixing different links in order to change the function of the given mechanism. Hence, he usefully classified mechanisms as chains of linked components. To demonstrate the value of his classification scheme, he correctly showed, for example, that several rotary steam engines that had been independently patented as distinct machines were actually based on the same kinematic chain.[10] By exposing such principles, Reuleaux expected that mechanists would better be able to invent new machines by using theory rather than by haphazard, laborious trials.

Like Betancourt, Reuleaux also assembled a collection of hundreds of models of various machines, his "kinematic collection," which was kept mainly at the Industrial Academy in Berlin. In 1879 he helped to found the Royal Institute of Technology in Berlin, where he remained until he retired in 1896. His approach to the kinematics of mechanisms is still employed today.

While mechanists studied mechanisms, mathematicians developed the abstract study of describing interrelations among moving geometric figures. Researchers on kinematics labored to expand the number of theorems available for the analytic description of motions irrespective of forces. We may mention a few such contributions. In 1841 Poncelet introduced the notion of "geometrical accelerations" for the description of curvilinear mo-

---

[10] For discussion, see Eugene S. Ferguson, "Kinematics of Mechanisms from the Time of Watt," in *Contributions from the Museum of History and Technology, United States National Museum, Bulletin 228*, paper 27 (Washington D.C.: Smithsonian Institution, 1962), pp. 185–230.

tions.[11] Abel Transon studied the movements of plane figures.[12] J. A. Charles Bresse developed additional ways of describing the curvilinear motion of plane figures.[13] M. Rivals established a theorem concerning the acceleration of the motion of a solid body around a fixed point.[14] These contributions demonstrated that Coriolis's theorem on the apparent force acting on a body viewed from a rotating system is only a kinematic effect, as clarified further by Henri Resal.[15] The latter also contributed theorems on the rolling and sliding of solid bodies, on the accelerations in the most general motions of an invariable system, on the relative motion of a point and a solid body, and generally about displacements of third order (measured by what he called *suraccélération*).

Henri Resal was a physicist and engineer of mines who invested much effort to the systematic development of kinematics. He was particularly interested in the more theoretical branch of kinematics, to which he gave the name *cinématique pure,* describing it as the science that deals with geometric movements independently of its application to the study of machines. In his *Treatise of Pure Kinematics* of 1862, he presented a collection of the principles and theorems of kinematics, formulated in a self-sufficient arrangement that would facilitate their employment in mechanics and geometry. Resal envisaged pure kinematics as the study of the possible motions of geometric figures, and he therefore conceived its employment in the study of material systems, mechanics, as a question of application.

---

[11] Henri Resal elucidated this notion further in his *Éléments de Mécanique* (Paris: MM. Didot frères and Mallot-Bachelier, libraires, 1851), additions.

[12] Abel Étienne Louis Transon, "Note sur les principes de la mécanique," *Journal de Mathématiques pures et appliquées* (*Journal de M. Liouville*), ser. 1, **10** (1845), 320–326.

[13] J. A. Charles Bresse, "Mémoire sur un théorème nouveau concernant les mouvements plans et l'application de la cinématique à la détermination des rayons de courbure," *Cahier du Journal de l'École Polytechnique,* vol. 20 (Paris: Imprimerie de la République, 1853), pp. 89–115. See also J. A. Charles Bresse, *Cours de mécanique et machines, professé à l'École Polytechnique: I, cinématique, dynamique du point matériel, statique; II, dynamique des systèmes matériels en général, mécanique spéciale des fluides, étude des machines à l'état de mouvement* (Paris: Gauthier-Villars, 1885).

[14] M. Rivals, in Bresse, *Cours de mécanique et machines* , appendix, p. 109.

[15] Resal, *Éléments de Mécanique,* additions. In 1835 Gaspard Gustave de Coriolis showed that the laws of motion apply in any rotating frame of reference if the equations of motion are supplemented with an acceleration term (now known as the Coriolis acceleration). Coriolis presented it as an additional force, though it is not actually a physical force but a descriptive formal by-product of the rotating reference frame.

Resal commented that the study of mechanics involved "serious difficulties" from the outset because of unverified hypotheses and posited principles upon which it was made to rest. Mechanics was based on the "considerably vague ideas" of mass and the causes of motion. Hence, in order "to not fall completely into obscurity," Resal explained, physicists recurred to "the fiction" of a single isolated particle of matter supposed to be devoid of dimensions, like a geometric point, while ascribing to it the physical properties of matter. In contradistinction, Resal argued that the laws of motion are really quite independent of the notions of force and mass, that some of the properties of motion of material systems can be disengaged from the specific related hypotheses. Accordingly, he argued that one should naturally divide mechanics into two parts: "one rational and purely geometric, where we study movement in itself; the other essentially physical, resting on the notions of force and mass." He ascribed this view to Newton himself and claimed that in teaching mechanics it is more logical to begin by studying the geometric properties of motion and then to pass on to the question of forces.

Newton had divided his *Mathematical Principles of Natural Philosophy* of 1687 into three "books." The first dealt with the motions of bodies in the absence of resistance, the second analyzed motions in resisting fluid media, and the third analyzed the motions of heavenly bodies. All throughout, Newton wrote about forces. But despite his verbal expressions, the first ten sections of Book 1 hardly involved his mathematical concept of motive force. Only accelerations were involved, that is, mainly accelerations directed toward fixed geometric centers, but with virtually no consideration of masses. In those first sections of his Book 1, Newton did not use accelerations to infer forces. Therefore, to an extent, Resal was right to claim that Newton first treated motion almost purely geometrically. Newton first analyzed rectilinear, curved, free, and constrained motions without calculating or inferring forces from such motions. Nonetheless, Newton yet alluded to the notion of forces all throughout; he even formulated the "laws of motion" in terms of the concept of force. In contradistinction, Resal and others sought to systematically formulate kinematics without relying on the notions, language, or mathematical concepts of force.

But not all physicists adopted the idea that motion was more fundamental than force. All throughout the 1800s numerous physicists continued to espouse the notion that force is prior to motion. An example commonly used to show the primacy of force was that of "muscular action impeded," as theorists claimed that this experience "gives us our primitive idea of force; our sense of muscular exertion itself is a primary one for which we have special nerves, and is not resolved into anything simpler."[16] Physical forces, at least as exerted by human beings, could be construed as resulting from the mind's will, and such forces could be counteracted without visibly producing motion, only a sense of difficulty and fatigue, for example, "as when we press our two hands together." Thus argued John Herschel in his *Preliminary Discourse on the Study of Natural Philosophy* of 1840. Herschel illustrated the idea of force without motion by writing: "Were there no such thing as motion, had we been from infancy shut up in a dark dungeon, every limb encrusted with plaster, this internal consciousness would give us a complete idea of *force*."[17] Nevertheless, Herschel and others realized that in the investigation of physical phenomena, even a phenomenon as simple as motion, the search for underlying causes presented grave difficulties. For instance, in the preceding example, it was clear that the inner workings of intentional human action remained quite mysterious. So, in place of a search for causes, some scientists advocated investing greater effort into the inquiry of natural laws, to at least resolve complex phenomena into simpler generalities.[18] In this connection, it seems remarkable that the concept of force had acquired such prominence that, despite its abstract character, it was used to explain an aspect of nature as elementary as motion.

Notwithstanding competing trends in the analysis of the nature of motion, Ampère's account gradually won acceptance among physicists. In England, it gained circulation owing to the influence of *Treatise on Natural Philosophy*, written by William Thomson and Peter Guthrie Tait, first published in 1867 and soon esteemed as an authoritative text. Like Ampère, Thomson and Tait distinguished between the study of motion with and without regard to the notion of force by using the terms *dynamics* and *kinematics*.[19] The science of dynamics was meant to investigate the relation-

[16] Oliver Lodge, *Elementary Mechanics; including Hydrostatics and Pneumatics* (London and Edinburgh: W. & R. Chambers, Ltd., 1896), p. 2.

[17] John Frederick William Herschel, *A Preliminary Discourse on the Study of Natural Philosophy*, "a new edition" (Philadelphia: Lea Blanchard, 1840), p. 66.

[18] See, e.g., ibid., pp. 67–69.

[19] See also the account by William Kingdon Clifford, in "Instruments Illustrating Kinematics, Statics, and Dynamics," in *Handbook to the Special Loan Collection of Scientific Apparatus* (1876); reprinted in Clifford, *Mathematical Papers*, ed. Robert Tucker, introd. H. J. S. Smith (Bronx, N.Y.: Chelsea, 1968), pp. 424–426.

ship between forces and motion, involving not only the study of how force produces or affects motion, *kinetics*, but also the study of how forces compel rest or prevent motion, *statics*.[20] By contrast, kinematics was conceived as the study of motion with regard neither to forces nor to any physical characteristics of the moving bodies (such as mass, chemical constitution, elasticity, and temperature). Kinematics appeared thus as the "pure" study of the geometry of motion.[21] Karl Pearson, for example, in his *Grammar of Science* of 1892, presented a classification of the sciences where he grouped kinematics with the "abstract sciences" of logic and mathematics. He argued that these sciences relied on "conceptual modes of discrimination," which in general were "narrowly defined and free from the infinite complexity of the contents of perception," such that their results "are absolutely valid for all that falls under their axioms and definitions."[22]

In Germany, increasingly many physicists also came to accept the idea that kinematics was the most fundamental science of physics. The word *kinematics* itself was not adopted as quickly in German-speaking countries as in France. In 1844 the mathematician Hermann Grassmann referred to "the pure theory of motion" as *phorometry* and claimed that it could be categorized "among the mathematical sciences with as much justice as geometry."[23] The word kinematics appeared in the title of Franz Reuleaux's book on mechanism, in 1875, as well as in a handbook written by Julius Petersen and published in German in 1884.[24] Reuleaux made considerable ef-

---

[20] William Thomson and Peter Guthrie Tait, *Treatise on Natural Philosophy*, vol. 1, new ed. (Cambridge: Cambridge University Press, 1879), pp. vi, 1, 219. See also William Thomson and P. G. Tait, *Elements of Natural Philosophy*, vol. 1 (Oxford: Clarendon Press, 1873).

[21] Thomson and Tait, *Elements of Natural Philosophy.*, vol. 1, p. 52.

[22] Karl Pearson, *The Grammar of Science*, part I: *Physical*, Contemporary Science Series, (London: Walter Scott, 1892); 3rd ed. (London: Adam and Charles Black, 1911), and also *Everyman's Library*, no. 939 (London: J. M. Dent & Sons, 1937), pp. 323–324. See also J. G. MacGregor, *An Elementary Treatise on Kinematics and Dynamics* (London, 1887).

[23] Hermann Grassmann, *Die lineale Ausdehnungslehre, ein neuer Zweig der Mathematik* (Leipzig: Otto Wigand, 1844); *A New Branch of Mathematics, The Ausdehnungslehre of 1844 and Other Works*, trans. Lloyd C. Kannenberg (Chicago/La Salle: Open Court, 1995), p. 25.

[24] Reuleaux, *Theoretische Kinematik*; Julius Petersen, *Kinematik*, German edition by R. von Fischer-Benzon (Copenhagen: Andr. Fred. Höst & Sohn, 1884). The book was originally written in Dutch: Julius Petersen, *Kinematik lehrbuch der statik fester kosper* (Copenhagen: Host, 1884). Another relatively rare German title is Ferdinand Wittenbauer's

forts to claim the term kinematics as proper for the theoretical study of machines. Meanwhile, he proposed that the word *phoronomy* (Phoronomie), an ancient term having roots even in Aristotle's works, be used to designate the geometric part of mechanics, that is, "the study of the measurement of the motions of bodies of every kind," especially those motions which occur through latent forces conditioned by the geometric properties of the pieces transmitting such forces.[25] He did not succeed in disseminating his definitions widely, yet the word phoronomy continued to be used in Germany partly because it was sometimes used at the Polytechnic schools to name the elementary study of the motions of a point. Later, some German theorists continued to avoid the word kinematics. For example, Ludwig Lange employed the term phoronomy when he divided the science of motion into the study of its pure geometry, and dynamics.[26]

At any rate, the ideas advanced in Ampère's plan were readily propagated. In 1868 Ernst Mach argued that the science of mechanics should be based essentially on the relation of measures of space and time, such that the concepts of force and mass would be reduced to purely mathematical expressions.[27] Gustav Kirchhoff, furthermore, advocated the idea that the *description* of motion is the fundamental aim of the whole science of mechanics.[28] Kirchhoff set forth his approach in definitive form in his monumental *Textbook on Mechanics,* first published in 1874, and again in 1876 as part of his *Lectures on Mathematical Physics.*[29] In his preface, Kirchhoff explained that he chose to deviate from the usual presentations of mechanics to remove from the subject the obscurities involved in the notions of force

Kinematik des Strahles (Graz, 1883). See also L. Burmester, *Lehrbuch der Kinematik,* vol. 1 (Leipzig, 1888).

[25] Reuleaux, *Theoretical Kinematics,* p. 56.

[26] Ludwig Lange, *Die Geschichtliche Entwickelung des Bewegungsbegriffes und Ihr Voraussichtliches Endergebniss. Ein Beitrag zur Historischen Kritik der Mechanischen Principien* (Leipzig: Wilhelm Engelmann, 1886), p. 72.

[27] Ernst Mach, "Ueber die Definition der Masse," *Carl's Repertorium der Experimentalphysik* 4 (1868); reprinted in Mach's *Erhaltung der Arbeit,* 2nd ed. (Leipzig, 1909). For discussion, see Max Jammer, *Concepts of Force: A Study in the Foundations of Dynamics* (New York: Harper and Brothers, 1962).

[28] Cf. Felix Klein, *Vorlesungen über die Entwickelung der Mathematik im 19 Jahrhundert,* Teil I (Berlin, 1928); translated by M. Ackerman, *Lie Groups: History, Frontiers and Applications,* vol. 9 (Brookline, Mass.: Math Sci Press, 1979), pp. 206–207.

[29] Gustav Kirchhoff, *Lehrbuch der Mechanik* (1874); *Vorlesungen über Mathematische Physik; Mechanik* (Leipzig, B. G. Teubner, 1876), 3rd ed. (Leipzig: B. G. Teubner, 1883).

and cause as well as the corresponding traditional definitions of mechanics as the science that investigates them.[30] Instead of presupposing forces as the cause of physical motions, Kirchhoff based his mechanics on the pure mathematical study of the spatial and temporal relations of matter.

Kirchhoff's approach was very influential in leading the German physics community to displace the concept of force from its traditional role in mechanics. Consider, for example, Heinrich Hertz's *Principles of Mechanics* of 1894. Hertz deemed it necessary to address the question of the causal relationship between force and motion. He recognized a difficulty in physicists' understanding of force as the cause of motion, on the one hand, and of motion as producing forces, on the other. If the idea that forces are the cause of motions was accepted, then, Hertz asked: "Can we, without confusing our ideas, suddenly begin to speak of forces which arise through motion, which are a consequence of motion?"[31] To resolve this apparent circularity in both a logical and causal way, Hertz proposed a novel statement as the "single fundamental law" of mechanics: "Every natural motion of an independent material system consists herein, that the system follows with uniform velocity one of its straightest paths."[32] He admitted that this statement condensed the ordinary law of inertia and Carl Friedrich Gauss's principle of least constraint into one law. From this proposal, he set out to derive the whole of mechanics in such a way that the concept of force appeared "not as something independent . . . but as a mathematical aid."[33] In this theoretical scheme, the motion of a body was determined by the motion of the other bodies in the given system. And a "force" entered simply as a matter of convenience in dividing the determination of the motion into two parts: "We thus say that the motion of the first body determines a force, and that this force then determines the motion of the second body."[34] This approach dissolved the dilemma of the causal relationship between force and motion, as any force could therefore "with equal justice be regarded as always being a cause of motion, and at the same time a consequence of mo-

---

[30] Kirchhoff, *Vorlesungen über Mathematische Physik; Mechanik,* 3rd ed. (1883), p. iii.

[31] Heinrich Hertz, *Die Prinzipien der Mechanik* (Leipzig: J. A. Barth, 1894), preface by Hermann von Helmholtz; authorized English translation by D. E. Jones and J. T. Walley as *The Principles of Mechanics, Presented in a New Form* (New York: Dover Publications, 1956), p. 6.

[32] Hertz, *Principles of Mechanics,* p. 27.

[33] Ibid., p. 28.

[34] Ibid.

tion," because "it is a middle term conceived only between two motions."[35] Hertz's interpretation served both to reorient the meaning of the concept of force toward Newton's original conception and to make the broad science of mechanics fundamentally dependent on the study of kinematics.

As physicists revised the elements of mechanics, the concept of forces instantaneously acting at a distance fell into disfavor. Thus, Kirchhoff elucidated his formulation of mechanics without fundamental reliance on the concept of force. He wanted to remove obscure questions of forces and causes from mechanics to conceive this science as the simplest and most unambiguous description of the motion of bodies.[36] When he later readopted the term force, he did so "merely as an abbreviation for certain algebraic expressions that constantly occur in the description of motion."[37] Subsequently, Hertz too renounced the concept of force more definitively, as he deemed impossible any motions that could propagate through space instantaneously. Hertz's formulation of mechanics thus involved the complete rejection of the idea of instantaneous action at a distance. For Hertz this approach was only natural, because he had obtained experimental proof that electric and magnetic forces do not act directly at a distance but instead are propagated through space like light, as Michael Faraday and James Clerk Maxwell had believed. It was thus reasonable to imagine that the force of gravity likewise propagated in time through the action of a medium in space, as Newton and others had believed.

Instead of a physics of force, Hertz structured mechanics in terms of only three "independent fundamental conceptions," namely, space, time, and mass. This approach was by no means new, as Hertz had, in the words of Hermann von Helmholtz, "reverted to the oldest theoretical conceptions, which may also be regarded as the simplest and most natural."[38] As an alternative to the Newtonian scheme, this ancient corpuscular conception reemerged at a time when other, more novel, alternative approaches were

---

[35] Ibid.

[36] Kirchhoff, *Vorlesungen über Mathematische Physik; Mechanik*, p. iii.

[37] Ludwig Boltzmann, "Über die Entwickelung der Methoden der theoretischen Physik in neuerer Zeit," in *Populäre Schriften*, Essay 14: Address to the meeting of natural scientists at Munich (22 September 1899); translated as "On the Development of the Methods of Theoretical Physics in Recent Times," in Boltzmann, *Theoretical Physics and Philosophical Problems: Selected Writings*, ed. Brian McGuiness (Boston: D. Reidel, 1974), pp. 77–100, quotation on p. 88.

[38] In Hertz, *Prinzipien der Mechanik* (1894); *Principles of Mechanics*, preface.

also being advanced. For example, the concept of energy conservation had been accepted and it became increasingly appealing for physicists because they now could fashion explanations of varied phenomena in terms of energy transformations. Likewise, the field concept attracted increased attention even though physicists failed to secure dynamical representations of the invisible physical structure of, say, electric and magnetic fields. Given the usefulness of the concepts of energy and field, it seemed reasonable for physicists to attempt to reformulate all of physics, including mechanics, by appeal to such concepts as fundamental notions. Yet mechanics was not readily reformulated in terms of such novel abstractions. Hence, as the concept of force became displaced from the foundations of mechanics, the analysis of motion was carried out on the basis of spatiotemporal relationships between definite physical bodies.

Physicists could hardly systematize physics in terms of any subdiscipline other than kinematics, as the fundamental concepts of other subdisciplines seemed hopelessly obscure. Because the underlying processes of heat, gravity, electricity, and magnetism, for example, were not amenable to direct observation, physicists had deemed it necessary to postulate invisible processes or abstract principles to make sense of such phenomena. By contrast, motion seemed sufficiently simple to allow straightforward analysis. As Ludwig Boltzmann argued, "Phenomena of motion are the ones that we observe most often and most directly, all other natural phenomena are more concealed. Besides we can cope with phenomena of motion by means of the least number of concepts, all we need for describing them are the concepts of position in space and its change in time, whereas for other phenomena many less clear concepts are required, such as temperature, luminous intensity and color, electric tension and so on."[39] Hence, for centuries scientists had sought to explain complex physical processes in terms of underlying motions of particles.

The idea that the study of motion was the simplest sort of physical analysis led physicists to imagine that kinematic knowledge enjoyed a degree of truth comparable to that of pure mathematics. Traditionally, many philosophers construed mathematics as a type of knowledge the truth of which was

[39] Ludwig Boltzmann, "Über die Prinzipien der Mechanik," *Populäre Schriften*, Essay 17: Inaugural lecture, Vienna (1902); translated as "On the Principles of Mechanics, Part II," in Boltzmann, *Theoretical Physics and Philosophical Problems*, pp. 146 158, quotation on p. 149.

known with certainty independently of any appeal to physical observations or experience. In his ordering of the sciences, Ampère ranked kinematics as the principal "physical-mathematical" science, immediately following the branches of pure mathematics. Thus conceived, it seemed reasonable to suppose that kinematics enjoyed the certainty of mathematical knowledge, as it only added to pure geometry the idea of motion. But Ampère abstained from drawing this inference, because in his opinion truth in geometry itself was dependent on experience. He deemed arithmetical truths to consist simply of the identity of different expressions for equal quantities, but the theorems of geometry were apparently true by virtue of the actual properties of space.[40] Because geometers could not justify the principles of geometry rigorously, Ampère argued that "geometrical truths therefore have an objective reality that is not found in those of arithmology."[41] For Ampère, only arithmology (arithmetic combined with algebra) enjoyed the status of true knowledge susceptible to *immediate intuition*.[42] Kinematics emerged as the physical science closest to this high degree of mathematical certainty. Accordingly, later theorists felt justified in surpassing Ampère in their assertions of the truthfulness of kinematics. For instance, Thomson and Tait understood the axioms of dynamics as derived from experience, while they deemed the axioms of kinematics to be independent of any such experience. Thomson and Tait construed kinematics to be "a branch of pure Mathematics."[43] In Germany, the same idea was endorsed by Hertz, who claimed at the outset of his analysis of the geometry and kinematics of material systems that he dealt with a subject "completely independent of experience," wherein "all assertions made are *a priori* judgments in Kant's sense."[44]

Like Hertz, other physicists expressed their utter certainty concerning the foundations of scientific knowledge by asserting the "*a priori*" character of such truths. This term had been advanced by the German philosopher Immanuel Kant to describe any intuitive knowledge deemed to

---

[40] André-Marie Ampère, *Essai sur la Philosophie des Sciences, ou exposition analytique d'une classification naturelle de toutes les connaissances humaines* (Paris: Bachelier, Imprimeur-Libraire pour les Sciences, 1834), p. 66.

[41] Ibid., p. 67; translation by A. Martínez.

[42] Ibid., p. 41.

[43] Thomson and Tait, *Elements of Natural Philosophy*, vol. 1, p. 52.

[44] Hertz, *Principles of Mechanics*, p. 45.

be necessarily true, the truth of which was established independently and prior to any experience. For example, for Kant the principles of mathematics were true independently of any human perceptions; these were *a priori* truths. The study of motion then appeared to have a high degree of certainty, for although the concept of motion presupposed empirical perceptions, it was based on the integration of the *a priori* truths of the notions of *space* and *time*.[45]

According to Kant, the notions of neither space nor time derived from experience. If the notions of space or time were derived from experience, then the truth of geometric knowledge would be limited because it would depend on the extent and precision of human observations.[46] Experience, even if refined by rigorous scientific methods, did not suffice for the universal validity and necessary truth of a proposition. For Kant, *a priori* truths were known by immediate intuition, as the notions of space and time seemed necessary conditions *underlying* the very possibility of any and all perceptions.[47] This was construed as the highest degree of truth that any knowledge could attain, because any absolute knowledge of nature independent of human cognition appeared as logically impossible. Newton had asserted the existence of "absolute space" and "absolute time" independent of any relationships to anything.[48] Kant denied such "absolute" or objective existence of space and time, in favor of admitting these notions solely as fundamental realities of *only* the human intellect.

It then might appear that Kant's ideas of space and time as subjective detracted certainty from scientific knowledge. But Kant argued that this "ideality" of space and time "leaves, however, the certainty of empirical knowledge unaffected, for we are equally sure of it, whether these forms necessar-

---

[45] Immanuel Kant, *Critik der reinen Vernunft* (Riga: Verlegts Johann Friedrich Hartknoch, 1781) [later spelled: *Kritik*]; *Critique of Pure Reason*, trans. and ed. Paul Guyer and Allen W. Wood, Cambridge Edition of the Works of Immanuel Kant (Cambridge: Cambridge University Press, 1998), see "Transcendental Aesthetics," pp. 166–167.
[46] Kant, *Critique of Pure Reason* (Cambridge Edition), "Transcendental Aesthetics," e.g., see p. 158.
[47] For discussion, see Henry E. Allison, *Kant's Transcendental Idealism: An Interpretation and Defense* (New Haven: Yale University Press, 1983), e.g., pp. 10, 87–96, 182–185.
[48] Isaac Newton, *Philosophia Naturalis Principia Mathematica* (1687), *Mathematical Principles of Natural Philosophy*, translated by Andrew Motte in 1729 and revised by Florian Cajori (Berkeley: University of California Press, 1946), p. 8.

ily inhere in things in themselves or only in our intuition of them."[49] In any case, like Newton, Kant based the truth of scientific knowledge on the fundamental notions of space and time. These "realities" served as "two sources of knowledge," argued Kant, "from which bodies of *a priori* synthetic knowledge can be derived. (Pure mathematics is a brilliant example of such knowledge, especially as regards to space and its relations.)" Hence, geometry seemed to depend on the notion of space, and the inclusion of time served for the development of what was later called kinematics: he elaborated that "our concept of time explains the possibility of that body of *a priori* synthetic knowledge which is exhibited in the general doctrine of motion, and which is by no means unfruitful."[50] Thus, many scientists and philosophers followed Kant in believing that the fundamental notions involved in the study of motion enjoyed the degree of certainty of the principles of mathematics.

Although the science of motion seemed to enjoy a high degree of mathematical certainty, this did not suffice to convince physicists that this field required no further study. In the first place, as remarked by Ampère, the field of kinematics had been neglected, as writers presumably assumed that the basic ideas of physics were quite self-evident and required no elucidation. Others came to share Ampère's feelings, judging from the subsequent proliferation of publications on the principles of physics.

Besides, the invention of new geometries and algebras may have inspired a few physicists to study the science of motion. As the traditional faith in mathematical truths began to weaken, it seemed necessary to identify the *specific* mathematical principles of physics. For example, Nikolai Lobachevsky, the mathematician, regarded the notion of space as derived

---

[49] Immanuel Kant, "Transcendental Aesthetics" (1781), translation by Norman Kemp Smith, *Critique of Pure Reason* (London, Macmillan, 1929), reprinted in J. J. C. Smart, *Problems of Space and Time* (New York: Macmillan, 1964), p. 115. Smith follows Ernst Laas, Erich Adickes and Hans Vaihinger in reading "Idealität" for "Realität," owing to the context. By contrast, Guyer and Wood give a literal translation: "This reality of space and time, further, leaves the certainty of experiential cognition untouched: for we are just as certain of that whether these forms necessarily adhere to the things in themselves or only to our intuition of these things" (Cambridge Edition of the *Critique*, p. 166).

[50] Immanuel Kant, "Transcendental Aesthetics" (1781), trans. Smith, reprinted in Smart, *Problems of Space and Time*, p. 111.

from the experience of observing motions of physical bodies (he opposed Kant's *a priori* conceptions of space and time). Hence, he argued that *physics* should decide the validity of conflicting geometric principles.[51] Nonetheless, physicists' pursuit of kinematics was scarcely motivated by any such concerns over the validity of mathematics, because most remained quite ignorant of any crisis in the foundations of mathematics at least until the late 1800s. Still, the study of kinematics was pursued by physicists who were unsatisfied with specific features of the broad science of mechanics.

Like Ampère, later physicists realized that the basic concepts involved in the study of motion had not been formulated exactly. For example, Newton had abstained from explaining the fundamental "well known" concepts of kinematics, beginning his *Mathematical Principles of Natural Philosophy* with the definitions of more-complex terms now called mass ("quantity of matter"), momentum ("quantity of motion arising from velocity and quantity of matter conjointly"), and force. The first section of his *Mathematical Principles* was followed by a scholium in which Newton explained: "Hitherto I have laid down the definitions of such words as are less known, and explained the sense in which I would have them understood in the following discourse. I do not define time, space, place, motion, as being well known to all."[52] Thus, Newton did not bother to elucidate the meaning of the basic concepts of kinematics: time, space, place, and motion. He discussed these only inasmuch as it seemed necessary to emphasize that the ordinary impressions or measurements of, say, motion and time, are actually distinct from the exact, "true," and "absolute" relations of physical processes in nature.

Other scientists followed Newton's approach. They did not define time, space, place, or motion. For example, Jean d'Alembert, in his *Essay on the Elements of Philosophy* of 1759, left the practical notions of time and space undefined, as obvious to "common sense." D'Alembert went so far as to claim that philosophical analysis of these notions served only to generate confusion. He claimed that the analysis of such concepts is "absolutely foreign and useless for mechanics." He argued, "This science does not as-

---

[51] Hans Wussing, *Die Genesis des abstrakten Gruppenbegriffes* (Berlin: Deutscher Verlag der Wissenschaften, 1969); translated by Abe Shenitzer with the editorial assistance of Hardy Grant, *The Genesis of the Abstract Group Concept* (Cambridge, Mass.: MIT Press, 1984), p. 31.

[52] Newton, *Principia*, p. 8.

sume anything other than the natural notions of space and time, as they are known to all men; notions that are very simple and very clear in themselves, and philosophy has only the privilege of obscuring and confounding them."[53] Likewise, decades later, William Whewell, when defining the elements of dynamics, noted, "The notions of *time, space, motion,* are taken for granted," that is, not defined.[54] Likewise, Siméon Denis Poisson noted, "One does not define either space or time," although he admitted that at least the procedures for measuring time required some explanation.[55] This slight concern for the procedures of measurement of the basic physical quantities led some physicists to analyze the concepts of space and time—despite any criticisms against "philosophical" analyses.

Once kinematics had been conceived as the independent basis of mechanics, physicists had to restructure the foundations of Newton's theory to distinguish between kinematic and dynamic principles. In his *Mathematical Principles,* Newton had formulated mechanics without treating questions of forces independently of the study of motion. One might say that his kinematics and dynamics were "inextricably mixed."[56] Newton had based mechanics on only three fundamental "Axioms, or Laws of Motion." For Newton, the laws of motion were more fundamental than other so-called laws of physics. Thus, he formulated the laws of motion at the beginning of his book, and thereafter capitalized the word "law" only when it referred to any of the three laws of motion. Of the three laws, only the first became admitted as a principle of kinematics. Newton had stated the first law as: "Every body continues in its state of rest, or of uniform motion in a straight line, unless it is compelled to change that state by forces impressed upon it."[57] Because this law describes the behavior of bodies in the absence of

---

[53] Jean d'Alembert, *Essai sur les élémens de Philosophie, ou sur les principes des connaissances humains, avec les éclaircissemens* (1759?); reprinted in *Oeuvres complètes de d'Alembert,* vol. 1, part 1 (Paris: A. Belin, et Bossange père, fils et frères, 1821), p. 316; translation by A. Martínez.

[54] William Whewell, *An Introduction to Dynamics, Containing the Laws of Motion and The First Three Sections of the Principia* (Cambridge and London: Cambridge University, J. and J. J. Deighton; Whittaker, Treacher, & Arnot, 1832), p. 14.

[55] S.-D. Poisson, *Traité de Mécanique,* 2nd ed., vol. 1 (Paris: Bachelier, 1833), p. 204.

[56] Roberto Torretti, *Relativity and Geometry* (1983; repr., New York: Dover, 1996), chap. 1; Torretti, *Creative Understanding: Philosophical Reflections on Physics* (Chicago: University of Chicago Press, 1990), p. 29.

[57] Newton, *Principia,* p. 14.

forces, it could serve as a principle of kinematics. By contrast, Newton's second and third laws of motion dealt directly with the behavior of bodies in the presence of forces. The second law stated that changes in a body's motion are proportional to motive forces impressed on it. The third law stated that to the action of every body upon another there is always opposed an equal reaction. These laws clearly pertain to dynamics. Hence, Newton's second and third laws of motion were not included in the new science of motion.

Furthermore, there were reasons to exclude the first law as well. By describing the behavior of bodies in the absence of forces, it yet incorporated the concept of force. Other philosophers before Newton had formulated essentially the same principle but without Newton's concept of force. Already in the 1650s, Christiaan Huygens had formulated an equivalent and fundamental statement: "Any body once moved continues to move, if nothing obstructs it, with a perpetual celerity and along a straight line."[58] Huygens called this a "Hypothesis." Subsequently, Newton too called it a hypothesis, as late as 1684, before deciding to call it a "Law of Motion" in his *Mathematical Principles of Natural Philosophy*.[59] In any case, formulations such as that of Huygens were more akin to the aims of kinematics in the 1800s. Furthermore, the "hypothesis," "principle," or "law" of inertia did not even need to be construed necessarily as a part of kinematics. Regardless of whether free bodies move in straight trajectories perpetually or not, the goal of kinematics was just to provide descriptions of motions, and kinematics would study motions irrespective of whether space is "empty."

Consider now the old reticence to "define" motion itself. Many writers simply characterized it as "the change of place," an ancient expression. For example, the geometer and natural philosopher John Playfair, in lectures at the University of Edinburgh, stated: "Motion, of which all bodies are

---

[58] Christiaan Huygens, *De Motu Corporum ex Percussione* (manuscript, mid-1650s), in Huygens, *Oeuvres complètes de Christiaan Huygens*, vol. 16 (La Haye: Société Hollandaise des Sciences, Martinus Nijhoff, 1929), pp. 29–91. See also Huygens, *Horologium Oscillatorium* (1673): *Christiani Hvgenii Zvlichemii Const. f. Horologivm oscillatorivm, sive, De motv pendvlorvm ad horologia aptato demonstrationes geometricae* (Paris: Apud F. Muguet, 1673); reissued in *Oeuvres*, vol. 18 (1934), pp. 69–368.

[59] Isaac Newton, *De Motu Corporum* (1684), in Newton, *The Preliminary Manuscripts for Isaac Newton's 1687 Principia, 1684–1686*, with an introduction by D. T. Whiteside (Cambridge: Cambridge University Press, 1989).

susceptible, is the continued change of place."[60] And, in another example among many, the *Encyclopædia Americana* stated, "The motion of a body is the change of its place in space."[61]

Some writers ascribed the shortcomings of the old theories of motion to the lack of attention paid by philosophers to the systematic analysis of observations. For example, Playfair criticized ancient definitions of motion as useless:

Aristotle's definition is highly characteristical of the vagueness and obscurity of his physical speculations. He calls motion "the act of a being in power, as far as in power;" words to which it is impossible that any distinct idea can ever have been annexed.

The truth is, however, that the best definition of motion can be of very little service in Physics. Epicurus defined it to be the "change of place," which is, no doubt, the simplest and best definition that can be given; but it must, at the same time, be confessed, that neither he, nor the moderns who have retained his definition, have derived the least advantage from it in their subsequent researches. The properties, or, as they are called, the laws of motion, cannot be derived from a mere definition; they must be sought for in experience and observation, and are not to be found without a diligent comparison, and scrupulous examination of facts. Of such an examination, neither Aristotle, nor any other of the ancients, ever conceived the necessity; and hence those laws remained quite unknown throughout all antiquity.[62]

[60] John Playfair, *Outlines on Natural Philosophy,* vol. 1 (Edinburgh and London: A. Neil & Co. for Archibald Constable and Company, and Longman, Hurst, Rees, Orme and Brown, Cadell and Davies, and John Murray, 1812), p. 8.

[61] *Encyclopædia Americana: A Popular Dictionary of Arts, Sciences, Literature, History, Politics and Biography* (on the basis of the Seventh edition of the German Conversations-Lexicon), ed. Francis Lieber, with the assistance of E. Wiggelsworth and T. G. Bradford, vol. 9 (Philadelphia: Carey and Lea, 1832), p. 69.

[62] John Playfair, *Dissertation second, exhibiting a general view of the progress of mathematical and physical science, since the revival of letters in Europe* (Boston: Wells & Lilly, 1817). Issued as the second dissertation of the supplement to the fourth, fifth, and sixth editions of the *Encyclopaedia Britannica.* Reissued in vol. 1 (1835) of the Seventh edition of the *Encyclopaedia Britannica,* as *Dissertations on the History of Metaphysical and Ethical and of Mathematical and Physical Science,* "Dissertation Third; Exhibiting a General View of the Progress of Mathematical and Physical Science, since the Revival of Letters in Europe," p. 450.

Likewise, various authors criticized Zeno, another famous philosopher of ancient times. For example, the *Encyclopædia Americana* briefly noted: "All changes in the material world consist of motion. The life of the organic creation, and the action of inorganic bodies, consists in motion: what we call rest, is only relative. Experience alone convinces us of the motion of bodies in space. Zeno of Elea, endeavored to prove this fundamental idea of motion to be contradictory in itself, in order to overthrow the testimony of experience."[63] Philosophers of nature saw the virtue of Newton's laws of motion as being grounded in a due regard for experience. The laws of motion did not seem to be deduced from metaphysical axioms. More generally, Playfair claimed, "It is from induction that all certain and accurate knowledge of the laws of nature is derived."[64]

Nonetheless, some writers variously proposed to reformulate Newton's laws. Some tried to reduce and consolidate Newton's laws and theorems. Others divided the same into multiples. Still, the majority of writers upheld Newton's formulation. William Whewell, for example, defended the notion that there are properly three fundamental laws of motion, against the idea that they can be condensed into only two. Still, in his *Introduction to Dynamics,* Whewell acknowledged, with disappointment, that most theorists at the time were hardly interested in subtleties or disquisitions about the foundations of mechanics:

> I am well aware that such discussions cannot be expected to excite much interest. It is a peculiar feature in the fortune of principles of such high elementary generality and simplicity as characterize the laws of motion, that when they are once firmly established, or supposed to be so, men turn with weariness and impatience from all questionings of the grounds and nature of their authority. We often feel disposed to believe that truths so clear and comprehensive are necessary conditions, rather than empirical attributes, of their subjects: that they are legible by their own axiomatic light, like the first truths of geometry, rather than discovered by the blind gropings of experience. And even when the experimental foundation of these principles is allowed, there is still no curiosity about the details of the induction by which they are established. The process of *deduction,* of

---

[63] *Encyclopædia Americana,* vol. 9, p. 69.
[64] Playfair, *Outlines on Natural Philosophy,* p. 2.

reasoning downwards from principles, fills the mind at every step with a confidence in its own workings, a consciousness of certainty, a distinctness of perception, a feeling of superiority to all more vague and doubtful impressions, which give a peculiar charm to this employment and have often tempted men to pursue it when the truth obtained was of no value, and even to apply it when it could not possibly lead to truth. But the process of *induction,* by which we arrive at principles, is one which, though at least as important to the progress of science, possesses no such fascination. To scrutinize this process after it has been successfully executed, we must put the mind in an attitude of doubt on doctrines not only certain, but the foundation of long trains of certainty; we must try to conceive ourselves ignorant of that which we most familiarly assume; in short we must attempt to guess a riddle of which we already know the answer. It is not surprising therefore that this should not be a favourite occupation with speculative men.[65]

Whewell was prepared for the indifference of readers who presumably would not care whether the doctrine of motion be founded on two laws or three.

Even in the mechanics of Newton, some scientists were unsatisfied with the inclusion of properties or principles of motion that seemed mere definitions. For example, the notion of absolute rest seemed unconnected with actual experience. In his *Elements of Kinematics,* the artillery general and engineer Arthur Morin commented that "for us on Earth there exists only *relative rest.*"[66] Likewise, in his *Pure Kinematics,* the physicist and engineer Henri Resal noted, "Absolute rest does not seem to exist in nature; one knows, in effect, that bodies placed on the surface of the earth participate of the double movement of the planet around its axis and around the sun, and that the sun itself is transposed through space as it drags with it the planets and satellites that accompany them."[67] If there was no evidence available for the existence of absolute rest, why would one want to found the sciences of dynamics or kinematics on that concept? If the first law of motion referred explicitly to bodies at rest, but the notion of rest lacked univocal meaning,

---

[65] Whewell, *An Introduction to Dynamics,* p. x.

[66] Morin, *Leçons de Mecanique Pratique, Élements de Cinématique* (1857), p. 7.

[67] Henri Resal, *Traité de Cinématique Pure* (1862), p. 1; translation by A. Martínez.

like that of force, some physicists thought that it would be preferable to formulate the theory of motion in more specific terms.

How History Came to Criticize Physics

Even though most physicists were perhaps confident in the principles of mechanics, some interest persisted in critical analysis. In April 1869 the faculty of philosophy of the University of Göttingen announced a prize competition for writers to compose "a critical history of the general principles of mechanics." There existed in Germany a well-known tradition of so-called critical history that pertained mainly to biblical exegesis. It was sometimes traced back to the kind of textual analysis of Hebrew scriptures advanced by the Dutch philosopher Baruch Spinoza in 1670. More recently, the tradition of critical history had gained impetus especially since the mid-1830s, owing to the works of Ferdinand Christian Baur and David Friedrich Strauss. They adopted ideas advocated by the philosopher Georg Hegel, such as the principle of evolution of all truths through the conciliation of contradiction, and applied such ideas to the analysis of the early development of the Christian church. By applying canons of "objective criticism," they and their followers came to challenge traditional orthodoxy about biblical scripture, arguing even that the Gospels might not all be authentic but may have been composed after the time of the apostles. Such scholars sought to apply current standards of scientific rationality to elucidate history and distinguish it from legend. Meanwhile, a kindred critical tradition had labored to reconstrue the history of antiquity. Likewise, as ancient and Christian authorities came under increased historical scrutiny, so too did the modern science of motion. Hence, in 1869 professors at Göttingen wanted to encourage the pursuit of "critical history" of mechanics. They wanted philosophers to distinguish to what extent its principles involved self-evident logical axioms, how they were connected to mathematical formulas and valid findings of experience, and to what extent they should be accepted in the totality of empirical knowledge.[68]

The prize competition at Göttingen had been facilitated by an endowment to the university by the Consistorialrath C. G. Beneke. Upon his

---

[68] E. Dühring, *Kritische Geschichte der allgemeinen Principien der Mechanik. Von der philosophischen Facultät der Universität Göttingen mit dem ersten Preise der Beneke-Stiftung gekrönte Schrift. Nebst einer Anleitung zum Studium mathematischer Wissenschaften* (Berlin: Theobald Grieben, 1873), 3rd ed. (Leipzig: Fues's Verlag/R. Reisland, 1887), p. ix.

death in 1864, his testament had willed that a sum of money be allocated to encourage the advancement of general scientific works dedicated to the remembrance of his brother. Friedrich Eduard Beneke had been a prolific writer on psychology and philosophy. In ethics, he advocated a relativist outlook, and thus he sharply opposed the philosophy of Hegel.[69] In 1822 he was abruptly barred from lecturing at the University of Berlin, owing to the influence of Hegel on the administration. Beneke's systematic approach was distinguished by the contention that all philosophy should be based on empirical psychology. Following the philosophical tradition of John Locke, Beneke denied the existence of innate ideas. Beneke insisted that psychology had to be treated as a natural science, based only on the critical examination of experience, albeit internal rather than external.[70] His approach was also distinguished by his strict treatment of mental phenomena by the "genetic method," that is, by investigation of the development and perfection of the "simple elements" of the mind.

Earlier, in Germany, the genetic method had been prominently advocated by Johann Gottfried von Herder, who had argued that it is important to pursue the "inner" history of the mind, to understand the origin and development of one's outlook, and to contrast it with those of other individuals, in order to recognize what is universal and invariant in it and what is distinctive and variable.[71] Herder advocated a philosophy based on empirical knowledge and concepts, in opposition to traditional metaphysics and to Kant's *Critiques*.

At any rate, Beneke's system was based on mechanical metaphors about the motions of (mental) elements and their combinations. He sought to explain the process of mental development in terms of the action and interactions of the powers of the soul with external stimuli, resulting in the formation of sensations and perceptions. Beneke lectured in Göttingen for a few years, and later he managed to return to Berlin in 1832, where he acted

[69] F. E. Beneke, *Neue Grundlegung zur Metaphysik; als Programm zu seinen Vorlesungen über Logik und Metaphysik* (Berlin: E. S. Mittler, 1822).

[70] F. E. Beneke, *Lehrbuch der Psychologie* (1832); and *Die neue Psychologie; Erläuternde Aufsätze zur zweiten Auflage meines Lehrbuches der Psychologie als Naturwissenschaft* (Berlin: E. S. Mittler, 1845).

[71] Johann Gottfried von Herder, *Auch eine Philosophie der Geschichte zur Bildung der Menschheit* (Riga, 1774); translated as *This Too; a Philosophy of History for the Formation of Humanity*, in *Johann Gottfried von Herder: Philosophical Writings*, trans. and ed. Michael N. Forster (Cambridge: Cambridge University Press, 2002), pp. 272–360.

as extraordinary professor until his mysterious death in 1854 (his body was found two years later in a canal).

Hence, owing to Beneke's legacy, the faculty of Göttingen announced its prize competition in 1869. Five contestants submitted entries. The winner was selected and announced in March 1872.[72] The prize was awarded to a 586-page manuscript on the history of mechanics, authored by Eugen Karl Dühring, a lecturer at the University of Berlin since 1864. Dühring was a philosopher and a political economist who had published works on capital and labor, the value of life, and natural dialectics, as well as, in 1869, a *Critical History of Philosophy*. His winning entry in the Göttingen contest was published in 1873 under the title *Critical History of the General Principles of Mechanics*.

Dühring's long book reviewed at length the history of mechanics, ranging mainly from the contributions of Galilei, Huygens, and Newton, to developments in the 1800s. He discussed the evolution of statics and dynamics. He discussed the origins and rise of the principle of virtual velocities, the principle of the equal speeds of falling bodies, the law of inertia, the principle of least action, the principle of the lever, Newton's law of gravitation, the principle of the conservation of living forces, the principle of the conservation of the moment of rotation, and much more. He also dedicated a chapter to discussing the relationship of mechanics to the philosophies of Kant, David Hume, Arthur Schopenhauer, and others. Dühring praised the analytical formulation of mechanics by Joseph-Louis Lagrange for being free of metaphysical notions. Dühring admired especially the relatively recent contributions of Robert Mayer that demonstrated the mechanical equivalence of heat.

Dühring dedicated the last chapter of his book to discussing the general importance of the principles of mechanics and their possible application to fields such as chemistry, electrodynamics, cosmic physics, and the study of living organisms. He there discussed the growing field of kinematics. He acknowledged Ampère's work but preferred to employ the old term phoronomy instead of kinematics. He stressed that this field could be considered a part of pure mathematics. He claimed that it had the same *a priori* certainty and consisted of the same kind of knowledge. He appreciated Lagrange's conception of mechanics as a geometry of four dimensions as "typical" of

---

[72] *Nachrichten der Königliche Gesellschaft der Wissenschaften zu Göttingen* **13**, no. 8 (13 March 1872).

this conception: "All that which can be represented in the three dimensions of space and in the fourth, corresponding to time, as appearances of motion or of rest, must become the subject of phoronomic formulation. . . . Disregarding metaphysical phantasms, this field will envelop the entire reality of spatial and temporal phenomena together and no more."[73] Throughout his writings, Dühring emphatically denounced all that which, like mysticism, seemed to veil reality. His philosophy was akin to that of Comte. Dühring rejected Kant's separation of the phenomenal from that which really exists, arguing that the human intellect is capable of grasping reality in its entirety. Within his materialistic perspective, the human mind was taken to be a by-product of the interrelations among matter, such that the laws of thought and life were, ultimately, the very laws of matter.

Incidentally, soon after Dühring's *Critical History* of mechanics appeared, the philosopher Friedrich Nietzsche independently published an essay "On the Use and Damage of History for Life," in 1874. It was the second installment of his "Untimely Mediations," which he had begun partly to criticize Strauss's recent work as superficial. Nietzsche identified three kinds of history, each with positive functions as well as possible dangers. So-called monumental history celebrated the great achievements of humanity and its heroes. By contrast, antiquarian history studied with reverence the customary aspects of culture. To counterbalance such reverence, critical history identified flaws and failures. Nietzsche commented: "Only the man whose chest is oppressed by a present stress and who craves to throw off his burden at any cost has a need for critical history, that is, history that sits in judgment and passes judgment."[74]

Traditionally, history of science had consisted prominently of celebrations of leading scientists and their achievements. In contradistinction, Dühring's work was tinted by irreverence. He sought not only to praise the monumental feats of scientific learning but also to undermine those who he did not hold in high esteem. Moreover, he was disrespectful of university authorities, though he was not even a professor. Dühring criticized university

---

[73] Dühring, *Kritische Geschichte der allgemeinen Principien der Mechanik,* 3rd ed. (1887); translation by A. Martínez.

[74] Friedrich Nietzsche, *Unzeitgemässe Betrachtungen. Zweites Stück: Vom Nutzen und Nachtheil der Historie für das Leben* (1874), part 2, translation by A. Martínez; translated fully in *Untimely Mediations,* by R. J. Hollingdale, ed. Daniel Breazeale (Cambridge: Cambridge University Press, 1997), pp. 57–124.

practices at Berlin and even accused Professor Helmholtz publicly of not being a competent scientist. The ensuing quarrel between Dühring and the faculty at Berlin resulted in the cancellation of his license to teach in 1877. Nonetheless, readers received his critical history of mechanics with interest. Hence, a second edition appeared in 1877, and a third in 1887.

In the 1870s Dühring's writings on national politics and economics were gaining much attention. He was known as a critic of *The Capital* of Karl Marx and, more generally, of Hegel's philosophy. The publication of Dühring's *Course on Philosophy* and the second edition of his *Critical History of National Economy and of Socialism* in 1875 earned him followers in the Social-Democratic Worker's Party—to the irritation of Marx and Friedrich Engels. The latter then invested two years writing articles to undermine Dühring's popularity and denounce him as a charlatan. The series of articles constituted a book, including a chapter by Marx, which was published in its entirety in 1878. In their book, which became widely known as *Anti-Dühring*, Engels did not comment on Dühring's *Critical History of Mechanics*, but his analysis of Dühring's attempt at a systematic philosophy included criticisms of views on mechanics which Dühring had expressed in other works. Dühring had criticized Hegel's method of dialectics by arguing that contradictions do not exist among physical things but only in thoughts.[75] He denounced the idea that contradictions are corporeal—"The reality of the absurd is the first article of faith in the Hegelian unity of the logical and the illogical"—which he ridiculed as inherited from theology and mysticism.[76] In response, Engels claimed that the motion of bodies evinces contradictions that exist in things and material processes: "Motion itself is a contradiction: even simple mechanical change of position can only come about through a body being at one and the same moment of time both in one place and in another place, being in one and the same place and also not in it. And the continuous origination and simultaneous solution of this contradiction is precisely what motion is."[77]

---

[75] Eugen Karl Dühring, *Cursus der Philosophie als streng wissenschaftlicher Weltanschauung und Lebensgestaltung* (Leipzig: E. Koschny, 1875), pp. 30–32.

[76] Eugen Karl Dühring, *Kritische Geschichte der Nationalökonomie und des Socialismus* (Berlin: T. Grieben, 1871), 2nd ed. (Berlin: T. Grieben, 1875), pp. 479–480.

[77] Friedrich Engels, *Herrn Eugen Dührings Umwälzung der Wissenschaft* (Leipzig, 1878); reissued in 1894 and translated in Karl Marx and Frederick Engels, *Collected Works*, vol. 25 (New York: International Publishers, 1987). *Anti-Dühring, Herr Eugen Dühring's Revolution in Science*, part I, chap. 12, p. 111.

With Marx's approval, Engels gave various other examples of contradictions in the material world and even in mathematics to justify the scope of dialectics. He also criticized Dühring's recurring tendency to conceive of statics as prior to dynamics and to claim that there is no bridge from the former to the latter. Engels rejected Dühring's contention that mechanical force was a state of matter that was originally identical with matter. Engels criticized such talk as meaningless mysticism, rejecting the idea of absolute immobility. Instead, he argued:

> *Motion is the mode of existence of matter.* Never anywhere has there been matter without motion, nor can there be. Motion in cosmic space, mechanical motion of smaller masses on the various celestial bodies, the vibration of molecules as heat or as electrical or magnetic currents, chemical disintegration and combination, organic life—at each given moment each individual atom of matter in the world is in one or other of these forms of motion, or in several forms at once. All rest, all equilibrium, is only relative, only has meaning in relation to one or other definite form of motion.[78]

He rejected Dühring's use of the concept of a motionless state of matter as nonsense. On the whole, Engels criticized Dühring's views as based on old metaphysics and devoid of substantive content.

Nevertheless, Dühring's irreverent approach helped to encourage some attention to highlighting deficiencies in the foundations of mechanics. Simultaneously and subsequently, other writers pursued similar lines. Soon after the faculty of the University of Göttingen announced its prize competition on the principles of mechanics, and before entries were received, similar lines were independently pursued in Leipzig.

In late 1869 Carl G. Neumann delivered an inaugural lecture at the University of Leipzig on the principles of classical mechanics, which was edited and published as *On the Principles of the Galilei-Newtonian Theory* in 1870. Neumann critically analyzed the principle of inertia, that is, Newton's first law of motion, which he called "Galileo's law of inertia." Following tradition, Neumann ascribed this law to Galileo Galilei. Yet Galilei had asserted no such thing. Actually, he had argued that free bodies in space move naturally in circular trajectories around certain centers. Still,

---

[78] Marx and Engels, *Anti-Dühring*, pp. 55–56.

Galilei had analyzed the motions of projectiles in terms of motions that deviate from rectilinear paths. Therefore, Newton had ascribed the principle of uniform rectilinear motion to him. However, Galilei referred only to rectilinear paths because such were the relatively small paths that were observed on the surface of the Earth, where the curvature of greater motions around the Earth could be neglected. Whereas Galilei had argued that bodies move naturally in circles, René Descartes and Christiaan Huygens expected instead that all free bodies move in straight lines. Newton likewise became convinced that free bodies move in rectilinear trajectories. Yet all four believed that such bodies move *at uniform rates,* that is, cover equal distances in equal intervals of time. Natural philosophers and mathematicians accepted this idea. For example, Leonhard Euler in the mid-1700s characterized it as an "unquestionable fact" that had been "certainly verified."[79]

Neumann expressed the law of inertia in a traditional formulation: "A material point that was set in motion will move on—if no foreign cause affects it, if it is entirely left to itself—in a *straight line,* and it will traverse in equal times equal distances."[80] Neumann then denounced this usual kind of formulation as unsatisfactory, repeatedly claiming that it was "incomprehensible" and "totally unintelligible," because it lacked definitions that should necessarily accompany it in order to give it meaning. In particular, one needed to establish precisely what is meant by the expressions "motion," "straight line," and "equal times." For example, you might imagine that it is easy to conceive of a rectilinear uniform motion, such as the motion of a bullet in free space, but notice, however, that this idea is insufficient, "because a motion which, e.g., is rectilinear when viewed from our Earth, will appear as curvilinear when viewed from the Sun."[81] Any motion that appears to be rectilinear when observed from one celestial body will appear as curvilinear if it is observed from another. Indeed, even an observer

---

[79] Leonhard Euler, "Réflexions sur l'espace et le tems," presented on 1 February 1748, *Mémoires de l'Académie des Sciences de Berlin* 4 (1750), 324–333; also in: *Opera Omnia,* ser. III, vol. 2, pp. 376–383; translated by Link M. Lotter, "Reflections on Space and Time," reprinted in *The Changeless Order: The Physics of Space, Time and Motion,* ed. Arnold Koslow (New York: George Braziller, 1967), secs. I–II, quotation on p. 116.

[80] Carl G. Neumann, *Ueber die Principien der Galilei-Newton'schen Theorie. Akademische Antrittsvorlesung gehalten in der Aula der Universität Leipzig am 3. November 1869* (Leipzig: Teubner, 1870), p. 14; translation by A. Martínez.

[81] Ibid.

on Earth will not observe rectilinear motion of free bodies in outer space but will find that bodies seem to move in curved trajectories because the Earth is spinning. Thus Neumann argued that certain definitions were needed to give meaning to the law of inertia and, moreover, that this law would better be formulated as not a single principle but several.

Neumann argued that to use Newton's law of inertia physicists have to assume the existence of a stationary reference point relative to which any motions are ascertained. He proposed a new fundamental principle underlying Newton's laws of motion, that "in some unknown place of the cosmic space there is an unknown body, which is an *absolutely rigid* body, a body whose shape and size are forever unchangeable."[82] This "Alpha-body" had not been observed by astronomers, but it could be assumed to correspond to some unknown object in space. Newton's law of inertia could then be deemed valid relative to this reference body. Moreover, the Alpha body need not be imagined as something material but simply as consisting of three mutually perpendicular geometric lines. Neumann thus advanced the concept of an *inertial frame of reference,* though he did not use that expression. Once the Alpha body was defined, any other body moving inertially relative to it would serve equally well as another reference frame where Newton's laws would be applicable.

Newton himself had considered the existence of a stationary reference body in space: "It is possible, that in the remote regions of the fixed stars, or perhaps far beyond them, there may be some body absolutely at rest; but impossible to know."[83] But Newton had not emphasized the need for such a reference point to carry out any analysis of motion. For Neumann the hypothesis of the concrete existence of the Alpha body served to substantiate the study of motion in the same way as the hypothesis of an electrical fluid underlay the theory of electricity, and the hypothesis of the luminiferous ether underlay the theory of light.

Neumann argued that the oblate shape of a celestial body such as a star could not depend on its motions relative to other bodies. He claimed that if all other bodies in the universe suddenly ceased to exist, the idea that motion is purely relative would entail that the star would then assume a spherical form as if it were at rest. Neumann claimed that this outcome would be a "contradiction so intolerable" that one just had to reject the assumption

---

[82] Ibid., p. 15.
[83] Newton, *Principia*, p. 10.

of the relativity of motion in favor of positing the body Alpha. Euler had employed essentially the same argument to try to prove the existence of absolute rest and motion.[84] Yet Neumann used it to substantiate also the existence of the fundamental body. Thus he rejected the attempts to define motion as something entirely relative.

Meanwhile, other theorists disagreed. Some maintained that all motion was purely relative. In the United States, the German-raised Johann Bernhard Stallo advocated a "principle" of the "relativity of all objective reality."[85] In the 1840s Stallo had extensively studied mathematics, physics, and philosophy, though he then chose a career in law, becoming a lawyer and judge in Cincinnati, as well as a political activist known for raising a regiment of Germans to fight in the Civil War to eliminate slavery. At first, Stallo had been a follower of German idealist philosophies, especially that of Hegel, but he eventually abandoned metaphysics in favor of practical American values. He became convinced that even the best theories of physics were infected with remnants of old metaphysical systems that should be repudiated. In 1881 he published a book titled *The Concepts and Theories of Modern Physics,* involving prominent features of critical history. In it, he denounced "radical errors" of metaphysics, such as the idea that things exist independently and prior to their relations, as though all relations are between absolute terms. The book was published simultaneously in England and then promptly in Paris in French translation. These and subsequent editions were reprinted several times.

The book was essentially a critique of contemporary physics in light of notions of cognition, of how knowledge develops, a nascent field that Stallo overestimated as a full-fledged scientific theory. He argued at length that the traditional atomic-mechanical conception, widely believed to underlie all the properties of matter, had been adopted by physicists owing more to metaphysical traditions than to practical experience. He wanted to distinguish between convenient hypothetical fictions and known realities. Stallo

---

[84] Leonhard Euler, *Theoria motus corporum solidorum seu rigidorum ex primis nostrae cognitionis principiis stabilita et ad omnes motus, qui in huiusmodi corpora cadere possunt, accommodata* (Rostochii et Gryphiswaldiae, 1765), reissued in *Opera Omnia,* ser. II, vol. 3 (1948), chap. 2, p. 31.

[85] Johann Bernhard Stallo, *The Concepts and Theories of Modern Physics* (New York: D. A. Appleton; London: Scientific Series/Kegan Paul, 1881); 3rd ed. (1888) reissued with an introduction by Percy W. Bridgman (Cambridge, Mass.: Belknap/Harvard University Press, 1960), p. 202.

despised the reification of concepts. Regarding the existence of things and relations presumed to be absolute, he argued:

> The real existence of things is coextensive with their qualitative and quantitative determinations. And both are in their nature relations, quality resulting from mutual action, and quantity being simply a ratio between terms neither of which is absolute. Every objectively real thing is thus a term in numberless series of mutual implications, and forms of reality beyond these implications are as unknown to experience as to thought. There is no absolute material quality, no absolute material substance, no absolute physical unit, no absolutely simple physical entity, no absolute physical constant, no absolute standard, either of quantity or quality, no absolute motion, no absolute rest, no absolute time, no absolute space. There is no form of material existence which is either its own support or its own measure, and which abides, either quantitatively or qualitatively, otherwise than in perpetual change, in an unceasing flow of mutations.[86]

Stallo argued that all properties of objects exist only in their relations to others, that we ascribe properties only by comparisons. He claimed that because all knowledge of reality stems from the recognition of relations, then key laws of physics, including the law of inertia, were based upon the principle of the "relativity of all objective reality." Stallo criticized physicists for, on the one hand, claiming that there are no *a priori* truths but only what is ascertained from variable experience, while on the other hand, assuming that phenomena are ruled by constant laws and that material constituents are absolutely invariable. He reviewed the positions of Descartes, Leibniz, Newton, Euler, and Kant and criticized them all for acknowledging the apparent relativity of motion without entirely abandoning notions of absolute motion. Stallo also rejected entirely the notions of absolute space and time:

> The same considerations which evince the relativity of motion also attest the relativity of its conceptual elements, space and time. As to space, this is at once apparent. And of time, "the great independent variable" whose supposed constant flow is said to be the ultimate mea-

---

[86] Ibid., p. 201.

sure of all things, it is sufficient to observe that it is itself measured by the recurrence of certain relative positions of objects or points in space, and that the periods of this recurrence are variable, depending upon variable physical conditions. This is as true of the data of our modern time-keepers, the clock and the chronometer, as of those of the clepsydra and hour-glass of the ancients, all of which are subject to variations of friction, temperature, changes in the intensity of gravitation, according to the latitude of the places of observation, and so on. And it is equally true of the records of the great celestial time-keepers, the sun and the stars. After we have reduced our apparent solar day to the mean solar day, and this, again, to the sidereal day, we find that the interval between any two transits of the equinoctial points is not constant, but becomes irregular in consequence of nutation, of the precession of the equinoxes, and of numerous other secular perturbations and variations due to the mutual attraction of the heavenly bodies. The constancy of the efflux of time, like that of the spatial positions which serve as the basis for our determination of the rates and amounts of physical motion, is purely conceptual.

On the whole, Stallo argued that philosophers and mathematicians had been misled by assuming that only what is absolute is real and that phenomena are merely appearances. He claimed that they had been deceived by the old metaphysical doctrine that the phenomenal is the opposite of the real. Instead, he advocated the reality of the relative. Many readers, teachers, and authoritative reviewers were annoyed by Stallo's brazen criticisms of the elements of physics.[87]

Furthermore, Stallo responded to Neumann's arguments. He objected that the hypotheses of the annihilation of all bodies and of the existence of an absolutely fixed and absolutely rigid body were "forbidden by the universal principle of relativity."[88] Stallo argued that the annihilation of all bodies but one would destroy not only the motion of the one star but also its very existence, because, he claimed, a body has being only in its system of relations. Like change of position, *presence,* for Stallo, was a relative property. He

---

[87] As noted by Josiah Royce in the introduction to *Science and Hypothesis* by Henri Poincaré, reissued in Poincaré, *Foundations of Science,* authorized translation by George Bruce Halsted (New York: Science Press, 1913), p. 10.
[88] Stallo, *The Concepts and Theories of Modern Physics* (1881), p. 215.

rejected the idea of the absolute fixity of any body, arguing that it involved the supposition of absolute finitude of the universe, a notion that he also denied. Thus, theorists disagreed over matters of principle.

Still, by referring all motions to the Alpha body, Neumann facilitated a picture that seemed easier to visualize than motion relative to space itself. For instance, Heinrich Streintz followed Neumann's lead by publishing in 1883 a book titled *The Physical Foundations of Mechanics*, which voiced his agreement that to make sense of any motions, the coordinates that appear in the equations of motion must be referred to some nonrotating, force-free "fundamental body."[89] Still, this idea was useless for practical purposes, as no such body was known to exist. Moreover, Neumann's predilection for hypothesizing the existence of the Alpha body as the embodiment of an arbitrary reference frame seemed pointless to some theoreticians. Boltzmann, for example, argued that the stationary Alpha body was "an absurd idea" and that even as a "mere mental object, however, I prefer to call it 'reference frame' than 'body α.' The name 'body' alone seems to me most inappropriate."[90] At any rate, physicists had relied previously on the idea that the motion of any body is detected only by reference to some other specific body, and they had used coordinate systems to map such motions. But Neumann expanded these ideas by stressing the need to define the conditions under which a reference frame would serve to substantiate Newton's law of inertia.

For the purpose of physical measurements Newton's laws of motion could be *observed* to be true only relative to *certain* bodies or physical systems. Only if an observer was positioned on an inertial reference body would this person see that free, nonaccelerated objects move in straight paths. Only certain bodies thus could be admitted to serve as "inertial" reference frames. The concept of an inertial reference frame was refined subsequently by James Thomson (1884) and by Ludwig Lange (1885) in their attempts to clarify the foundations of mechanics. Whereas Neumann had used the notion of Alpha body, Lange wrote of *inertial systems*, and Thomson used the term *frame* to mean the same thing.

[89] Heinrich Streintz, *Die Physikalischen Grundlagen der Mechanik* (Leipzig: Teubner, 1883).
[90] Ludwig Boltzmann, *Vorlesungen über die Principe der Mechanik*, Part II (1904), §88; translated as "The Law of Inertia," in Boltzmann, *Theoretical Physics and Philosophical Problems*, quotation on p. 262.

Neumann's hypothesis of the existence of a reference body would seem rather useless because he did not say how to identify a similar body in a state of rest or inertial motion. Streintz attempted to remedy this deficiency by arguing that a gyroscope could be used to verify that a reference body is not rotating or accelerating. But this argument did not satisfy theorists, who realized its circularity: like Neumann, Streintz required a fundamental body at rest relative to which Newton's laws were valid, but Streintz claimed that such a body could be identified on the basis of a spinning device that itself was supposed to follow Newton's dynamics.[91] At any rate, the deficiency in Neumann's account was remedied by Ludwig Lange.

In a paper of 1885, Lange built upon Neumann's work, and he extended his arguments in a historical and critical review of the development of the concept of motion in his booklet of 1886: *The Historical Development of the Concepts of Motion and their Probable Destination: A Contribution to the Historical Critique of the Mechanical Principles.*[92] Lange did not assert the existence of an Alpha body, as he understood that it was not needed as a foundation for the theory of motion. Indeed, Lange did not rely on the concept of absolute rest at all. Instead, he proposed an ideal construct that would serve to identify any "inertial system" of reference. Lange argued that any three material points simultaneously projected from any location and left to move freely would constitute an inertial system. He assumed the uniform rectilinear motion of these points as a matter of "mere convention," for it seemed impossible to ascertain the motion of these points, because they were meant to function precisely as a standard by which to measure other motions. He expected that any objects moving freely through space should move uniformly in straight paths, and the elementary three points of any reference system then served as a means to verify that any other free particles move inertially. Any inertial system, though established by convention, would serve to verify that other material objects move according to Newton's first law of motion.

In short, the main advantage of the concept of inertial reference frame

---

[91] For criticism see, e.g., Lange, *Geschichtliche Entwickelung des Bewegungsbegriffes* (1886), p. 115. See also Ludwig Boltzmann, "Über die Grundprinzipien und Grundgleichungen der Mechanik," *Populäre Schriften,* Essay 16: Lecture delivered at Clark University (1899); translated as "On the Fundamental Principles and Equations of Mechanics," in Boltzmann, *Theoretical Physics and Philosophical Problems,* p. 103.

[92] Lange, *Die Geschichtliche Entwickelung des Bewegungsbegriffes.*

was that it allowed physicists to imagine conditions under which they could *test the validity* of Newton's first law of motion. Although Lange assumed the uniformity of motion of the three basic reference particles *as a matter of convention,* the motions of any other bodies could be measured relative to these, and thus physicists would be able to verify whether any such free bodies *actually* obey Newton's law of inertia. Though Lange's construction was essentially abstract, it constituted a step toward the elucidation of Newton's laws of motion in terms of physical processes. As Lange admitted, "This construction is, as indicated, fully ideal, as nobody is in a position to directly bring it about. But its value is not thus diminished. The real and practical (approximate) construction-method of the inertial system flows from its primary ideal construction in the exact same way as, for example, all real methods of (approximately) measuring electrical force arise from a fully ideal fundamental method, which now more than ever find immediate application."[93] Lange emphasized that necessary recourse to stipulated conventions in the definition of fundamental concepts "makes possible the immediate transition from theory to praxis," otherwise lacking in the theory of motion.[94]

Because physicists traditionally had held mechanics in the highest esteem, it came as something of a surprise that critical inspection of its elements eventually generated considerable debate. In hindsight, Ludwig Boltzmann characterized the situation as follows:

Analytical mechanics is a science worked out by its very founder Newton with a precision and perfection almost unrivalled in the whole field of human knowledge. The great masters that succeeded him have further strengthened the structure he had erected, so that it seemed quite inconceivable that there could be a creation of the human spirit more perfect and uniform than the foundations of mechanics as they confront us in the works of Lagrange, Laplace, Poisson, Hamilton and so on. Especially the establishment of the first principles seemed to have been carried out by these enquirers with a precision and logical consistency that has always furnished the paradigm according to which people sought to fashion the foundations of other branches of knowledge, if not always with the same success. It

[93] Ibid., p. 140; translation by A. Martínez.
[94] Ibid., p. 108.

long seemed quite impossible to expand or modify those foundations in any way.

It is the more noticeable and unexpected that at present there have arisen, especially in Germany, fairly lively controversies precisely about the fundamental principles of analytical mechanics.[95]

Although physicists had long entertained a high level of confidence in the validity of Newton's mechanics, some were troubled by what they found at the base of the system. Boltzmann commented that progress in physics had "provided a certain leisure for looking at the construction of the Newtonian edifice through a magnifying glass as it were, and lo and behold, this yielded many difficulties."[96] Such difficulties did not appear as problems in the practical pursuit of experimental physics, for the most part. Instead, they were conceptual; they seemed to consist mainly in clearly explaining basic ideas or justifying their validity. Heinrich Hertz, for example, undertook the task of investigating the foundations of mechanics not because this field in any way clashed with experience but "solely in order to rid myself of the oppressive feeling that to me its elements were not free from things obscure and unintelligible."[97] By contrast, he commented that "to many physicists it appears simply inconceivable that any further experience whatever should find anything to alter in the firm foundations of mechanics."[98] Most physicists seemed content to work within Newton's conceptual framework. Meanwhile, others who tried to interrelate results obtained in various subfields had to wrestle with the various basic concepts.

Owing to great advances in the fields of optics, thermodynamics, electricity, and magnetism, physicists were driven to analyze the foundations of their science to secure a coherent theoretical scheme that would accommodate a broad variety of phenomena. It was one thing to claim that the mathematical description of one class of phenomena, say, electricity, could in principle be derived from Newton's laws of mechanics; it was another to carry out such a derivation. If the concept of an electric field was to be explained in terms of Newtonian forces, then it and the basic elements

---

[95] Boltzmann, "On the Fundamental Principles and Equations of Mechanics" (1899), p. 101.
[96] Ibid., pp. 101–102.
[97] Hertz, *Principles of Mechanics*, p. 33.
[98] Ibid., p. 9.

of Newtonian mechanics would have to be defined precisely. As physicists tried to establish such connections, they encountered difficulties that made them focus more closely on the logical foundations of Newton's theory. Moreover, whereas most physicists agreed that all natural phenomena should be traced back to mechanical laws, there was disagreement as to what those laws actually were. In some cases, propositions that had been originally introduced and widely received as "principles" of physics had been reduced to more fundamental statements and still continued to be referred to as "principles." The increasing complexity in physics demanded greater precision in the formulation of its basic concepts. In turn, concepts that long had rested on intuitive notions were analyzed critically and reconceived in terms of exact physical procedures.

In particular, theorists increasingly pondered the exact meaning of the law of inertia. They then realized that even if the truth of Newton's first law of motion is assumed, one still needs to establish the procedures by which the uniformity of a given motion is actually *ascertained*. Likewise, physicists acknowledged the need to specify the circumstances in which the constancy of the direction of any inertial motion, that is, its rectilinear trajectory, can be observed. Some physicists became increasingly concerned with the interpretation of the laws of physics in terms of plausible physical processes and operations.

Engineers Complained about the Abstract Principles of Physics

While Neumann and Lange advanced their arguments in Germany, similar lines of reasoning were pursued in Great Britain. In early 1881 a leading article in the journal the *Engineer* called attention to the need for some treatise that would effectively clarify the definitions and principles of mechanics and carefully explain how they may be applied to actual experience. The anonymous author, perhaps an editor of the *Engineer*, complained about what seemed to be a deplorable lack of consensus on the definitions of even basic concepts, along with a widespread reliance on arguments from authority: "There is still too great a tendency to use a great name as sufficient excuse for putting forward a startling statement, or an inconsequent conclusion. To the objection that such and such a thing is perhaps not true, or that it seems to involve a contradiction, it is deemed answer sufficient, 'Newton said it,' or 'It will be found in Rankine.' It is high time that this kind of argument was displaced for ever, and that, in its stead, we should have accurate definitions which might be generally accepted as sound."[99]

The writer complained that fundamental terms such as matter, momentum, and especially motion were defined carelessly and ambiguously by many or even most authors. He commented, "If we ask twenty men of average education to define what motion is in a very few words, they will break down. If we turn to the text books, which are supposed to put dynamical facts before us in the most precise shape, we shall fare very little better." The author referred to Stephen Parkinson's *Treatise on Mechanics* of 1874 (first published in 1855) as a representative text, among many, involving poorly formulated definitions. The dynamical principle of the equality of action and reaction, usually known as Newton's third law of motion, seemed to be vaguely formulated, presenting difficulties for students. The law or principle of inertia, also suffered from unnecessary ambiguity by being formulated often by use of the "especially unhappy" words "tendency to motion" and even the concept of force: "As to the word 'force,' we confess we approach it with the utmost hesitation. It is continually used by every one, and almost invariably in a vague and purposeless kind of way, which is very wearisome."[100] The anonymous author mentioned that even excellent writers, such as Maxwell, usually defined force superficially without really explaining the concept, whereas some others, such as Tait, by contrast seemed even to deny the very existence of forces as things. On the whole, the anonymous author drew attention to the lack of unanimity on fundamental matters, and pleaded for the use of accurate and unambiguous language: "Ideas should be definite, and the words into which they are crystallized should be as translucent as the diamond. Clear thinking makes clear writing; but in the world of science there can be neither the one nor the other, in the fullest sense of the term, until much that is now vague has been accurately defined."

The article in the *Engineer* promptly triggered two series of publications in the same journal. On one side, several individuals wrote letters to the editor, which were published in several issues under the heading "The Foundations of Mechanics." These discussions were centered mainly on the question of debating the meaning and validity of the law of action and reac-

[99] Anonymous, "The Language of Physical Science," *Engineer* **51** (25 February 1881), 143–144; quotation on p. 143.
[100] Ibid., p. 144.

tion. The other series of publications consisted of a series of articles, each published also under the title "The Foundations of Mechanics," authored by Walter Raleigh Browne, a civil engineer and Fellow of Trinity College in Cambridge. The articles consisted of sequential sections of a single long monograph, which was finally reprinted separately as a whole in 1882. In response to the leading article in the *Engineer,* Browne argued:

> It cannot, I believe, be denied that amongst those who have to apply science to practice, and especially therefore amongst engineers, there exists considerable confusion as to the meaning of the fundamental definitions and fundamental principles on which the science of Mechanics is built. . . . the explanations are not always as full and as clear as would seem requisite to prevent confusion; nor is it always easy to reconcile, at least at first sight, the definitions and explanations given in one work with those given in another.[101]

To clarify the principles of mechanics, Browne explicitly decided not to divide them sharply in any necessary accord with the "in some ways unfortunate" divisions of the fields of statics, dynamics, and kinematics. He sought to clarify the elements of mechanics "to give an accurate plan of the foundations—a part of the structure which architects and householders are both somewhat inclined to neglect."[102]

By critically reviewing several widespread definitions of elementary concepts, Browne provided his own definitions of mechanics, kinematics, motion, force, matter, and more. For him, kinematics was the study of pure motion, the science of motions considered apart from forces. He characterized motion as an ultimate conception or elementary fact that cannot be further clarified, especially absolute motion, but he defined relative motion as any change of place of a thing recognized by an observer with reference to some other thing supposed to be fixed:

---

[101] Walter R. Browne, *The Foundations of Mechanics,* reprinted from the *Engineer* (London: Charles Griffin and Co., 1882), p. 1. Browne's work was originally published as a series of articles in the *Engineer,* in parts: I (30 September 1881), II (4 November 1881), III (2 December 1881), IV (16 December 1881), V (20 January 1882), VI (17 February 1882), VII (24 March 1882), VIII (12 May 1882).
[102] Browne, *Foundations,* p. 2.

If there were anything which we knew to be absolutely fixed in space, we might perceive absolute motion by change of place with reference to that thing. But as we know of no such thing, it follows that all motion, as tested and measured by us, must be relative—must relate, that is, to something which we assume to be fixed for the moment. Hence the same thing may often be properly said to be at rest and in motion at the same time; for it may be at rest with regard to one thing, and in motion with regard to another.[103]

Like Carnot and others, Browne argued that motion cannot be separated from matter, as an independent thing. He therefore cautioned against language concerning the transference and the conservation of motion, notions that imply that motion is a thing in itself rather than a condition or relation. Despite his emphasis on relative motions, Browne asserted the existence of the condition of absolute motion, though he considered it impossible to recognize or to measure it. Likewise, he accepted the old notion of force as an ultimate fact of consciousness, following Whewell, Thomson and Tait, and Herbert Spencer. He did not deny the existence of forces as things or agencies that may be independent of matter. Following tradition, Browne defined force as the cause of motion; yet he focused on what he claimed to be the more useful definition of force as an abstract mathematical term. He then based mechanics on a "universal principle of Conservation," which he characterized with two words: "Effects live," and upon this principle he based the first law of motion, which he stated as *"a body, under the action of no external force, will remain at rest, or move uniformly in a straight line."*[104] Browne claimed that this law and other fundamental principles also derive from the joint definitions of motion, force, and the principle of conservation. Thus, we see that unlike writers who sought to base the foundations of mechanics on kinematics, Browne chose to base mechanics on the definition of force and its implications. This approach led him to wrestle with old problems concerning the relationship between force and matter and the question of the action of alleged forces emanating from the centers of bodies at a distance from one another, across empty space.

Afterward, Browne proceeded to highlight the importance of physical

---

[103] Ibid., pp. 7–8.
[104] Ibid., pp. 13–14.

measurement, especially measurements of motion, matter, and forces, and then he discussed Newton's laws of motion and other principles of mechanics. Like other followers of Newton (such as d'Alembert, Poisson, and Whewell), Browne too explicitly chose not to attempt to define the notions of space and time, though he considered them to be things rather than abstractions.[105] At least he gave some brief indication, as usual, of how artificial and convenient standards of time, length, and weight are established, such as the standard of length established as the distance between the marks on a particular bar of metal. He then skipped ahead quickly to what seemed to be the more important subject of establishing measures of motion, matter, and force, and discussing the laws of mechanics. In closing, despite his interest in clarifying ambiguities in the works of others, he claimed that there existed no major differences of opinion, at least among the great writers of mechanics, but that the main questions were merely about proper terminology.

Also in Great Britain, arguments similar to those advanced by Ludwig Lange in Germany were advanced also by James Thomson, independently. Thomson was a civil engineer and physicist, professor at Glasgow University, and president of the Institution of Engineers and Shipbuilders in Scotland. Earlier, he had published papers on a wide variety of subjects, including one in 1878 on the lack of clarity in the nomenclature and notation used in units of measure in texts such as Maxwell's *Treatise*. Early in March 1884, James Thomson presented a paper, "On the law of inertia, the principle of chronometry and the principle of absolute clinural rest, and of absolute rotation," at a meeting of the Royal Society of Edinburgh, in the presence of such influential physicists as his younger brother William Thomson and Peter Guthrie Tait. In this 1884 paper, he elucidated Newton's law of inertia in terms of physical processes and procedures.

Thomson argued that to draw "a perfectly intelligible conception" of the law of inertia, scientists were "confronted with the preliminary difficulty or impossibility as to forming any perfectly distinct notion of a meaning in respect to a single body, for the phrase '*state of resting or of moving uniformly in a straight course.*'"[106] Newton's appeal to absolute space did not

---

[105] Ibid., p. 25.

[106] James Thomson, "On the law of inertia, the principle of chronometry and the principle of absolute clinural rest, and of absolute rotation," *Royal Society of Edinburgh*

serve to clarify the notion of inertia, because "it involves in itself the whole difficulty of our inability to form a distinct notion of identical points or places in unmarked space at successive times, or, of our inability to conceive any means whatever of recognizing afterwards in any one point of space, rather than in any other, the point of space which, at a particular moment of past time, was occupied by a specified point of a known body."[107] Thomson stressed that scientists had "no means of knowing, nor of even imagining" distinctions between places, motions, or directions in "unmarked" space, because any point occupied by a material body "at any specified past moment is utterly lost to us as soon as that moment is past."[108] Likewise, Kant had remarked on the absence of any criteria to distinguish "the different parts and diverse places that are not occupied by anything corporeal."[109]

Newton had known this, admitting that "because the parts of space cannot be seen, or distinguished from one another by our senses, therefore in their stead we use sensible measures of them."[110] But Newton had not specified the means of establishing sensible measures of space exactly. Newton's aims were distinct from those of later researchers, because at least in the *Mathematical Principles* he was concerned not with *practical mechanics* but with the exact philosophy of "rational mechanics."[111] Thus, Newton was concerned less with the practical questions of how to establish approximately exact measurements of distance and length than with the theoretical analysis of understanding the "absolute" relations among objects in space. For Newton, space existed independently of any measurements or relations among physical bodies. He had stressed the importance of this notion of "absolute space" not to facilitate any experimental verification of the law of inertia but to distinguish between *apparent* motions and *true* motions. "It is indeed a matter of great difficulty to discover, and effectually to distinguish,

---

Proceedings **12** (1884), 568–578; quotation on p. 572; reprinted as document no. 57, in James Thomson, *Collected Papers in Physics and Engineering* (Cambridge: Cambridge University Press, 1912); emphasis in the original.

[107] Thomson, "On the law of inertia, the principle of chronometry," p. 573.

[108] Ibid., p. 568.

[109] Immanuel Kant, *Neuer Lehrbegriff der Bewegung und der Ruhe und der damit verknüpften Folgerungen in den ersten Gründen der Naturwissenschaft* (Königsberg: Johann Friedrich Driest, 1758).

[110] Newton, *Principia*, p. 10.

[111] Ibid., preface.

the true motions of particular bodies from the apparent; because the parts of that immovable space in which those motions are performed, do by no means come under the observation of our senses. Yet the thing is not altogether desperate."[112] Newton provided a few *dynamical* arguments to substantiate the existence of absolute motions. His arguments were brief and subsequently were criticized by many theorists. Yet the distinction between true and apparent motions was essentially the purpose for which Newton had composed the *Principia:* "But how we are to obtain the true motions from their causes, effects, and apparent differences, and the converse, shall be explained more at large in the following treatise. For to this end it was that I composed it."[113]

In contradistinction, in the 1800s some of the writers who reformulated Newton's law of inertia were more concerned with *practical* mechanics. The application of Newton's laws had become a matter of such exactitude that physicists and engineers desired explicit and detailed specifications of how Newton's basic concepts should be interpreted in terms of mathematical constructions and physical operations. Newton and others could substantiate absolute *dynamical* motion because certain effects of forces could be interpreted to suggest that some motions were not simply the results of any particular *point of view.* However, the *kinematic* concept of uniform rectilinear motion had not been distinguished empirically from any "absolute" state of rest, because any body could be construed to be in motion precisely as a function of the places from which it was observed. Therefore, including the distinction between rest and uniform motion in the law of inertia did not seem essential. Hence, James Thomson, in particular, formulated the law of inertia without appeal to the concept of rest.[114] Like Neumann and Lange, Thomson argued that one could know or deal with motions of bodies only relative to specific points of reference. He noted that three points marked on a physical body sufficed to construct rectangular coordinate axes or planes to which to refer the positions of any other bodies. He named any such changeless configuration of points, lines, or planes a "reference frame." Consequently, he reformulated Newton's law of inertia in terms of the *relations* between any moving body and a reference frame *and* a clock (a "reference dial-traveler").

[112] Ibid., p. 10.
[113] Ibid., p. 13.
[114] Thomson, "On the law of inertia, the principle of chronometry," p. 576.

Perceptions versus Metaphysics

The trend to formulate the foundations of physics in terms of relationships between perceptible bodies received much support owing to the works of Ernst Mach. A teacher of physics with strong interests in physiology and psychology, Mach emphasized the importance of measurement procedures in the definition of the physical concepts of science, claiming that "our ideas must be based directly upon sensations."[115] He argued that measurement consists of "the derivation of one portion of a phenomenon from some other."[116]

For two centuries already, theorists' growing interest in sensations had prominent roots in the philosophy of John Locke. According to Locke, all ideas originated from empirical sensations, on which the mind then operates according to its faculties (such as memory). His followers, David Hume and Étienne Bonnot de Condillac, rejected his claim that some ideas arise from internal reflection and sought to trace the growth of consciousness as a succession of sensations. In the mid-1700s, for example, Condillac denied the existence of innate faculties and argued that all mental operations, like all ideas, derive from simple outer sensations. For example, he claimed that the ideas of extension, figure, and motion originate directly from sensations. Hence, he tried to ground philosophy on the "certainty" of perceptions, and he advanced the study of how ideas and the mind are generated.[117] Consequently, François-Pierre-Gonthier Maine de Biran further advocated the genetic method, rejecting Condillac's notion of the passive receptivity of the mind, to argue that the human mind should be understood as growing in connection with external conditions. In a distant but kindred vein, Mach investigated the development of physical concepts psychologically and historically.

At an early age, Mach had been interested in the writings of Kant. When

---

[115] Ernst Mach, address delivered on the anniversary meeting of the Imperial Academy of Sciences, at Vienna (25 May 1882); translated by T. J. McCormack, "The Economical Nature of Physical Inquiry," in Mach, *Popular Scientific Lectures* (1898; repr., La Salle, Illinois: Open Court, 1986), p. 206.

[116] Ibid.

[117] Étienne Bonnot abbé de Condillac, *Essai sur l'Origine des Connoissances Humaines* (Amsterdam: P. Mortier, 1746).

he was about fifteen years old, Kant's *Prolegomena to Any Future Metaphysics* made a profound impression on him.[118] Yet just a few years later he abruptly became convinced that Kant's notion of "the thing-in-itself," that is, of what exists independently of our sensations, is superfluous in science. Mach increasingly became aware that theories of physics involved superfluous or false metaphysical notions along with the useful parts. By pursuing systematic research in physics, physiology, and history, Mach tried to resolve the traditional apparent conflict between the physical and the psychological. He argued that the knowledge of the senses belonged to both the physical and the mental, connecting them. His interests were stimulated by contributions to the study of "psychophysics" by leading writers such as Gustav Theodor Fechner (who, incidentally, had been influenced by Beneke).[119] Mach cultivated a kind of approach that he deemed "antimetaphysical."[120]

Mach critically studied the development of the principles of mechanics and concluded, "All our principles of mechanics are, as we have shown in detail, experimental knowledge concerning the relative positions and motions of bodies."[121] His arguments were published in his book of 1883, *Mechanics: Historically-Critically Presented in its Development*. He was aware of Dühring's critical history and called it an "estimable work," but he commented that it did not particularly influence him because he had developed many of his ideas earlier and even published some.[122] Instead, he noted that he had been especially stimulated by "the beautiful introductions" of

[118] Ernst Mach, *Beiträge zur Analyse der Empfindungen* (Jena: G. Fischer, 1886), translated as *Contributions to the Analysis of the Sensations,* trans. C. M. Williams (Chicago: Open Court, 1897), p. 23.

[119] G. T. Fechner, *Elemete der Psychophysik* (Leipzig: Breitkopf and Hartel, 1859); Ernst Mach, *Vorlesungen über Psychophysik* (Vienna: Sommer, 1863). Fechner used to be a professor of physics at Göttingen in the 1840s, but he abandoned that position to investigate the quantified measurement of sensations, in the nascent field that he came to call psychophysics.

[120] Mach, *Analysis of the Sensations,* p. 1.

[121] Ernst Mach, *Die Mechanik in ihrer Entwickelung historisch-kritisch dargestellt* (Leipzig: F. A. Brockhaus, 1883); 2nd ed. (1889), translated by T. J. McCormack, *The Science of Mechanics: A Critical and Historical Account of Its Development* (1893); revised in 1942 to include additional alterations up to the ninth German edition of 1933 (repr., La Salle, Illinois: Open Court, 1960), p. 127.

[122] Mach, *Die Mechanik* (1883), preface.

Lagrange to chapters of his *Analytical Mechanics* and also by "the lucid and lively" book of Philipp Jolly, *The Principles of Mechanics.*[123]

Like d'Alembert, Mach argued that the motion of a body could be substantiated only by comparison to the motion of another body: "A motion is termed uniform in which equal increments of space described correspond to equal increments of space described by some motion with which we form a comparison, as the rotation of the earth. A motion may, with respect to another motion, be uniform. But the question whether a motion is *in itself* uniform, is senseless."[124] Thus, Newton's principle of inertia seemed to require reformulation, as suggested also by Neumann, Thomson, and Lange. Mach argued that it was not necessary to refer the law of inertia to an absolute space, and he dismissed the Alpha body as a "fiction." Accordingly, in his analysis the traditional concepts of absolute space and time appeared to be meaningless.

Mach sought to develop in physics a point of view that would not change when regarding other domains of science. Hence, in his *Analysis of Sensations,* he argued that the ultimate elements that make up all phenomena are colors, sounds, spaces, times, tactile pressures, odors, and temperatures. He argued that all bodies are but thought symbols that stand for complexes of such elementary sensations. Science would then consist of ascertaining systematically the connections between sensations. Mach rejected any dichotomy between bodies and mind. He rejected the idea of the ego as "a *real* unity," and he also rejected the notion of "a world of unknowable entities (which would be quite idle and purposeless)."[125] For him, the human body, "like all others, is part of the world of sense; the boundary-line between the physical and the psychical is purely practical and conventional."[126] The dividing line could be traced, in order to pursue *physics,* by leaving the human body out of account, but the line could also be moved or erased for other purposes of science.

Regarding time and space, Mach argued on physiological grounds that these notions were essentially sensations, not *a priori* forms of pure intuition as Kant had argued. Most sensations could present themselves

---

[123] Philipp Johannes Gustav von Jolly, *Die Principien der Mechanik* (Stuttgart: Franck, 1852).

[124] Mach, *Science of Mechanics*, p. 127.

[125] Mach, *Analysis of the Sensations*, p. 21.

[126] Ibid., p. 152.

without others, for example, as odors can be perceived without regard to sensations of space. Yet he highlighted "time-sensation" as always accompanying all other sensations, inseparably.[127] Nevertheless, this apparent inseparability did not give time an absolute standing. Mach argued: "With just as little justice, also, may we speak of an 'absolute time'—*of a time independent of* change. This absolute time can be measured by comparison with no motion; it has therefore neither a practical nor scientific value; and no one is justified in saying that he knows aught about it. It is an idle metaphysical conception."[128] Thus, Mach criticized Newton for admitting the concepts of absolute time, space, and motion into the science of mechanics. Likewise, he criticized Kant for not rejecting entirely Newton's ideas of absolute space and time.[129] In short, Mach stressed, "No one is competent to predicate things about absolute space and absolute motion; they are pure things of thought, pure mental constructs that cannot be produced in experience."[130]

Likewise, it made no empirical sense to say that an object moves through absolute space. The notion of space seemed useful only as an expression of mutual relationships among material bodies. Mach discussed Newton's arguments for absolute space and found them unsatisfactory. As such arguments were essentially of a dynamical nature, it is not necessary to review them here. Suffice it to say that Mach argued that the idea of absolute space could not be substantiated experimentally and that scientists therefore should restrain their expressions to deal only with the facts of relative motions. He suggested that Newton's idea of motion relative to absolute space should be substituted by the idea of motion relative to a hypothetical *medium* filling all of space.[131] Although Mach was skeptical of hypothetical entities, he allowed that such entities could be admitted into science, as long as we clearly acknowledge that they have to be regarded as more artificial and dubitable than the "unconsciously constructed mental symbol" known as ordinary matter.[132]

---

[127] Ibid., p. 109.
[128] Mach, *Science of Mechanics*, p. 127.
[129] Mach, "The Economical Nature of Physical Inquiry" (1882), translated in *Popular Scientific Lectures*, p. 204.
[130] Mach, *Science of Mechanics*, p. 127.
[131] Ibid., p. 128.
[132] Mach, *Analysis of the Sensations*, p. 152.

Other theorists shared some of the views popularized by Mach. Lange wrote his critical review of the historical development of the concept of motion to show that the traditional distinction between "true" and "apparent" motions was "entirely groundless and superfluous."[133] In this interpretation, the Copernican description of heavenly motions around the Earth as "apparent" was erroneous, for this was valid as the description of the Earth rotating on its axis and traveling around the Sun.[134] Lange argued that the ideas of absolute space and absolute motion should be removed from physics as "incomprehensible and superfluous fictions entirely in the spirit of metaphysics."[135] By contrast, in the mid-1700s, Leonhard Euler and Colin Maclaurin had characterized the ideas of absolute space and motion as correctly maintained by "mathematicians," whereas they criticized "metaphysicians" for dismissing such concepts in favor of statements expressing relations to reference bodies.[136]

Instead of conceiving of motion as relative to absolute space, scientists such as Mach suggested that motion be conceived as relative to the luminous ether. The success of the wave theory of light, owing to the works of Augustin Fresnel and others, had led physicists to posit the existence of an ethereal substance filling all space, through which light propagated as transverse undulations. Most scientists of the time believed in the existence of the ether, but it could not serve to determine motion because no way had been found to detect it directly. Nonetheless, the ether concept helped to stimulate the discussion of ideas of relative motion in the area of electrodynamics.

At the time when Mach wrote his *Mechanics,* he was unaware of Stallo's *Concepts and Theories of Modern Physics.* When Mach finally became familiar with the work, he saw a kindred approach in Stallo's skepticism against metaphysical aspects of the principles of mechanics and in his interests in the cognitive development of physical notions. The two began a brief cor-

---

[133] Lange, *Geschichtliche Entwickelung des Bewegungsbegriffes,* p. 119.

[134] Ibid., pp. 120–122.

[135] Ibid., p. 116.

[136] Euler, "Reflections on Space and Time" (1748), trans. Lotter, in Koslow, *The Changeless Order,* p. 119; Colin Maclaurin, *An Account of Sir Isaac Newton's Philosophical Discoveries* (1748); facsimile of the first edition, with a new introduction by L. L. Laudan, Sources of Science, no. 74 (New York: Johnson Reprint Corporation, 1968); book II, p. 102.

respondence, which soon ended with Stallo's death in 1900. Still, Mach then instigated the publication of a German translation of Stallo's book in 1901.

The tradition of ideas that stressed the relational character of motion led physicists to conceive of a so-called "principle of relative motion," or "principle of relativity," as inherent in physics. This notion had existed in various forms long before it became known by such names.[137] For centuries, many scientists had asserted that all states of uniform nonaccelerated motion could be detected only by reference to other arbitrary bodies. To early theorists such as Nicole Oresme and Galileo Galilei, physical phenomena seemed to transpire identically whether they took place on the Earth or on a vessel traveling uniformly relative to the ground. Subsequently, this notion was inherent in the mechanics of Newton and his followers. In the mid-eighteenth century, the necessity of always describing motions as relations was emphasized by Kant, who claimed, "I would never say a body is at rest, without adding in relation to what thing is it immobile; I would never say that it moves without making explicit the objects in relation to which it changes its position."[138] Accordingly, Ludwig Lange, in his historical review of the science of motion of 1886, commented on what he called Kant's "law of relativity."[139]

The idea of the relativity of motion was borrowed from mechanics by physicists researching electricity and magnetism. James Clerk Maxwell, in 1876, three years after publishing his *Treatise on Electricity and Magnetism*, published his book *Matter and Motion*, which he conceived as "an introduction to the study of Physical Science in general."[140] Maxwell recognized the importance of relative measures of space and time, arguing that "we cannot describe the time of an event except by reference to some other body. All our knowledge, both of time and place, is essentially relative."[141] He also deemed motion a relative phenomenon. In addition to mechanics, Maxwell's views

[137] Marie-Antoinette Tonnelat, *Histoire du principe de relativité* (Paris: Flammarion, 1971), part 1, pp. 5–77; part 2, pp. 262–272.
[138] Kant, *Neuer Lehrbegriff der Bewegung und Ruhe* (1758); reissued in *Immanuel Kants Werke*, ed. Ernst Cassirer (Berlin: Bruno Cassirer, 1922), vol. 2, pp. 15–28.
[139] Lange, *Geschichtliche Entwickelung des Bewegungsbegriffes*, p. 102.
[140] James Clerk Maxwell, *Matter and Motion* (London: Society for promoting Christian knowledge, 1876; repr., New York: Dover, 1991); preface.
[141] Ibid., p. 12.

on relativity derived from a particular philosophical source. In the mid-nineteenth century, as a student in Edinburgh, Maxwell had learned of the doctrine of the "relativity" of human knowledge from its proponent, the philosopher William Hamilton (not to be confused with William Rowan Hamilton). Maxwell hence emphasized the relativity of motion in his book, *Matter and Motion*. While endorsing Newton's basic scheme, he laid more stress on the need to recognize the relativity of physical measures. He emphasized the need to establish explicit reference frames when discussing motion, saying that expressions concerning position, displacement, velocity, rest, and acceleration had little meaning outside of a relativized context. For instance, he argued that acceleration "is a relative concept and cannot be interpreted absolutely": "If every particle of the material universe . . . were at a given instant to have its velocity altered by compounding therewith a new velocity, the same in magnitude and direction for every such particle, all relative motions of bodies . . . would go in perfectly continuous manner, and neither astronomers nor physicists, though using their instruments all the while, would be able to find out anything had happened."[142] Maxwell concluded: "It is only if the change of motion occurs in a different manner in the different bodies of the system that any event capable of being observed takes place." He likewise argued that forces, and dynamical knowledge in general, are relative conceptions, because we cannot detect forces directly but can detect only the *differences* between the forces acting on one thing and those acting on another.[143]

Like Maxwell, other nineteenth-century physicists became increasingly aware that the propositions of physical science required specifications of reference to possess definite meaning. For example, in their *Treatise on Natural Philosophy*, Thomson and Tait declared, "All motion that we are, or can be, acquainted with, is *Relative* merely."[144] Likewise, in his paper of 1882 Walter R. Browne affirmed that all motion that can be recognized is relative. Also, Oliver Heaviside, in his *Electromagnetic Theory*, asserted bluntly that "all motion is relative."[145] A number of other late nineteenth-century theo-

---

[142] Ibid., p. 25.
[143] Ibid., pp. 80–82.
[144] Thomson and Tait, *Treatise on Natural Philosophy*, vol. 1, part 1, p. 32.
[145] Oliver Heaviside, *Electromagnetic Theory*, vol. 1 (London: Benn Brothers, Ltd., The Electrician Printing and Publishing Company, 1893; repr., New York: Chelsea, 1971), p. 182.

rists, including Heinrich Hertz, August Föppl, and Henri Poincaré, also laid stress on the relativity of motion. By 1903, consensus was so widespread that Bertrand Russell commented, "Not only other philosophers, but also men of science, have been nearly unanimous in rejecting absolute motion, the latter on the ground that it is not capable of being observed, and cannot therefore be a datum in an empirical study."[146]

Despite its currency, however, the idea of the relativity of motion was questioned because the ether could be interpreted as a unique frame of reference. Maxwell's equations of electricity and magnetism seemed to be valid exactly only relative to the ether. The question of relativity increasingly gained importance as physicists fruitlessly attempted to detect the motion of the Earth relative to the invisible ethereal medium purported to transmit electromagnetic waves, such as light. Thus, in 1904 Henri Poincaré (a former student of Henri Resal at the Polytechnique and at the School of Mines) wrote about "the principle of relativity, according to which the laws of physical phenomena must be the same for a stationary observer as for an observer carried along in a uniform motion of translation; so that we have not and can not have any means of discerning whether or not we are carried along in such a motion."[147] In his writings Poincaré used the expressions "principle of relative motions," "law of relativity," and "principle of relativity." His papers were reprinted and translated widely.

In textbook presentations of kinematics, authors decreasingly used the concept of absolute motion. For example, consider the works of Paul Appell, professor of the Faculté de Sciences of Paris. In textbooks of 1893 and 1902, he began his treatment of kinematics by discussing absolute motion, though he qualified it as a "pure abstraction," noting that "one observes only *relative motions*. Nonetheless, we can imagine three axes of coordinates absolutely fixed: the motion of a body, by reference to those axes, will be called the *absolute* motion of the body. Absolute motion is therefore a pure abstraction; but the relative motions can always be referred to absolute motions, and those being subject to the simplest laws, it is convenient to begin

---

[146] Bertrand Russell, *The Principles of Mathematics* (Cambridge: University Press, 1903), p. 489.

[147] Henri Poincaré, "L'état actuel et l'avenir de la Physique mathématique" (1904); "The History of Mathematical Physics," *The Value of Science*, trans. George Bruce Halsted, reprinted in *Foundations of Physics* (New York: Science Press, 1913), p. 300.

by the study of *absolute motion*."[148] Yet in later lessons and in a textbook of 1905, Appell changed his outlook. Alongside his coauthor J. Chappuis, professor of the École Centrale, they presented a new plan: "In the principles of mechanics, we have conveyed, by numerous examples, the *relativity* of the notion of *motion*, and we have completely suppressed the words *absolute motion*."[149]

The changing character of the study of motion may now be summarized. The pure science of motion gained prominence in the work of Ampère, who realized that although this science had been widely neglected, it nonetheless should constitute the fundamental study underlying all other branches of physics. That viewpoint gradually superseded others that focused on statics, or on agencies such as forces or causes as more fundamental than motion. Meanwhile, many theorists soon forgot that Ampère had characterized kinematics as the study of motion *as it appears to observation*. Because kinematics thus became construed as the "pure geometry of motion," theorists to varying extents construed the truth of kinematic propositions as independent of experimental or observational knowledge. Indeed, physicists imagined that the propositions of kinematics had the highest degree of truth in physical science, comparable to that of the propositions of pure mathematics.

As theorists elucidated the central concepts of the neglected science of motion, however, they realized increasingly that the propositions of kinematics actually depended on empirical notions. Physicists surmised that the connection between general theoretical statements and definite experiences had to be explained by reformulating such general statements in terms of plausible physical operations or processes. To give clear meaning to Newton's laws of motion, physicists and engineers proposed to reconstrue them in the form of observable relations. Hence, defined concepts such as "uniform velocity" could acquire a presumably unambiguous meaning. Yet this line of analysis showed that even if such concepts were

---

[148] Paul Appell, *Traité de Mécanique Rationelle; Cours de Mécanique de la Faculté de Sciences*, vol. 1: *Statique – Dynamique du Point* (Paris: Gauthier-Villars et Fils, 1893), chap. 2: Cinématique, p. 44; Paul Appell, *Cours de Méchanique à l'usage des Candidats à l'École Centrale des Artes et Manufactures* (Paris: Gauthier-Villars, 1902), part 1: Cinématique du Point, p. 14; translation by A. Martínez.

[149] Paul Appell and James Chappuis, *Leçons de Mécanique Élémentaire à l'usage des Élèves des Classes de Première C et D, conformément aux Programmes du 31 mai 1902* (Paris: Gauthier-Villars, 1905), p. vii; translation by A. Martínez.

empirically grounded, some of them had to be defined by means of conventions. Lacking any nonconventional means to distinguish true and apparent motions, several physicists rejected any distinction between the two. Thus, the idea of relativity became a prominent and fundamental principle of kinematics, as physicists ascribed meaning to the concepts of space, time, and motion only as measures relative to specific frames of reference.

On the whole, the science of motion evolved from a stage in which physicists regarded it as a virtually perfect science, with firm foundations and self-evident principles, to one that required careful elucidation of its basic concepts. Belief in the abstract mathematical truth of kinematics was replaced by belief in the necessity of grounding concepts on perceptions or on feasible or idealized procedures of measurement. The transition from statics and dynamics to kinematics, as the new foundation of mechanics, corresponded to physicists' growing interest in describing rather than explaining phenomena. Gradually, the "pure" science of motion grew to be progressively empirical.

CHAPTER FOUR

Debates over Language
Coordinates versus Vectors

As we encounter them nowadays, textbook discussions on kinematics involve the languages of coordinates and vectors. But how did vector methods come to accompany coordinates in kinematics? Students learn these methods as if they are entirely equivalent, just languages that describe physical subjects. Yet when vectors entered physics, they were presented in opposition to the old coordinate methods, which in turn were denounced as misleading and obsolete. The history of physicists' disagreements and conflicts over mathematical methods, though, is entirely hidden by physics textbooks.

Before the mid-1600s, mathematicians commonly practiced geometry without associating points with *numbers*. They studied the relationships among geometric figures without necessarily measuring the lengths of lines in diagrams. Yet a new approach to geometry was being invented, precisely by quantifying lengths and by designating points by numbers.

The Legacy of Descartes

Analytic geometry was one of the names given to the system of algebraic procedures developed to analyze geometric problems. This branch of mathematics emerged in 1637 in a work authored by René Descartes, under the title *The Geometry*, as an appendix to his philosophical work, *Discourse on the Method for Rightly Conducting Reason, and for Finding Truth in the Sciences.*[1]

---

[1] René Descartes, *Discours de la Méthode pour Bien Conduire Sa Raison, et Chercher la Verité dans les Sciences; Plus La Dioptriqve. Les Meteores. Et La Géométrie. Qui sont des essais de cete methode* (Leyde: l'imprimerie de 1. Maire, 1637).

Descartes turned to mathematics mainly as a foundation upon which to base a philosophical worldview. At the time, geometric investigations had been pursued much in the tradition of the ancient Greeks. For instance, contemporary mathematicians, such as François Viète, Willebrord Snell, and Marino Ghetaldi, were attempting to reconstruct parts of the ancient lost works of the Greek geometer Apollonius.[2] Meanwhile, Descartes authored a new approach that eventually led other mathematicians to neglect the traditional geometric approaches for almost two centuries.[3] Whereas geometers had relied traditionally on methods that essentially employed diagrams, Descartes' method allowed geometers to investigate problems chiefly by means of symbols and, in the long run, even without visualizing figures. Descartes believed that a symbolic mathematics such as algebra should occupy a fundamental place in the system of knowledge in general.[4]

In *The Geometry*, Descartes established simple rules by which to associate geometric curves and figures with algebraic equations.[5] Coordinates served as an aid in the analysis of problems. Following earlier mathematicians, Descartes established a correspondence between arithmetical operations and the construction of geometric figures with straightedge and compass. He thus justified the use of arithmetic in geometry.[6] Otherwise, in the eyes of mathematicians and philosophers, the mixture of arithmetic and geometry presumably would detract certainty from geometric deductions. The representation of geometric figures in algebraic equations consisted, in

[2] Morris Kline, *Mathematical Thought from Ancient to Modern Times*, vol. 1 (New York: Oxford University Press, 1990), p. 285.

[3] Carl B. Boyer, *History of Analytic Geometry* (New York: Scripta Mathematica, 1956), p. 83.

[4] Jakob Klein, "Die griechische Logistik und die Entstehung der Algebra," in *Quellen und Studien zur Geschichte der Mathematik, Astronomie und Physik*, Abteilung B: *Studien*, vol. 3, fasc. 1 (Berlin, 1934), pp. 18–105 (part I); fasc. 2 (1936), pp. 122–235 (part II); translated by Eva Brann as *Greek Mathematical Thought and the Origin of Algebra* (New York: Dover, 1992), p. 184.

[5] René Descartes, *La Géométrie* (1637); facsimile of the first edition, with translation and annotations by David Eugene Smith and Marcia L. Latham, *The Geometry of René Descartes* (New York: Dover, 1954). Also published as "First Book: 'Of Problems That Can Be Constructed Using Only Circles and Straight Lines,'" in *Discourse on Method, Optics, Geometry, and Meteorology*, trans. Paul J. Olscamp (Indianapolis: Bobbs-Merrill, 1965).

[6] Boyer, *History of Analytic Geometry*, p. 84.

the first place, of expressing ideas of situation (position) in terms of ideas of magnitude.[7] Descartes began by considering curves that could be generated by simple motions and then proceeded to ascertain symbolic expressions to describe such curves, to operate with them algebraically.[8] He used a single reference axis upon which to construct geometric figures. In his *Geometry*, algebra was subordinated to geometry as a tool for the analysis of constructions. Nonetheless, Descartes granted a certain superiority to algebra by designating as "geometric curves" only those that could be formulated algebraically.[9]

Descartes was not the first to analyze geometric figures by tracing distances from reference lines. A French contemporary, Pierre de Fermat, independently developed an equivalent system of analytic geometry.[10] Furthermore, in ancient times, Greek mathematicians already had analyzed the properties of curves by reference to other lines. Given a curve under investigation, they would superimpose certain lines upon it to describe it verbally (or symbolically) by reference to its intersections with the lines or, say, to its distance from them. Fermat reversed this process. By *beginning* with an algebraic equation, he showed that it could be interpreted as describing a curve (a locus of points) by referring the terms of the equation to points (coordinates) along reference lines.[11] The fundamental principle of analytic geometry, that an equation in two unknowns is an algebraic expression of the properties of a curve, was expressed by Fermat more clearly than by Descartes.[12] Whereas Descartes thought of curves as generated by motions,

---

[7] As noted, e.g., by Auguste Comte, *Cours de Philosophie Positive*, translated by W. M. Gillespie as *The Philosophy of Mathematics* (New York: Harper & Bothers, 1851), p. 233.

[8] Descartes, *La Géométrie*, book I: "Des problesmes qu'on peut construire sans y employer que des cercles & des lignes droites."

[9] Henk J. M. Bos, "The Structure of Descartes's *Géométrie*," in *Lectures in the History of Mathematics, History of Mathematics*, vol. 7 (Providence, R.I.: American Mathematical Society and London Mathematical Society, 1993), pp. 37–53, pp. 46–49. Originally published in *Descartes: il metodo e i saggi; Atti del convegno per il 350° anniversario della publicazione del Discours de la Méthode e degli Essais*, 2 vols. (Florence: Paoletti, 1990), pp. 349–369.

[10] See, e.g., Gino Loria, "Da Descartes e Fermat a Monge e Lagrange. Contributo alla storia della geometria analitica," *Atti della R. Accademia Nazionale dei Lincei*, Anno 120, 5th ser., *Memorie della Classe di Scienze Fisiche, Matematiche e Naturali*, vol. 14 (Rome: Accademia Nazionale dei Lincei, 1923), pp. 777–845.

[11] Boyer, *History of Analytic Geometry*, p. 75.

[12] Ibid., p. 101.

Fermat conceived them as determined by algebraic equations. Most of Fermat's work remained unpublished and unknown until 1679, fourteen years after his death, and forty-two years after Descartes' *Geometry* first appeared. By then, Fermat's contributions did not seem novel, and analytic geometry was commonly known as Cartesian geometry. This name was justified, even in retrospect, because this new branch of mathematics had grown primarily under the influence of Descartes.[13]

Algebraic geometry also came to be known as analytic geometry, in contradistinction to the traditional geometry known as synthetic geometry. Since antiquity, mathematicians had distinguished between the processes of analysis and synthesis in the following way.[14] *Analysis* was any procedure by which mathematicians initially assumed as true a proposition about the relationships among lines in a geometric construction to find whether the proposition can be broken down to others previously accepted. *Synthesis* was the process by which a proposition was proved as true by showing that a specific series of deductions from accepted principles suffice to derive it.[15] Whereas analysis served for the *investigation* of the validity of propositions, synthesis served as an expository device, a means to *display* convincingly the validity of propositions. Although the meanings of the terms analysis and synthesis varied throughout the centuries, it remained clear that most of the geometric writings of the ancient Greeks were synthetic, because they consisted of sequences of increasingly complex propositions, deduced from simple principles, without even hinting about the procedures used to *discover* the propositions and their proofs.[16] Centuries later, when European mathematicians during the 1600s developed the methods of algebraic

[13] Ibid., p. 82.

[14] These notions apparently began with Plato and later were refined by Theon of Alexandria, Pappus, and others. See Klein, *Greek Mathematical Thought and the Origin of Algebra,* pp. 154–155.

[15] E.g., see Aristotle, *Metaphysics,* Θ. 9. 1051^{a}21; in *Mathematics in Aristotle,* trans. and ed. Thomas L. Heath (Oxford: Clarendon Press, 1949), pp. 270–273. See also Thomas L. Heath, *A History of Greek Mathematics,* vol. 2 (Oxford: Clarendon Press, 1921), p. 400.

[16] E.g., as Heath indicates, despite Euclid's use of arguments involving the reductio ad absurdum, in the *Elements,* analysis has otherwise virtually "no place in the exposition"; in Heath, *History of Greek Mathematics,* vol. 1, p. 371. One extant ancient text that discloses methods of mathematical discovery is the *Geometrical Solutions Derived from Mechanics,* also known as "The Method," by Archimedes, trans. J. L. Heiberg (Chicago: Open Court, 1909).

geometry and calculus, some thought that they had rediscovered the lost methods of analysis of the ancients.

Descartes himself did not refer to his method as analytic geometry. The name emerged afterward from a variety of sources. It appeared in print "perhaps for the first time, in a sense analogous to that of today" in writings of Michel Rolle in 1709.[17] Rolle used the term to refer to the transformation of geometric questions into algebra. Likewise, the term *co-ordinates* acquired currency during the early eighteenth century, being first used by Gottfried Wilhelm Leibniz in the early 1690s.[18] Meanwhile, the designation of Cartesian for algebraic geometry was advanced by Jean Bernoulli in 1692.[19]

In the early 1600s, algebra was not a well-defined subject. So it was hardly regarded as an independent branch of mathematics. Since its introduction, algebraic geometry was viewed as a *method* of analysis of geometric problems, rather than as an integration of two branches of mathematics.[20] At the time, the terms mathematics and geometry were practically synonymous. Algebra worked as a practical shorthand to describe generally valid arithmetical relationships. Still, arithmetic itself was a subject lacking logical foundations.[21] Ever since the days of Euclid, at least by 300 BC, most geometric knowledge had been organized effectively in a systematic scheme where all true propositions were derived sequentially from fundamental principles. By contrast, arithmetic and algebra had developed as makeshift collections of rules, convenient mathematical procedures that served to generate practical results, but without definite logical structure. Thus, in the 1630s Descartes commented that "algebra indeed, as it is usually taught, is so restricted by definite rules and formulas of calculation, that it seems rather a confused kind of an art, by the practice of which the mind is in a certain manner disturbed and obscured, than a science by which it is cultivated and made acute."[22]

---

[17] Boyer, *History of Analytic Geometry*, p. 155.

[18] See, e.g., Florian Cajori, *A History of Mathematics*, 2nd ed. (New York: Macmillan, 1926), p. 175.

[19] Boyer, *History of Analytic Geometry*, p. 133.

[20] Descartes, *Discourse on Method, Optics, Geometry, and Meteorology*, trans. Olscamp, pp. 206–207, 251, 258. Dover facsimile of original edition: pp. 340–341, 401–402, 412–413.

[21] Morris Kline, *Mathematics: The Loss of Certainty* (Oxford: Oxford University Press, 1980), p. 125.

[22] Descartes, quoted by Augustus De Morgan in *A Budget of Paradoxes*, vol. 1 (1872), 2nd ed., ed. David Eugene Smith (Chicago: Open Court, 1915), pp. 204–205.

To mathematicians accustomed to the logical rigor and structure of classic synthetic geometry, the methods of analytic geometry also appeared as a makeshift admixture of geometry and algebra. Neither of the originators of analytic geometry was a professional mathematician. Fermat was a lawyer, and Descartes was a philosopher. From the outset, some leading mathematicians, such as Blaise Pascal and Isaac Barrow, sought to keep geometry free of algebraic methods because the latter lacked logical foundations. Leibniz, for instance, described the work in algebra as a "mélange of good fortune and chance," though he recognized its effectiveness.[23] Likewise, Isaac Newton, though contributing much to the rise of algebra, stated, "Algebra is the analysis of bunglers in mathematics."[24] Thomas Hobbes, the philosopher, criticized the "whole herd of them who apply their algebra to geometry."[25] Traditional synthetic geometry continued to be widely used, and a number of mathematicians rejected analytic geometry entirely. For example, Leibniz, like other seventeenth-century mathematicians, construed Descartes' work as a continuation of the work of the ancients. It seemed that Descartes, like Viète before him, had merely succeeded in applying equations to curves that had not yet been algebraically represented. As mentioned in chapter 2, Leibniz sought to formulate a more profoundly novel system of symbolic analysis. Thus he complained: "algebra is nothing but the characteristic of indeterminate numbers or *quantities*. But it does not directly express *situation, angles,* and *motion,* whence it is often difficult to reduce what is in a figure to calculation, and even more difficult to find sufficiently convenient demonstrations and constructions, even after the algebraic calculation is completely done."[26]

Still, analytic geometry drew sufficient attention for its steady development. The circulation of Descartes' *Geometry* became comparable to that of other major mathematical texts once Frans van Schooten translated the text into Latin, the scholarly language of the time, and published it in several editions between 1649 and 1695. Van Schooten's editions and

---

[23] Kline, *Mathematics*, p. 125.

[24] Ibid., p. 124.

[25] Kline, *Mathematical Thought from Ancient to Modern Times,* vol. 1, p. 318.

[26] Leibniz, letter to Christiaan Huygens, 8 September 1679, in *Christiaani Hugenii aliorumque seculi XVII. virorum celebrium exercitationes mathematicae et philosophicae,* ed. Uylenbroek (1833), fasc. I, p. 10; in Hermann Grassmann, *A New Branch of Mathematics, The Ausdehnungslehre of 1844 and Other Works,* trans. Lloyd C. Kannenberg (Chicago: Open Court, 1995), p. 404.

commentaries served to popularize analytic geometry throughout Europe. Likewise, in England, Cartesian geometry was popularized mainly through the contributions of John Wallis, who thoroughly developed Descartes' program to arithmetize geometry. For the most part, such early commentators "were more attracted by the algebra of *La geometrie* than by the geometry."[27] Descartes had used algebra as a method with which to analyze geometric problems. He successfully exhibited the versatility of a symbolic approach for the evolution of mathematical truths. Algebra emerged as a powerful method for obtaining results of great generality.

Geometry helped to establish the validity of algebra, but eventually algebra was acknowledged as a legitimate mathematical system in its own right. By the mid-1700s, "the reluctance to use algebra had been overcome."[28] For one, the mathematician Colin Maclaurin acknowledged that algebra "may have been employed to cover, under a complication of symbols, abstruse doctrines, that would not bear the light so well in a plain geometrical form; but, without doubt, obscurity may be avoided in this art as well as in geometry, by defining clearly the import and use of the symbols, and proceeding with care afterwards."[29] The association of arithmetic and algebra to geometry facilitated not only the advance of geometric knowledge but also the growth of algebra. Following Descartes' work, many mathematicians were apparently so impressed by the usefulness of algebraic analysis in geometry that they pursued algebra as an end in itself. Commentators looked upon Descartes' work not just as geometry but as a significant contribution to algebra.

Commentators continued to build upon Descartes' work with a greater focus on algebraic expressions than on diagrams of figures. Whereas Descartes and Fermat had developed a mixture of geometry and algebra, their most prominent followers proceeded to emancipate algebra from geometric constructions. Leonhard Euler praised algebra without reservations as being far superior to traditional synthetic geometry.[30] In the late 1700s, mathematicians led by Gaspard Monge labored to elucidate a system of coordi-

---

[27] Boyer, *History of Analytic Geometry*, p. 109.

[28] Kline, *Mathematics*, p. 125.

[29] Colin Maclaurin, *A Treatise of Fluxions*, vol. 2 (Edinburgh, 1742); quoted in Florian Cajori, *A History of Mathematical Notations*, vol. 2, *Notations Mainly in Higher Mathematics* (Chicago: Open Court, 1929), p. 330.

[30] Leonhard Euler (1748), quoted in Kline, *Mathematics*, p. 125.

nate geometry that was purely arithmetical. Joseph-Louis Lagrange, though not a professional geometer, devoted much time to replacing geometric figures with analytic formulas, especially in the field of solid geometry, and he suggested that geometers should build a completely analytic form of geometry. His student Sylvestre François Lacroix carried out this program, arguing that algebra and geometry "should be treated separately, as far apart as they can be; and that the results in each should serve for mutual clarification, corresponding, so to speak, to the text of a book and its translation."[31] Hence, the methods of analytic geometry were developed to the extent that algebraic techniques could stand independently of diagrammatic constructions. In the 1800s the tradition initiated by Monge was continued prominently in Germany by Julius Plücker, while in France Augustin-Louis Cauchy advanced "the so-called 'arithmetization' of all mathematics."[32]

As mathematicians and natural philosophers relied increasingly on algebraic formulas without reference to geometric diagrams, the practice of physics became increasingly abstract. Physicists came to commonly use analytic geometry to analyze physical problems. Like their predecessors, they believed that Euclidean geometry could rightly describe the lengths and shapes of objects and the spaces between them. They also believed that the algebra that described this geometry was likewise appropriate for the analysis of phenomena. Physicists could use algebra to study motion because they could use variables in equations to represent the changing positions of objects. With the proliferation of analytic representations, theorists became increasingly convinced that all notions of quality could be reduced to relations of quantity. Many came to believe that "every phenomenon is logically susceptible of being represented by an *equation*."[33] This conviction grew to the extent that traditional geometry came to be viewed as superfluous to the complete analysis of nature.

Hence, in 1788 Lagrange published a purely analytic formulation of

[31] Sylvestre François Lacroix (1797), quoted in Boyer, *History of Analytic Geometry*, p. 212.

[32] Felix Klein, *Vorlesungen uber die Entwickelung der Mathematik im 19 Jahrhundert*, part I (Berlin: Springer-Verlag, 1928); translated by M. Ackerman in *Lie Groups: History, Frontiers and Applications*, vol. 9, *Development of Mathematics in the 19th Century* (Brookline, Mass.: Math Sci Press, 1979), pp. 110, 76.

[33] Comte, *Cours de Philosophie Positive,; The Philosophy of Mathematics*, trans. Gillespie, p. 37.

mechanics, titled *Analytical Mechanics,* in which he boasted that he had developed the subject completely without reference to a single diagram: "One will not find any figures at all in this work. The methods that I present here require neither constructions nor geometrical or mechanical reasoning, but only algebraic operations, carried out in a regular and uniform march."[34] Lagrange's work served as a demonstration that algebra and calculus sufficed for the analysis of physical problems without appeal to the methods of synthetic geometry. Coordinate algebra had developed more quickly in the study of solid rather than plane geometry (it circumvented the difficulties of pictorial representation and interpretation of three-dimensional figures).[35] Because algebra was not originally invented for the description of motion, though, it was not a foregone conclusion that Cartesian algebra, from its inception, would be a precisely suitable mathematical tool for physics. Moreover, some of the principles of algebra were hardly understood in physical terms. For instance, what would it mean to say that the multiplication of two negative motions results in a positive magnitude?

Lagrange's analytic formulation of mechanics appeared before any texts on pure analytic geometry had been written. By the 1790s, especially following the work of Lacroix, "Elementary coordinate geometry, as now usually taught in a first course, had indeed reached its definitive form, with only details to be added here and there."[36] By the 1800s Cartesian geometry was sometimes referred to as Cartesian algebra. It appeared as a highly effective system, though still lacking logical foundations. Analytic formulas served well for physics but relied on ambiguous principles and often entailed complicated calculations. Because algebra had been used more often for the analysis of three-dimensional problems than for plane problems, the solutions to simple plane problems seemed more readily obtained by geometric construction than by algebra. Historian Carl Boyer noted that "among students of the École Polytechnique there were those who asserted that analytic geometry often failed where synthesis afforded short and elegant solutions."[37] Soon enough, however, analytic methods were employed

---

[34] J. -L. Lagrange, *Mécanique Analytique* (1788), 3rd ed., annotated by M. J. Bertrand, vol. 1 (Paris: Mallet-Bachelier, Gendre et Successeur de Bachelier (École Polytechnique), 1853), advertissement de la première édition, pp. I–II.

[35] Boyer, *History of Analytic Geometry,* p. 200.

[36] Ibid., p. 224.

[37] Ibid., p. 226.

to facilitate simple solutions to problems in plane as well as solid geometry. In this line, Joseph-Diaz Gergonne labored in favor of analytic geometry:

> Even those who are most familiar with the advantages which are presented by analytic geometry properly so-called, knowing full well that it alone affords the privilege of leading us constantly to the end of our researches without any kind of uncertainty, reproach the subject pretty generally for furnishing in the solution of problems only very complicated constructions, and of demonstrating theorems only by a calculation of which the prolixity often is repulsive. I have always thought that, most often, these inconveniences pertain perhaps less to the nature of the instrument than to the manner in which it is employed.[38]

Although the early development of analytic geometry was slow, the pace accelerated in the nineteenth century, owing to the publication of new periodicals such as the *Journal de l'École Polytechnique,* and the *Annales de mathématiques pures et appliqeés,* edited by Gergonne. The range of applicability and elegance of analytic geometry was extended, eventually exceeding that of traditional geometry, yet there remained ambiguity as to the validity of its foundations.

As some mathematicians focused on the study of algebra, independently of geometry, other mathematicians likewise contributed to the separation of algebra and geometry by employing only purely geometric methods. In the early 1800s there occurred a revival of interest in traditional geometry, owing partly to the works of Lazare Carnot. Geometers such as Jean-Victor Poncelet "scorned algebraic methods as alien and impenetrable to the insights and values which geometry proper afforded."[39] But even Poncelet occasionally employed algebraic methods as a basis or test in the process of ascertaining mathematical propositions. Moreover, the revival of interest in the methods of synthetic geometry did not suffice to diminish the growing role played by algebra in the minds of mathematicians. The importance of algebra in mathematics only continued to grow as mathematicians realized that the geometry of Euclid actually lacked the degree of certainty that it had been assumed to possess for more than two thousand years.

[38] Joseph Diaz Gergonne (1812), quoted in Boyer, *History of Analytic Geometry,* p. 230.
[39] Kline, *Mathematics,* p. 162.

The traditional faith in the validity of synthetic geometry was gradually undermined when mathematicians invented new geometries. In the early 1800s Carl Friedrich Gauss, Nicolai Ivanovich Lobachevsky, and János Bólyai independently devised new geometries based on a departure from the principles ordinarily assumed to be true. Gauss developed his "anti-Euclidean" geometry before the others, but he did not publish his results, in the belief that such work would not be well received.[40] Lobachevsky and Bólyai both published systematic deductive formulations of non-Euclidean geometry, and thus they became widely known as the creators of the new mathematics. Still, their works were preceded by earlier advances made by other mathematicians. Gauss eventually became known not only for having developed a fundamentally new geometry but especially for asserting that non-Euclidean geometry was applicable to the world, and hence that "we could no longer be sure of the *truth* of Euclidean geometry."[41] This realization slowly led to an upheaval in the understanding of mathematics. The non-Euclidean geometries were not just additional approaches to the study of space. They constituted also mutually conflicting accounts that could not be simultaneously true when used to describe the behaviors of free-moving bodies.

Once there existed multiple geometries, algebra emerged as a singular alternative upon which mathematical truth could be grounded. The results of arithmetic were readily acceptable as true, whereas geometric claims appeared as uncertain. Thus, for example, as Gauss became increasingly convinced that the ultimate validity of Euclidean geometry in physical space could not be proved, he argued that "we must not place geometry in the same class with arithmetic, which is purely *a priori*, but with mechanics."[42] Likewise, in his comprehensive ordering of the fields of human knowledge, André-Marie Ampère presented "arithmology," the integration of arithmetic and algebra, as more fundamental than geometry.[43] Traditional geometric knowledge appeared as less certain because its validity seemed to depend on tacit appeals to experience. Auguste Comte also came to hold ge-

---

[40] Klein, *Entwickelung der Mathematik im 19 Jahrhundert*, pp. 54–55.

[41] Kline, *Mathematics*, p. 87.

[42] Carl Friedrich Gauss (1817), quoted in Kline, *Mathematics*, p. 87.

[43] André-Marie Ampère, *Essai sur la Philosophie des Sciences, ou Exposition analytique d'une classification naturelle de toutes les connaissances humaines* (Paris: Bachelier, 1834), p. 66.

ometry in a lesser standing than algebra. In the *Course of Positive Philosophy*, Comte discussed at length the "immense necessary superiority" of modern analytic geometry over synthetic geometry.[44] Comte was so impressed by the arithmetization of geometry that he regarded analysis alone as abstract mathematics, whereas he regarded pure geometry and mechanics as concrete mathematics. Thus, "he called analytic geometry the most decisive step in mathematical education."[45]

In antiquity, Aristotle and others had granted arithmetic a more fundamental standing than geometry. But following the problems faced by those who attempted to express all mathematical ideas in number form, especially the problem of representing irrational quantities, the ancients came to value geometry as more certain than arithmetic. This interpretation was embodied in Euclid's thoroughly geometric formulation of the elements of mathematics. Mathematicians and philosophers followed this tradition. But after more than two thousand years, arithmetical knowledge once again came to be esteemed as superior to geometric propositions, owing to the maturation of algebra and the invention of non-Euclidean geometries. At any rate, mathematicians did not quickly surrender their traditional views. For all practical purposes, though, by the mid-1800s the part of mathematics based on arithmetic "was far more extensive and vital for science than the several geometries."[46]

Despite its growing acceptance, algebra developed with little theoretical footing. Even as algebra became commonly employed by physicists, there remained aspects of algebra that were only vaguely understood, aspects that had not been interpreted clearly or justified in geometric or logical terms. For example, the concepts of negative and imaginary numbers and the use of basic arithmetical operations upon them were practically devoid of justification.[47]

Strange New Algebras of Time and Space

In the early 1800s, a growing number of mathematicians regarded algebra as a symbolic system devoid of meaning. Like others, William Rowan

[44] Comte, *Cours de Philosophie Positive*, p. 207.
[45] Auguste Comte (1843), quoted in Boyer, *History of Analytic Geometry*, p. 267.
[46] Kline, *Mathematics*, p. 89.
[47] Alberto A. Martínez, *Negative Math* (Princeton: Princeton University Press, 2005).

Hamilton was frustrated by the lack of logical foundations of algebra. Hamilton was the Astronomer Royal at Dunsink Observatory and Andrews Professor of Astronomy at Trinity College in Dublin, Ireland. He rejected the view of algebra as a meaningless system of symbols. In contradistinction, he sought an account that would give true meaning to algebra. He sought a kind of meaning that would transcend even physical usefulness in specific applications. Hamilton required that the symbols of algebra must stand for something *intuitively real*.[48] Working in isolation, he wished to know what *real things* are represented by numbers and symbols.[49]

In 1827 Hamilton proposed to reconceive algebra as "the Science of Pure Time," in a similar way as geometry could be understood as the science of space.[50] Previously, Kant had argued that the concepts of number and arithmetic originate from the notion of time.[51] Hamilton apparently arrived at similar ideas before being acquainted with Kant's writings, yet he found support in Kant's *Critique of Pure Reason* when he read this work in the 1830s.[52] Hamilton noted that he had been "encouraged" to pursue and publish his ideas upon reading Kant's *Critique*.[53] Kant—and, to a greater

[48] William Rowan Hamilton, "Theory of Conjugate Functions, or Algebraic Couples; with a Preliminary and Elementary Essay on Algebra as the Science of Pure Time," *Transactions of the Royal Irish Academy* 17 (1837), 293–422; reprinted in *The Mathematical Papers of Sir William Rowan Hamilton*, vol. 3, ed. A. W. Conway and J. L. Synge (Cambridge: Cambridge University Press, 1967), p. 5. See also Thomas L. Hankins, *Sir William Rowan Hamilton* (Baltimore: Johns Hopkins University Press, 1980), p. 250.

[49] Hamilton, "Theory of Conjugate Functions," and Hankins, *Sir William Rowan Hamilton*, p. 254.

[50] W. R. Hamilton, Abstract to "Account of a System of Rays," presented to the Royal Irish Academy on 23 April 1827; reprinted in Robert Perceval Graves, *Life of William Rowan Hamilton*, 3 vols. (Dublin: Hodges, Figgis, & Co., 1882), vol. 1, pp. 228–229.

[51] Immanuel Kant, *Prolegomena zu einer jeden künftigen Metaphysik die als Wissenschaft wird auftreten können* (Riga: Johann Friedrich Hartknoch, 1783), sec. 10; *Kant's Prolegomena to Any Future Metaphysics*, trans. and ed. Paul Carus (Chicago: Open Court; London: Kegan Paul, Trench, Trübner & Co., 1902). See also Norman Kemp Smith, *A Commentary to Kant's Critique of Pure Reason*, 2nd ed. (London: Macmillan, 1923); p. 129.

[52] Thomas L. Hankins, "Algebra as Pure Time: William Rowan Hamilton and the Foundations of Algebra," in *Motion and Time, Space and Matter*, ed. Peter K. Machamer and Robert G. Turnbull (Ohio: Ohio State University Press, 1976), pp. 335–339.

[53] W. R. Hamilton, *Lectures on Quaternions: containing a systematic statement of a new mathematical method; of which the principles were communicated in 1843 to the Royal Irish Academy; and which has since formed the subject of successive courses of lectures, delivered in 1848 and subsequent years, in the halls of Trinity College Dublin: with numerous illustra-*

extent, Arthur Schopenhauer afterward—posited the pure intuition of time as the foundation of arithmetic.[54] Hamilton placed the emphasis on *algebra* as based on the notion of time.

The notion that events have a specific order in time served as a foundation for the ordinal character of real numbers. Hamilton hence attempted to justify negative numbers as corresponding to steps backward in time. Furthermore, he proposed an algebraic definition of complex numbers that avoided the concept of imaginary numbers. He interpreted complex numbers as "couples," ordered pairs of real numbers, which he saw as corresponding to sets of moments in time. Hamilton's treatment of algebra was independent of geometric or spatial considerations. Indeed, he construed the major advances in the history of analysis, "the most remarkable discoveries," those of Newton, Lagrange, John Napier and others, as being dependent on the notion of time, say, in the form of ideas about "continuous progression," "fluxions," "change," or "variability."[55]

Although Hamilton had not appealed to geometry in his analysis of algebra published in 1837, he believed that the "sciences of Space and Time" were "intimately intertwined and indissolubly connected with each other."[56] Cartesian algebraic geometry did not appear as a satisfactory integration of these systems, especially for physics. Accordingly, he labored to elucidate a new framework for the analysis of three-dimensional space. But Hamilton, though he struggled for more than a decade to find three-part

---

tive diagrams, and with some geometrical and physical applications (Dublin: Hodges and Smith, 1853), preface.

[54] Kant, *Prolegomena*, sec. 10; and *Kritik der reinen Vernuft* (1781), *Critique of Pure Reason*, trans. and ed. Paul Guyer and Allen W. Wood, Cambridge Edition of the Works of Immanuel Kant (Cambridge: Cambridge University Press, 1998), pp. 144, 633–636. Arthur Schopenhauer, *Vierfache Wurzel des Satzes von zureichenden Grunde*, translated by Karl Hillebrand as "On the Fourfold Root of the Principle of Sufficient Reason," in *Two Essays by Arthur Schopenhauer*, rev. ed. (London: George Bell and Sons, 1891), chap. 6, sec. 38, p. 156. Arthur Schopenhauer, *Die Welt als Wille und Vorstellung*, vol. 2 (1844); chap. 4: "Von der Erkenntniß a priori," translated by E. F. J. Payne, in *The World as Will and Representation*, vol. 2 (New York: Dover, 1966), pp. 34–35, 45, 51, 179, 379; Schopenhauer explicitly includes algebra with arithmetic as constituting the "doctrine of the ground of being in time," only in chapter 7, p. 127.

[55] Hamilton, "Theory of Conjugate Functions," pp. 5–6.

[56] Hamilton, "Account of a Theory of Systems of Rays" (1827), reprinted in Graves, *Life of William Rowan Hamilton*, vol. 1, p. 229.

numbers analogous to complex numbers, failed to find any such triplets upon which he could apply all the ordinary operations of algebra. In 1843 he finally found a system that satisfied most of the algebraic properties that he deemed indispensable. Surprisingly, the new sort of "number" did not consist of three parts but of four. Hence, Hamilton named them *quaternions*.

Hamilton conceived of quaternions as hypercomplex numbers of the form

$$w + ix + jy + kz.$$

He distinguished the final three-part expression, $ix + jy + kz$, from the real number w, by calling the three-part expression a *vector*. He further designated the term w as the *scalar* part of the quaternion, as it may receive any real number values "contained on the one *scale* of progression of numbers from negative to positive infinity."[57] The operations Hamilton defined for his quaternions were mathematically consistent but did not follow all of the ordinary laws of algebra. The single outstanding deviation from ordinary algebra was that quaternions did not obey the commutative law for multiplication. Ordinarily, mathematicians had believed that the product of two numbers is the same regardless of the order in which they are multiplied, such that

$$ab = ba.$$

By contrast, Hamilton abandoned the commutative law to multiply quaternions. He made the i, j, k units obey the following laws:

$$ij = k \quad jk = i \quad ki = j$$
$$ji = -k \quad kj = -i \quad ik = -j.$$

Here, it is clear that $ab \neq ba$. Furthermore, Hamilton also established that

$$ii = jj = kk = ijk = -1,$$

[57] William Rowan Hamilton, "On Quaternions; or on a New System of Imaginaries in Algebra," *Philosophical Magazine* 29 (1846), 26–31, 113–122, 326–328; quotation on p. 26; reprinted in *Mathematical Papers*, vol. 3, quotation on p. 236; emphasis in the original.

in accordance with the ordinary rule for squaring imaginary numbers.[58] Notwithstanding the violation of the commutative law for multiplication, the other traditional laws of arithmetic applied for quaternions as for other numbers. Hamilton had made a new kind of algebra. Thus, it became evident that ordinary algebra was not unique, as alternative schemes could be conceived.

To represent quaternions geometrically, Hamilton eventually proposed that the three imaginary terms of any quaternion be interpreted in analogy to mutually perpendicular lines in space. Meanwhile, he argued that the fourth, real-number, part of quaternions be construed as corresponding to a *scale* of values in *time*. Thus, Hamilton introduced the concept of a scalar as a line that is outside of three-dimensional space, a line without any direction in space.[59] Later, this concept became used to characterize any numerical magnitude, such as ordinary positive and negative numbers, that do not specify any particular orientation in space. But even Hamilton's early interpretation of the real part of a quaternion as a one-dimensional scale hardly served to clarify how this sort of term could be *added* to the imaginary terms representing vectors in three-dimensional space. In geometry, mathematicians did not allow the addition of quantities of different dimensions. For example, it seemed meaningless to add a line to a volume, or a number to a line. Accordingly, the sum of a real number w and the expression $ix + jy + kz$ seemed geometrically meaningless to Hamilton's contemporaries.

While Hamilton developed his theory of quaternions in Ireland, another mathematician working in Germany independently conceived of a similar system. Hermann Grassmann arrived at his theory by attempting to represent negative quantities and the arithmetical operations upon them geometrically. Apparently, Grassmann was unaware of the various attempts made by others to represent complex numbers geometrically. His investigations in geometry were motivated by his conviction that ordinary approaches to geometry lacked a definite and valid foundation. Grassmann believed that "geometry can in no way be viewed, like arithmetic or combination

---

[58] William Rowan Hamilton, manuscript notebook 24.5, entry for 16 October 1843: published in *Mathematical Papers*, vol. 3, pp. 103–105. See also Hamilton, "On a New Species of Imaginary Quantities Connected with the Theory of Quaternions," *Proceedings of the Royal Irish Academy* 2 (1844), 424–434; reprinted in *Mathematical Papers*, vol. 3, pp. 111–116.

[59] Hamilton, "On a New Species," pp. 424–434.

theory, as a branch of mathematics; instead, geometry relates to something already given in nature, namely, space."[60] Hence, he labored to elucidate an algebraic system that would secure geometric truths independently of any appeals to physical experience. Indeed, Grassmann's researches were guided by an interest in an abstract notion of space, a space of indefinitely many dimensions.

Nonetheless, Grassmann established the fundamental principles of his system in accordance with physics. In 1839 he applied his new method of geometric analysis to the physical study of tides. While reformulating the mathematics of tidal theory, he realized that the whole science of mechanics could be simplified by means of his methods. He systematically replicated many results presented in Lagrange's *Analytical Mechanics*, realizing that he could do so "in such a simple way that the calculations often came out more than ten times shorter" than in Lagrange's work.[61] Impressed by the usefulness of his approach, Grassmann extended it in accord with his interest in pursuing a general theoretical basis for the study of geometry. He sought to develop "a purely abstract branch of mathematics" that would consist of laws of greater generality than those of Euclidean and Cartesian geometry.

Grassmann devised a "Theory of Extension," an algebraic scheme meant to underlie all branches of mathematics. It allowed, for example, the analysis of spaces not only of three dimensions but of many, in terms of the relationships among directed line segments. In 1844 Grassmann published his theory in *The Lineal Theory of Extension*.[62]

Although Grassmann's book included methods that were practically identical to the vector methods advanced by Hamilton, Grassmann's ideas remained almost unknown. His ideas were abstract and far removed from the concepts ordinarily used by mathematicians. Few copies of Grassmann's book were sold, and skilled contemporaries, such as August Ferdinand Möbius, Heinrich Richard Baltzer, and even Hamilton himself found

---

[60] Hermann Grassmann, *Die lineale Ausdehnungslehre ein neuer Zweig der Mathematik dargestellt und durch Anwendungen auf die übrigen Zweige der Mathematik, wie auch auf die Statik, Mechanik, die Lehre vom Magnetismus und die Krystallonomie erläutert* (Leipzig: Otto Wigand, 1844), foreword, VII.

[61] Ibid.

[62] A second, much revised edition appeared in 1862: Grassmann, *Die Ausdehnungslehre* (Berlin: Adolph Enslin, 1862). Both were reissued in *Hermann Grassmanns Gesammelte Mathematische und Physikalische Werke*, ed. Friedrich Engel (Leipzig: B. G. Teubner, 1896).

the book difficult to understand.[63] Grassmann used directed magnitudes continually to exemplify the content of his system, but his arguments were complicated and idiosyncratic. When Hamilton finally began to understand Grassmann's work and to gauge its similarity to his own, he developed a great admiration for it: "If I could hope to be put in rivalship with Des Cartes on the one hand, and with Grassmann on the other, my scientific ambition would be fulfilled!"[64] Hamilton's admiration did not remain at such a high point, for he eventually concluded that Grassmann had failed to discover *quaternions*, though he had come quite close. Moreover, Grassmann's approach left his results practically inaccessible. For instance, Hamilton acknowledged that he and Grassmann shared "the interpretation of *B − A*, where *A* and *B* denote *points*, as the *directed line AB*," but he noted that Grassmann "comes to this, in his page 139 of the *Ausdehnungslehre*, after *long* preparations, and ostrich-stomach-needing iron previous doses. I . . . STARTED with the same."[65] Despite the difficulties that plagued the reception of Grassmann's work, many of his ideas were more general and thorough than Hamilton's and were grasped better when they were rediscovered later by other researchers.

Grassmann's contributions exerted practically no influence on the propagation of the ideas of vectorial methods.[66] Nonetheless, his work serves as an example of mathematical approaches that were being developed concurrently with Hamilton's work. For various reasons, a few theorists in the 1800s attempted to develop new algebraic methods for the analysis of space that would be preferable to coordinate geometry. A common characteristic of these pursuits was that the new methods distinguished clearly between the length and direction of lines, such that arithmetical operations upon lines would yield distinct results whether only the magnitude of the lines, or their relative orientations, or both were involved in each operation. This approach was useful in physics because it facilitated the study of the motions of objects. It elucidated the distinction between physical quantities

---

[63] Michael J. Crowe, *A History of Vector Analysis: The Evolution of the Idea of a Vectorial System* (1967; repr., New York: Dover, 1994), pp. 69, 80.

[64] Hamilton, letter to Augustus De Morgan, 31 January 1853; published in Graves, *Life of William Rowan Hamilton*, vol. 3, p. 441.

[65] Hamilton, letter to Augustus De Morgan, 9 February 1853, in ibid., vol. 3, p. 444; emphasis in the original.

[66] Crowe, *History of Vector Analysis*, p. 94.

that involve only magnitude, such as length, and physical quantities that also involve direction, such as velocity.

At first, however, the works of Hamilton and Grassmann did not seem easily applicable to the study of physics. Both Hamilton and Grassmann had formulated their schemes as "pure" mathematical systems. Nonetheless, their guiding interests in spatial analysis and physics served to produce a useful physical mathematics, though clothed in abstract algebraic and metaphysical arguments. It took decades before the physical value of Grassmann's work became appreciated. Yet Hamilton's work generated considerable interest relatively quickly.

Hamilton's system ultimately yielded a simple method for the analysis of three-dimensional space. Thus, Peter Guthrie Tait, Hamilton's successor as the leading advocate of quaternions, claimed emphatically, "*From the most intensely artificial of systems arose, as if by magic, an absolutely natural one!*"[67] Quaternions and vectorial methods were eventually advocated by physicists who wished to replace traditional Cartesian algebra. The discovery that multiple and distinct algebras could be established logically led a few physicists to inspect critically the mathematical methods ordinarily used, to find whether any of the new algebras were actually superior or preferable to ordinary algebra.

Hamilton conceived of the method of quaternions as an alternative to the traditional methods of mathematics and sought a system independent of the procedures of Cartesian algebra: "I regard it as an inelegance and imperfection in this calculus, or rather in the state to which it has hitherto been unfolded, whenever it becomes, or *seems* to become, necessary to have recourse . . . to the resources of ordinary algebra, for the *solution of equations in quaternions.*"[68] Hamilton knew that instead of relying on the x, y, z expression of a vector as in

$$ix + jy + kz,$$

---

[67] Peter Guthrie Tait, "On the Intrinsic Nature of the Quaternion Method," *Proceedings of the Royal Society of Edinburgh*, 2 July 1894; reprinted in P. G. Tait, *Scientific Papers*, 2 vols. (Cambridge: Cambridge University Press, 1900), vol. 2, doc. CXVI, quotation on pp. 394–395; emphasis in the original.

[68] Peter Guthrie Tait. "Quaternions," *Encyclopaedia Britannica* (1886); reprinted in Tait, *Scientific Papers*, vol. 2, doc. CXXIX, quotation on p. 453; emphasis in the original.

so reminiscent of the Cartesian expression of a line segment,

$$\sqrt{x^2 + y^2 + z^2},$$

he could equate the vector to a single letter,

$$ix + jy + kz = \alpha,$$

so that this letter (Hamilton used the Greek alphabet) would suffice to designate the vector. Then, a quaternion could be expressed as the sum

$$a + \alpha,$$

or by a letter q, by letting $q = a + \alpha$. Although Hamilton often used the more complex notation, Tait later emphasized the immense gain in simplicity.[69] Given the abridged notation, the imaginary terms i, j, k would disappear from the face of the calculus, such that Hamilton's novel insight into complex numbers would become invisible along with the older Cartesian traces.

Owing to mathematicians' familiarity with Cartesian algebra and their interest in the study of complex numbers, Hamilton's use of the more complex notation seemed "indispensable to the reception of his method by a world steeped in Cartesianism."[70] Nonetheless, in Tait's opinion, the benefits of using the method of quaternions would be exhibited better by "removing from its formulae the fragments of their Cartesian shell."[71]

> The quaternion exists, as a space-reality, altogether *independent of* and *antecedent to* i, j, k, or x, y, z. *It* is the natural, *they* the altogether artificial, weapon. And I venture further to assert (1) that if Descartes, or some of his brilliant contemporaries, had recognised the quaternion,

---

[69] Ibid.

[70] Tait, "On the Intrinsic Nature of the Quaternion Method" (1894), in Tait, *Scientific Papers*, vol. 2, doc. CXVI, quotation on p. 395.

[71] Peter Guthrie Tait, "On the Importance of Quaternions in Physics," *Philosophical Magazine* (January, 1890); reprinted in Tait, *Scientific Papers*, vol. 2, doc. XCVII, quotation on p. 303.

(and it is quite conceivable that they might have done so), science would have then advanced with even more tremendous strides than those which it has recently taken; and (2) that the wretch who, under such conditions, had ventured to introduce i, j, k, would have been justly regarded as a miscreant of the very basest and most depraved character: possibly subjected to "brave punishments," the *peine forte et dure* at the very least![72]

Tait, at Edinburgh University, helped develop Hamilton's theory as a research tool for physicists. By studying Hamilton's *Lectures on Quaternions* in the 1850s, Tait formulated many new theorems that exploited the use of quaternion (or vectorial) methods in physics, as he regularly wrote articles, texts, and letters to propagate their employment. He published approximately seventy papers on quaternions, and he authored, coauthored, edited, and translated eight books on the subject.

Already physicists employed pictorial geometry less and less. They regarded analytical methods as being more versatile. The emergence of non-Euclidean geometries was taken to suggest that the principles of ordinary geometry were based on physical experience. Traditional formulations of such principles seemed to lack rigorous and purely logical justifications. By contrast, analytic geometry seemed to eliminate such uncertainty. Because algebra served to derive equalities from other equalities, its results appeared as truisms, as they necessarily had to be equivalent to one another. This is what philosophers called "tautologies": statements that are logically true simply because they assert an identity. Hence, analytic geometry was construed as a valid means to escape difficulties involved in the traditional formulation of the principles of geometry, as argued by Hermann von Helmholtz: "We escape them, if in our investigation of basic principles we employ the analytic method developed in modern calculative geometry. The calculation is wholly carried out as a purely logical operation. It can yield no relationship between the quantities subjected to the calculation which is not already contained in the equations forming the starting point of the calculation. For this reason, the mentioned recent [geometric] investigations have been pursued almost exclusively by means of the purely abstract method

[72] Tait, "On the Intrinsic Nature of the Quaternion Method," *Scientific Papers*, vol. 2, quotation on p. 397; emphasis in the original.

of analytic geometry."[73] Analytic methods served both to substantiate the many geometries *and* to formulate physical theory. This high degree of effectiveness had to be confronted by anyone advocating new mathematical methods for physics.

The early interest in developing a version of quaternion methods for physical applications is well illustrated by the works of one of Tait's close friends, the prominent physicist James Clerk Maxwell. He crucially helped to appraise the value of quaternions for many physicists. By 1873, the year his *Treatise on Electricity and Magnetism* was published, Maxwell's reputation as one of the greatest physicists of the nineteenth century had been established. In the late 1860s he became acquainted with quaternion methods partly owing to his friendship with Tait.[74] Both believed that the mathematical methods used in physics should describe physical processes closely as they appear in nature. Maxwell's interest in quaternions thus can be traced to his views on the function of mathematics in physics.

Maxwell appreciated Hamilton's systematic treatment of quantities as consisting of only magnitude, direction, or both. Maxwell ascribed great importance to "the systematic classification of quantities."[75] In 1870, in an address to the Mathematical and Physical Sections of the British Association, he publicly drew attention to quaternions. Subsequently, in a paper titled "On the Mathematical Classification of Physical Quantities," Maxwell emphasized his claims:

> A most important distinction was drawn by Hamilton when he divided the quantities with which he had to do into Scalar quantities, which are completely represented by one numerical quantity, and Vectors, which require three numerical quantities to define them.

[73] Hermann von Helmholtz, "Über den Ursprung und die Bedeutung der geometrischen Axiome" (1870); *Vorträge und Reden*, 5th ed., vol. 2, pp. 1–31; reissued in: Helmholtz, *Epistemological Writings*, trans. Malcolm F. Lowe and ed. Robert S. Cohen and Yehuda Elkana, Boston Studies in the Philosophy of Science, vol. 37 (Boston: D. Reidel, 1977), quotation on p. 5.

[74] Crowe, *History of Vector Analysis*, p. 132.

[75] James Clerk Maxwell, "Address to the Mathematical and Physical Sections of the British Association," *British Association for the Advancement of Science Report* 40 (1870); reprinted in *The Scientific Papers of James Clerk Maxwell*, ed. W. D. Niven, vol. 2 (Cambridge: Cambridge University Press, 1890; repr., New York: Dover, 1965), quotation on p. 218.

> The invention of the calculus of Quaternions is a step towards the knowledge of quantities related to space which can only be compared, for its importance, with the invention of triple coordinates by Descartes.[76]

Maxwell believed that if physicists only paid closer attention to tracing necessary distinctions between physical quantities of different sorts, then it would be easier to visualize physical processes and advance scientific understanding.

Maxwell's interest in quaternions may have been motivated by reservations about the ordinary methods of analysis. Thus, for example, he wrote to Tait that he was trying "to sow 4nion seed at Cambridge." In Maxwell's opinion, "Algebra is very far from o. k. after now some centuries, and diff. calc. is in a mess and $\iiint$ is equivocal at Cambridge with respect to sign. We put down everything, payments, debts, receipts, cash credit, in a row or column, and trust to good sense in totting up."[77] These words, written in late 1872, reveal that Maxwell viewed algebra and calculus with significant dissatisfaction. From his perspective, quaternions may have appeared as a promising alternative formalism. Still, he chose not to thoroughly employ quaternions because he was not entirely satisfied with Hamilton's formulation of its methods.

By the time Maxwell wrote his *Treatise*, he was not inclined to formulate electromagnetic theory in quaternion operations, nor did he think it would help the book's accessibility to readers. However, he believed that "the value of Hamilton's idea of a vector is unspeakable," as he wrote to Tait in correspondence of 1870.[78] Thus he argued in the *Treatise* that

> for many purposes of physical reasoning, as distinguished from calculation, it is desirable to avoid explicitly introducing the Cartesian coordinates, and to fix the mind at once on a point of space instead of

[76] James Clerk Maxwell, "On the Mathematical Classification of Physical Quantities," *Proceedings of the London Mathematical Society* 3 (1871), 224–232; in *Scientific Papers*, vol. 2, quotation on p. 259.

[77] J. C. Maxwell, letter to P. G. Tait, 9 October 1872, in Cargill Gilston Knott, *Life and Scientific Work of Peter Guthrie Tait* (Cambridge: Cambridge University Press, 1911), p. 151.

[78] J. C. Maxwell, letter to P. G. Tait, 14 November 1870, in ibid., p. 144.

its three coordinates, and on the magnitude and direction of a force instead of its three components. This mode of contemplating geometrical and physical quantities is more primitive and more natural than the other, although the ideas connected with it did not receive their full development till Hamilton made the next great step in dealing with space, by the invention of his Calculus of Quaternions.[79]

Hence, Maxwell chose to employ the *notation* of vectors, convinced that it would be "of the greatest use in all parts of science."[80]

One example of Maxwell's dissatisfaction with quaternion methods is Hamilton's prescription for the operation of squaring vectors. In Hamilton's account, the square of any given vector is a negative quantity. Hamilton had established this fundamental convention in accordance with the common understanding of imaginary numbers. Ordinarily,

$$i^2 = ii = -1,$$

just because that was why the term i had been introduced in the first place: to serve as a solution for the operation of extracting the square root of a negative number. Accordingly, when Hamilton conceived of the hypercomplex terms of a quaternion, namely, i, j, k, he expected that

$$ii = jj = kk = -1.$$

This idea served well to establish a consistent system of quaternion multiplication. When only vectors are considered, however, for the purpose of physical science, this convention seemed rather unnatural. In the expression

$$w + ix + jy + kz,$$

the terms i, j, k would be considered as vectors directed along the x-, y-, and z-axes respectively (their sum, $ix + jy + kz$, of course, is also a vector). Given any such vector, its square would equal a quantity with a negative

[79] James Clerk Maxwell, *A Treatise on Electricity and Magnetism*, vol. 1 (1873), 2nd ed. (Oxford: Clarendon Press, 1881), p. 9.

[80] Maxwell, "On the Mathematical Classification of Physical Quantities" (1871), in *Scientific Papers*, vol. 2, quotation on p. 259.

sign prefixed. Consider now a vector meant to represent a physical quantity, such as a velocity, having both magnitude and direction. The square of this vector would be negative, an odd result to interpret in physical terms. For instance, the kinetic energy of an object, $\frac{1}{2}mv^2$, would yield a negative result. In contradistinction, the square of a velocity analyzed in Cartesian algebra would be positive. Thus, Maxwell complained that "it is troublesome, to say the least, to find the square of AB is always positive in Cartesians and always negative in 4nions, and that when the thing is mentioned incidentally you don't know which language is being spoken. . . . It is also awkward when discussing, say, kinetic energy to find that to ensure its being +*ve* you must stick a – sign to it."[81] Therefore, Maxwell abstained from employing quaternion operations for the purpose of mathematical calculations, but he did employ the *language* of quaternions to make some mathematical results appear more suggestive and expressive.

Maxwell did not rely on Hamilton's *methods* for the advancement of electromagnetic theory, but he viewed Hamilton's *ideas* as a useful instrument for physical reasoning, that is, for the interpretation of symbols as visualizable entities. Maxwell detailed his position in a review of Philip Kelland and Tait's *Introduction to Quaternions* of 1873: "Now Quaternions, or the doctrine of Vectors, is a mathematical method, but it is a method of thinking, and not, at least for the present generation, a method of saving thought. . . . It calls upon us at every step to form a mental image of the geometrical features presented by the symbols, so that in studying geometry by this method we have our minds engaged with geometrical ideas."[82] For Maxwell, the quaternion approach was more than just a system to abbreviate Cartesian expressions; it was a means of understanding physical relations in space, though it fell short of being necessarily a labor-saving device. The historian Michael Crowe characterized Maxwell's stance on quaternions, pointing out that "by means of the vectorial approach the physicist attains to a direct mathematical representation of physical entities and is thus aided in seeing the physics involved in the mathematics."[83] Hence, Maxwell had even encouraged Tait to defend quaternions on the grounds that "the virtue of the 4nions lies not so much as yet in solving hard questions, as in

---

[81] J. C. Maxwell, letter to P. G. Tait, 7 September 1878, in Knott, *Life and Scientific Work of Peter Guthrie Tait*, pp. 151–152.

[82] Maxwell, quoted in Crowe, *History of Vector Analysis*, pp. 133–134.

[83] Crowe, *History of Vector Analysis*, p. 134.

enabling us to see the meaning of the question and of its solution, instead of setting up the question in $x\, y\, z$, sending it to the analytical engine, and when the solution is sent home translating it back from $x\, y\, z$, so that it may appear as A, B, C to the vulgar."[84]

Maxwell's *Treatise* motivated some physicists to study quaternions. Given Maxwell's praises and appeals to quaternion expressions, plus the reasonable impression that he had perhaps abstained from relying more fully on quaternions throughout simply because prospective readers were not quite ready for such methods, "it seems probable that not a few readers of his *Treatise* left it with the impression that Maxwell had rather strongly recommended Quaternions."[85] Yet Maxwell's reservations against Hamilton's methods continued throughout the remainder of his career. In none of the four books he wrote after 1873 did he even once mention quaternions, and only once did he briefly discuss the idea of vectors, in *Matter and Motion*, his elementary treatment of mechanics first published in 1876. In sum, Maxwell's appraisal of quaternions was not as enthusiastic as that of other advocates. Tait agreed with Maxwell that quaternion expressions were more representative of physical relations than the corresponding Cartesian expressions. But Tait and others further believed that using quaternions as a method of investigation was superior to the ordinary forms of analysis.

To its leading advocates, quaternions were an extraordinary contribution to the methods of physics. To Hamilton, the advent of the new method was comparable to the invention of the calculus by Newton and Leibniz. He wrote that "this discovery appears to me to be as important for the middle of the nineteenth century as the discovery of fluxions was for the close of the seventeenth."[86] Tait, in turn, insistently argued the superiority of quaternions over coordinate geometry. He claimed that "quaternions form as great an advance relatively to Cartesian methods as the latter, when first propounded, formed relatively to Euclidean geometry."[87] His assessment of the advantages of quaternions over Cartesian algebra warrants special attention, for he was the most prominent physicist who critically argued

---

[84] J. C. Maxwell, letter to P. G. Tait, 2 November 1871, in Knott, *Life and Scientific Work of Peter Guthrie Tait*, p. 101.

[85] Crowe, *History of Vector Analysis*, p. 137.

[86] Graves, *Life of William Rowan Hamilton*, vol. 2, p. 445.

[87] Peter Guthrie Tait, "Hamilton," *Encyclopaedia Britannica* (1880); reprinted in Tait, *Scientific Papers*, vol. 2, doc. CXXVIII, quotation on p. 443.

against the use of coordinate methods in physics. Maxwell had written his *Treatise* in what he called a "bilingual" approach,[88] whereas Tait construed quaternions as the only legitimate language of space analysis.

Beyond advocating quaternions as the most valuable language for physics, Tait claimed that physicists should abandon Cartesian methods. According to him, the days of coordinate geometry should end: "The intensely artificial system of Cartesian coordinates, splendidly useful as it was *in its day*, is one of the wholly avoidable encumbrances which now retard the progress of mathematical physics."[89] Tait repeatedly denounced coordinates as utterly "artificial," while characterizing quaternions as easily intelligible and natural. He deemed it necessary to advocate Hamilton's approach *at the expense of* coordinate geometry because quaternions were designed for the solution of the same sorts of problems commonly solved analytically. He claimed that the new method was "expressly fitted for the symmetrical evolution of truths which are usually obtained by the ordinary Cartesian methods only after great labour of calculation, and by modes so indirect, and at first sight so purposeless, as to bewilder all but a very small class of readers."[90] He claimed that quaternions served to formulate physical results in expressions that were simpler than, or at least equally simple as, any other approach. In particular, he argued that quaternions "enable us to exhibit in a singularly compact and elegant form, whose meaning is obvious at a glance on account of the utter inartificiality of the method, results which in the ordinary Cartesian coordinates are of the utmost complexity."[91]

To illustrate, consider the trajectories of two bodies (see figure 19). The two depart from one point; relative to us, one takes a horizontal path, while the other takes a diagonal path. In coordinate geometry the horizontal tra-

[88] E.g., J. C. Maxwell, letter to Tait, 7 September 1878, in Knott, *Life and Scientific Work of Peter Guthrie Tait*, p. 151.

[89] Tait, "On the Importance of Quaternions in Physics" (1890), in Tait, *Scientific Papers*, vol. 2, doc. XCVII, quotation on p. 299; emphasis in the original.

[90] Peter Guthrie Tait, "On the Rotation of a Rigid Body about a Fixed Point," *Transactions of the Royal Society of Edinburgh* 25; received on 13 October, read on 21 December 1868; reprinted in Tait, *Scientific Papers*, vol. 1, doc. XV, quotation on p. 86.

[91] Peter Guthrie. Tait, "Address to Section A of the British Association," *British Association Report*, Edinburgh (3 August 1871); reprinted in Tait, *Scientific Papers*, vol. 1, doc. XXIII, quotation on p. 164.

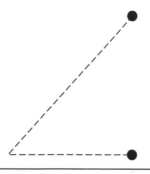

Figure 19. Two moving bodies

jectory would be represented by a simple expression such as *x*, whereas the diagonal trajectory would be represented by an expression such as

$$\sqrt{x^2 + y^2 + z^2}.$$

By contrast, vector methods facilitated equally simple symbols to designate the two trajectories, such as α and β, respectively. Therefore, Tait judged coordinate representations to be cumbersome and artificial. He noted that vector concepts "enable us by a mere mark to separate the ideas of length and direction without introducing the cumbrous square roots of sums of squares which are otherwise necessary."[92]

Tait explained that thanks to quaternions he had been able to understand, "with a thoroughness and vastness of comprehension which I had despaired of attaining (at least with Cartesian processes)," the many applications of Laplace's equation.[93] He used historical hyperboles in his criticisms of Cartesian methods. Thus he said of "the puzzling machinery" of coordinates that it "ought never to have been introduced, or that its use was an indication of a comparatively savage state of mathematical civilization."[94]

---

[92] Ibid.

[93] Peter Guthrie Tait, "On Green's and Other Allied Theorems," *Transactions of the Royal Society of Edinburgh* 26, received on 29 April, read on 16 May 1870; reprinted in Tait, *Scientific Papers*, vol. 1, doc. XIX, quotation on p. 137.

[94] Tait, "Address to Section A of the British Association" (1871), in Tait, *Scientific Papers*, vol. 1, doc. XXIII, quotation on p. 166.

He praised the advantages of quaternions and modern quantitative systems over outdated arbitrary schemes:

> Comparing a Quaternion investigation, no matter in what department, with the equivalent Cartesian one, even when the latter has availed itself to the utmost improvements suggested by Higher Algebra, one can hardly help making the remark that they contrast even more strongly than the decimal notation with the binary scale or with the old Greek Arithmetic, or even the well-ordered subdivisions of the metrical system with the preposterous no-systems of Great Britain, a mere fragment of which (in the form of Tables of Weights and Measures) forms perhaps the most effective, if not the most ingenious, of the many instruments of torture employed in our elementary teaching.[95]

Other advocates of quaternions, such as Alexander McAulay and Alexander Macfarlane, regarded Hamilton's work as "the greatest mathematical work of the century."[96] They felt that quaternions embodied the culmination of the long quest for an efficient method of space analysis. Thus, Macfarlane claimed that Hamilton's researches "contain what was long sought-after—a veritable extension of algebra to space: I do not say *the,* for I believe that there is more than one. The Cartesian analysis is also an extension of algebra to space, but it is fragmentary and incomplete."[97] Following Tait's criticisms of Cartesian methods, a few other theorists also entertained the replacement of ordinary analytic geometry in physics. McAulay, for one, argued that physics would advance more rapidly if quaternions were "introduced to serious study to the almost total exclusion of Cartesian Geometry, except in an insignificant way, as a particular case of the former."[98]

Hamilton's theory of quaternions generated considerable interest. By the time of his death in 1865, Hamilton had published 109 papers on quaternions, and fifteen other authors had contributed a total of 41 additional

---

[95] Ibid., p. 167.
[96] Alexander Macfarlane, "[Review of] *Utility of Quaternions in Physics.* By A. McAulay," in *Physical Review* 1 (1893), 389.
[97] Ibid.; emphasis in the original.
[98] Alexander McAulay, *Utility of Quaternions in Physics* (London: Macmillan, 1893), p. 2.

papers to the subject.[99] By 1890 the number of articles in print exceeded 400, and no less than twenty-seven books had been published. Notwithstanding this significant number of publications, quaternion advocates often felt that their system received very little attention, especially from physicists. Thus, McAulay, for example, complained, "It is a curious phenomenon in the History of Mathematics that the greatest work of the greatest Mathematician of the century which prides itself upon being the most enlightened the world has yet seen, has suffered the most chilling neglect."[100] McAulay described the great difficulties he experienced in trying to generate interest in Hamilton's method: "To break down the solid and well-nigh universal scepticism as to the utility of Quaternions in Physics seemed too much like casting one's pearls—at least like crying in the wilderness."[101] Such remarks seem peculiar, in retrospect, considering that few other new mathematical or physical systems generated as much interest in the nineteenth century.[102] Nonetheless, the concerns of quaternion advocates were justified. Not only did Hamilton's system fail to acquire the currency that some expected, but there also appeared a variation of his system, an alternate approach, that eventually came to replace the use of quaternions by physicists.

Which Algebra Is Fictitious and Unnatural?

Vector theory emerged from quaternions as a simplification tailored to the needs of physicists. It was chiefly the contribution of two individuals, Josiah Willard Gibbs and Oliver Heaviside, working independently of each other. Gibbs worked in the United States as a professor of mathematical physics at Yale University. Heaviside lived in London and did not work for a university nor had he even received a university education. Although Gibbs and Heaviside developed vector methods almost simultaneously, Gibbs's efforts began slightly earlier.

Gibbs first became acquainted with quaternions by reading Maxwell's *Treatise on Electricity and Magnetism* of 1873. Gibbs decided to pursue Hamilton's methods to master electromagnetic theory but soon realized that "the

---

[99] Crowe, *History of Vector Analysis,* p. 41.
[100] McAulay, *Utility of Quaternions in Physics,* p. 1.
[101] Ibid., p. lx.
[102] Crowe, *History of Vector Analysis,* p. 219.

idea of the quaternion was quite foreign to the subject," because Maxwell actually did not employ quaternions but vectors.[103] Hence, Gibbs formulated a systematic approach in which the vector concept was fundamental and quaternions did not appear. He privately printed, in 1881 and 1884, a pamphlet titled *Elements of Vector Analysis* and personally distributed copies to many leading scientists. After devising his vectorial system, he became acquainted with Grassmann's work and realized that Grassmann's methods were essentially identical to his own. Thus, he became more sympathetic to Grassmann than Hamilton. Nonetheless, Gibbs's vector theory arose from his revisions of the work of Hamilton and Tait.[104] The substance of the vector equations advanced by Gibbs was not original, but their presentation, in the form of a new symbolism, helped many physicists to subsequently adopt vector methods.

In addition to circulating his pamphlet on vector analysis, Gibbs encouraged physicists to employ vector methods by lecturing on the subject and publishing various articles on it. In an address to the Section of Mathematics and Astronomy of the American Association for the Advancement of Science, Gibbs argued that modern geometry had evolved to a stage in which new algebraic methods were necessary for the treatment of spatial magnitudes, "and therefore, that a certain logical necessity calls for throwing off the yoke under which analytical geometry has so long labored."[105] In addition to the application of vector ideas to the science of electricity and magnetism, Gibbs advocated the adoption of vectors in other branches of science, such as astronomy. Moreover, his arguments sought to respond to the needs of any physicist who, like Maxwell, had sympathized with quaternion ideas but had found quaternion methods unsatisfactory. Thus, in a paper demonstrating the usefulness of vectors in the field of astronomy, Gibbs stated that his objective "could best be obtained, not by showing, as I might have done, that much in the classic methods could be conveniently and perspicuously represented by vector notations, but rather by showing

[103] J. W. Gibbs, letter to Victor Schlegel, 1 August 1888, in Lynde Phelps Wheeler, *Josiah Willard Gibbs: The History of a Great Mind* (New Haven: Yale University Press, 1962), p. 107.

[104] Crowe, *History of Vector Analysis*, p. 155.

[105] Josiah Willlard Gibbs, "On Multiple Algebra," *Proceedings of the American Association for the Advancement of Science* 35 (1886), 37–66; quotation on p. 49; reprinted in *The Scientific Papers of J. Willard Gibbs*, vol. 2 (New York: Dover, 1961), quotation on p. 103.

that these notations so simplify the subject, that it is easy to construct a method for the complete solution of the problem."[106]

While Gibbs was developing vector algebra in America, Oliver Heaviside, a retired telegraph operator and engineer, was carrying out practically the same program in England. Like Gibbs, Heaviside first became acquainted with quaternions through Maxwell's *Treatise*. In his early studies, Heaviside realized that operations with quaternions did not yield results in full agreement with Cartesian mathematics. Moreover, the quaternion approach did not appear sufficiently convenient when applied to electrical theory. Heaviside realized the utility of vectors, as Maxwell and Tait had employed them, but he did not understand why the laws of vector algebra were justified in terms of quaternions: "I am not sure that anyone has ever understood this establishment. . . . I never understood it."[107] He argued, for example, that in Hamilton's system "some of the properties of vectors professedly proved were wholly incomprehensible. How could the square of a vector be negative?"[108] Quaternions appeared as a hindrance to physical understanding, a poor way to establish the laws of vectors: to Heaviside, "the quaternionic is an undesirable way of beginning the subject, and impedes the diffusion of vectorial analysis in a way which is as vexatious and brain-wasting as it is unnecessary."[109]

Heaviside decided to simplify the method of quaternions and deal only with vectors. In correspondence with Tait, Heaviside disclaimed "any idea of discovering a new system," as he admitted to having but "derived my system from Hamilton and Tait by elimination and simplification," to devise "a thoroughly practical system."[110] Heaviside desired a vector system that would "harmonise" with ordinary mathematics.[111] After a series of preliminary papers, Heaviside published his first unified presentation of his vector

[106] Josiah Willard Gibbs, "On the Determination of Elliptic Orbits from Three Complete Observations," *Memoirs of the National Academy of Sciences* 4 (1889), part. II, 79–104; reprinted in *The Scientific Papers of J. Willard Gibbs*, vol. 2, quotation on p. 104.

[107] Oliver Heaviside, *Electromagnetic Theory*, vol. 1, chap. III: "The Elements of Vectorial Algebra and Analysis" (London, 1893); repr., New York: Chelsea, 1971), quotation on p. 137.

[108] Heaviside, *Electromagnetic Theory*, vol. 3 (London: Benn, 1912), p. 135.

[109] Heaviside, *Electromagnetic Theory*, vol. 1, chap. III: "The Elements of Vectorial Algebra and Analysis," quotation on p. 137.

[110] Heaviside, *Electromagnetic Theory*, vol. 3, p. 137.

[111] Ibid., p. 136. See also Crowe, *History of Vector Analysis*, p. 163.

system in 1885. In his paper "On the Electromagnetic Wave Surface," he explained: "Owing to the extraordinary complexity of the investigation when written out in Cartesian form (which I began doing, but gave up aghast), some abbreviated method of expression becomes desirable. I may also add nearly indispensable, owing to the great difficulty in making out the meanings and mutual connections of very complex formulae. . . . I therefore adopt, with some simplification, the method of vectors."[112] In 1891 Heaviside further characterized his system: "It rests entirely upon a few definitions, and may be regarded (from one point of view) as a systematically abbreviated Cartesian method of investigation, and be understood and practically used by any one accustomed to Cartesians, without any study of the difficult science of Quaternions. It is simply the elements of Quaternions without the quaternions, with the notation simplified to the uttermost, and with the very inconvenient *minus* sign before scalar products done away with."[113] A year later, in 1893, Heaviside published a lengthy treatment of the subject as chapter 3 of the first volume of his *Electromagnetic Theory*. Because there did not exist any other introductory texts on vector methods "suitable for mathematical physics," Heaviside offered this chapter of his book "useful as a stopgap."[114]

Heaviside presented vector mathematics as a "language" equivalent to Cartesian analysis but better suited to represent physical magnitudes with "natural" simplicity and compactness. He acknowledged that it was *sometimes* useful to express vector magnitudes in the Cartesian components, x, y, z, as physicists ordinarily did. "[In] the usual treatment of physical vectors, there is an avoidance of the vectors themselves by their resolution into components. That this is a highly artificial process is obvious, but it is often convenient. More often, however, the Cartesian mathematics is ill-adapted to the work it has to do, being lengthy and cumbrous, and frequently calculated to conceal rather than to furnish and exhibit results in a useful manner."[115] He added that, "in the Cartesian method, we are led away from

---

[112] Oliver Heaviside, *Electrical Papers*, vol. 2 (London: Macmillan, 1892), p. 3.

[113] Oliver Heaviside, "On the Forces, Stresses, and Fluxes of Energy in the Electromagnetic Field," Royal Society: Abstract in *Proceedings* 50 (1891); *Transactions*, A. (1892); reprinted in Heaviside, *Electrical Papers*, vol. 2, quotation on p. 529.

[114] Heaviside, *Electromagnetic Theory*, vol. 1, quotation in the preface, [p. ii], see also p. 305.

[115] Heaviside, *Electromagnetic Theory*, vol. 1, chap. III: "The Elements of Vectorial Algebra and Analysis," quotation on p. 133.

the physical relations that it is so desirable to bear in mind, to the working out of mathematical exercises upon the components. It becomes, or tends to become, blind mathematics."[116] Opposed to the "bulky inanimateness" of the Cartesian system, Heaviside suggested that symbolic expressions should represent their physical counterparts clearly, such that "the mere sight of the arrangement of symbols should call up an immediate picture of the physics symbolised."[117]

Gibbs and Heaviside realized independently that, although Hamilton and his followers stressed the utility of quaternions in physics, in actuality only vectors were involved in most physical applications of quaternion methods. For Hamilton and Tait, any vector was itself a quaternion,

$$w + ix + jy + kz,$$

where $w = 0$. By contrast, Gibbs and Heaviside saw no practical need to mix the "imaginary" part of the expression with any real number w, and thus they accepted the vector, $ix + jy + kz$, as a fundamental entity, as a concept independent of any reference to quaternions. Furthermore, the expression was further simplified by omitting any reference to imaginaries and by denoting the vector by a single letter, as the proponents of quaternions already had suggested. Thus, instead of beginning the analysis of space with the quaternion concept, or with the terms i, j, k, or with the coordinates x, y, z, Gibbs and Heaviside established the operations of vectors simply in terms of single characters to represent each vector. Gibbs and Heaviside placed the emphasis on vectors directly, understanding that, despite the use of the word "quaternion" in the writings of Hamilton, Maxwell, Tait, and others, what were really involved in most physical applications were simply vectors. Heaviside argued:

> For the word "quaternion," we may read vector or vectorial here, because a vector is considered by Hamilton and Tait to be a quaternion, or is often counted as one. This practice is sometimes confusing. Thus the important operator ∇ is called a quaternion operator. It is really a vector. It is as unfair to call a vector a quaternion as to call a man a quadruped; although, four including two, the quadruped

[116] Ibid.
[117] Ibid., p. 134.

might be held (in the matter of legs) to include the biped, or, indeed, the triped, which would be more analogous to a vector. It is also often inconvenient that the name of the science, viz., Quaternions, should be a mere repetition of the name of the operator. There is some gain in clearness by preserving the name "quaternion" for the real quaternion—the quadruped, that is to say.[118]

Likewise, in 1891 Gibbs published an article in *Nature* in which he indicated that even Tait, for the most part, had used vectors rather than quaternions in his *Elementary Treatise on Quaternions* of 1890.[119]

In accordance with Maxwell's criticisms of some of the principles of quaternions, Gibbs and Heaviside tailored the principles of vectors to maximize their efficiency and practicality in physical analysis. They avoided appeals to any quaternion expressions, as quaternions simply appeared as sums of scalar and vector terms, a mixture that went against Maxwell's prescription to distinguish between physical quantities. Furthermore, Gibbs and Heaviside both defined the square of a vector as *positive*, thus abandoning Hamilton's original association of vectors with imaginary quantities. Likewise, Macfarlane, though endorsing the concept of quaternion, diverged from Hamilton and Tait by arguing that the squares of vectors should be positive. This move, though physically convenient, served to further distance vector methods from ordinary algebra, as some advocates of quaternions readily pointed out.

By the late 1800s, however, various algebras existed as independent and consistent systems, such that the departures of Gibbs, Heaviside, and others from the parameters of traditional algebra could just as well be accepted, at least in the interest of practicality in physical mathematics.

Of central importance in the applicability of vector algebras to physics was the operation of combining vectors. The "addition of velocities" was the name given by vector theorists to the concept of compounding velocities. Directed magnitudes, such as the velocities w' and v, were said to be "added" when placed end to end and represented by the resultant vector, v' (see figure 20). For example, a person inside a train car walks at a velocity v with respect to the car. Meanwhile, the train moves relative to the ground

---

[118] Ibid., quotation on p. 172.
[119] Josiah Willard Gibbs, "Quaternions and the 'Ausdehnungslehre,'" *Nature* 44 (28 May 1891), 79–82; reprinted in *The Scientific Papers of J. Willard Gibbs*, vol. 2.

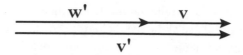

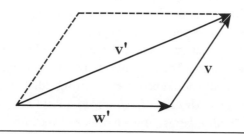

Figure 20. Two velocities, added together

Figure 21. Two velocities with distinct directions, added together

with a velocity **w′**. Hence one could say that the person moved relative to the ground at a velocity **v′**. This vector addition would be expressed symbolically by the equation

$$\mathbf{v} + \mathbf{w}' = \mathbf{v}'.$$

This expression was fully in accord with traditional algebraic and arithmetic ideas. However, this same equation was also employed by vector theorists to describe the composition of velocities in any arbitrary direction, such as shown in figure 21.

Here the concept of "addition" no longer carried the same meaning as in ordinary arithmetic.[120] This was because the lengths of the lines **w′** and **v** are not added *numerically* to produce the resultant length of **v′**.

The rules for the composition of directed magnitudes existed long before the creation of vector theory. For centuries earlier, the rule for adding directed magnitudes, such as velocities or forces, had existed and had been expressed by means of diagrammatic and algebraic geometry. Geometers since antiquity, as well as during the Renaissance, occasionally employed diagrams showing that the result of compounding two rectilinear motions

[120] E.g., see Bertrand Russell, *The Principles of Mathematics* (1903; 2nd ed., New York: W. W. Norton, 1938), article 451, p. 477.

is the diagonal line of the parallelogram determined by the two motions.[121] Despite the widespread use of the parallelogram construction, not until the nineteenth century was it designated as the "addition" of lines.[122]

To gain support for their system, vector theorists used a different approach from that used by the advocates of quaternions. Both camps had to contend with the problem that the areas of physics served by their methods were already routinely covered by the methods of Cartesian coordinate geometry. In response, the most vocal advocates of quaternions chose to argue that physicists should *abandon* the older approach to benefit from the new method. For example, as mentioned earlier, Tait argued that Cartesian methods should be rejected entirely, that they should not be mixed with quaternions, because "recourse to quasi-Cartesian processes is fatal to progress."[123] By contrast, the advocates of vector theory argued that their method should be adopted *alongside* the methods of Cartesian coordinate geometry.

Vector theorists were quite aware that most physicists had not substituted their reliance on ordinary analytic methods for the substance of Hamilton's theory. The attacks of critics, such as Tait, against the ordinary tools of mathematical physics had not swayed physicists away from their established methods. For the most part, physicists disregarded quaternions and construed them as obscure. In turn, Heaviside and Gibbs employed the strategy of dissociating their system from that of Hamilton.[124] In their arguments, vectors appeared simply as *a convenient shorthand* for coordinate geometry. Thus, Gibbs argued, "If I wished to attract the student of any of these sciences to an algebra for vectors, I should tell him that the fundamental notions of this algebra were exactly those with which he was daily conversant . . . [those] which he who reads between the lines will meet on every page of the great masters of analysis."[125] In this approach, vec-

---

[121] See, e.g., René Dugas, *A History of Mechanics*, translated from the French by J. R. Maddox (New York: Dover, 1988), e.g., pp. 21, 151, 208; Max Jammer, *Concepts of Force: A Study in the Foundations of Dynamics* (1957); repr., New York: Harper and Brothers, 1962), pp. 128–133.

[122] Crowe, *History of Vector Analysis*, p. 2.

[123] Peter Guthrie Tait, *An Elementary Treatise on Quaternions*, 3rd ed. (Cambridge: Cambridge University Press, 1890), p. vii.

[124] Crowe, *History of Vector Analysis*, p. 196.

[125] Josiah Willard Gibbs, "Quaternions and the Algebra of Vectors," *Nature* 47 (16 March 1893), 464; reprinted in *Scientific Papers of J. Willard Gibbs*, vol. 2, quotation on p. 171.

tor theory was vector *algebra* and, in broader terms, vector *analysis*. Much like Gibbs, Heaviside contrasted the approach of the quaternion advocates to his own. In ending his chapter on "Elements of Vectorial Algebra and Analysis," Heaviside wrote: "The quaternionists want to throw away the 'cartesian trammels,' as they call them. This may do for quaternions but with vectors would be a grave mistake. My system, so far from being inimical to the cartesian system of mathematics, is its very essence."[126] From early in his investigations, Heaviside had sought a vector approach that would be "in harmony with the Cartesian mathematics (a matter to which I attach the greatest importance)."[127] In his early papers, Heaviside's approach appeared as a middle ground between quaternions and Cartesian algebra, though in its fully developed form, Heaviside presented vectors as independent of quaternions and equivalent to Cartesian coordinate expressions: "The vector analysis I use may be described either as a convenient and systematic abbreviation of Cartesian analysis; or else, as Quaternions without the quaternions, and with a simplified notation harmonising with Cartesians."[128] This tactic worked, as physicists more readily incorporated the use of vector theory into their researches, while quaternions remained rather neglected.

Physicists' criticisms of Cartesian methods in the 1890s emerged within a broader struggle between the advocates of quaternions and of vectors. Perhaps vector and quaternion advocates opposed one another mostly for tactical reasons, because it was clear that vector theory originated in Hamilton's work and was, in its original form, an integral part of quaternions. Because quaternions included vectors, many criticisms against vectors would count also against quaternions, whereas the converse was not necessarily true. Despite debates about methods, foundations, notations, and so forth, their approaches were quite similar: "Gibbs and Heaviside were indeed heretics, but their cause was 90 percent in harmony with the Hamilton-Tait orthodoxy."[129] Moreover, the two approaches could be made to harmonize with Cartesian methods. This, for example, was the path taken

---

[126] Heaviside, *Electromagnetic Theory*, vol. 1, chap. III: "The Elements of Vectorial Algebra and Analysis," quotation on p. 305.
[127] Ibid., preface.
[128] Ibid., quotation on p. 135.
[129] Crowe, *History of Vector Analysis*, p. 219.

by Macfarlane in his treatment of the subject: "By 'Vector Analysis' is meant a space-analysis in which the vector is the fundamental idea; by 'Quaternions' is meant a space-analysis in which the quaternion is the fundamental idea. They are in truth complementary parts of one whole; and . . . they will be treated as such, and developed so as to harmonize with one another and with the Cartesian Analysis."[130] In the long run, a focus on quaternions was just not convenient for the needs of most physicists. Heaviside spoke of physicists' lack of interest in quaternions as a sign of their lack of naturalness and usefulness: "They don't want it. They have said so by their silence. Common sense of the fitness of things revolts against the quaternionic doctrines about vectors. Nothing could be more unnatural."[131] Meanwhile, Heaviside's researches in electricity and his contributions to electromagnetic theory, cast in vectorial form, did much to encourage physicists to employ vector theory.

Regardless of the similarities or equivalence between the various approaches, practicing physicists had to decide whether to replace their use of Cartesian methods and, if so, to what extent. After all, the mere equivalence of analytic geometry and a vectorial system did not necessarily warrant the abandonment of one for the other. In this context, and given the diverse attacks upon Cartesian methods, it appears odd that few arguments were published on the fundamental necessity to pursue any vector mathematics. Michael Crowe studied the debate between the advocates of the different systems from 1890 to 1894 and found it "surprising that only two papers in the debate directly discussed the fundamental question of whether any vectorial system should be adopted, especially since vectorial systems were at the time still something of an innovation."[132] Only the prominent mathematician Arthur Cayley cared enough to argue in print that no vector system should be adopted. Presumably, however, "a large number of readers of the debate must, at least initially, have been of that persuasion."[133] Consider, for example, that William Thomson, "probably the most influential British physicist of the late nineteenth century," strongly opposed the use of quater-

[130] Alexander Macfarlane, *Vector Analysis and Quaternions* (1896); 4th ed. (New York: John Wiley & Sons; London: Chapman & Hall, Ltd., 1906), p. 7.
[131] Heaviside, *Electromagnetic Theory*, vol. 1, chap. III: "The Elements of Vectorial Algebra and Analysis," quotation on p. 303.
[132] Crowe, *History of Vector Analysis*, p. 218.
[133] Ibid.

nions and vectors in physics for decades, although he never argued the matter in print.[134] Thomson (who since 1892 became known as Lord Kelvin) collaborated with Tait in authoring a *Treatise on Natural Philosophy*, an authoritative treatment of the subject of mechanics, but because Thomson did not share Tait's mathematical sympathies, he prevented any employment of quaternion methods in it. Given this sort of silent rejection of vectorial methods, Cayley's public and critical participation in the discussions of the relative values of vectorial and Cartesian methods warrants attention.

Cayley was the only professional mathematician who contributed a major article to the debate, while most of the other participants were physicists. Cayley was the first person after Hamilton to publish a paper on quaternions. Subsequently, after years of research, Cayley claimed to have "the highest admiration" for quaternions as a pure mathematical subject, but he did not endorse their application in physics.[135] In an article of 1895 titled "Coordinates versus Quaternions," he argued that quaternion formulas had to be translated into coordinates to be understood. Cayley summarized his position by stating that, "while coordinates are applicable to the whole of geometry, and are the natural and appropriate basis and method in science, quaternions seem to me a particular and very artificial method for treating such parts of the science of three-dimensional geometry as are most naturally discussed by means of the rectangular coordinates x, y, z."[136] Of course, Cayley's arguments also applied to vectors. He was encouraged to read his paper before the Royal Society of Edinburgh by Tait, with whom he had corresponded frequently. In response to Cayley's objections, Tait argued that quaternions were not merely "a species of Analytical Geometry," as Cayley would have it, but a natural "Mode of Representation." This was essentially the line of defense that Maxwell had advised Tait to employ. But Tait argued further:

> To me Quaternions are primarily a Mode of Representation:—immensely superior to, but of essentially the same kind of usefulness as, a diagram or a model. They *are*, virtually, the thing represented:

[134] Ibid., p. 120.
[135] Arthur Cayley, "Coordinates Versus Quaternions," *Proceedings of the Royal Society of Edinburgh* 20 (1895), 271–275; read on 2 July 1894; reprinted in *The Collected Mathematical Papers of Arthur Cayley*, vol. 13 (Cambridge: Cambridge University Press, 1897), doc. 962, pp. 541–544; quotation on p. 541.
[136] Ibid., 544.

and are thus antecedent to, and independent of, coordinates: giving, in general, all the main relations, in the problem to which they are applied, without the necessity of appealing to coordinates *at all*. Coordinates may, however, easily be *read into* them:—when anything (such as metrical or numerical detail) is to be gained thereby. Quaternions, in a word, *exist* in space and we have only to recognize them:— but we have to *invent* or *imagine* coordinates of all kinds.[137]

Notice that Tait's penchant for the naturalness of quaternions led him to claim the fundamental existence of quaternions in the world. Likewise, Heaviside claimed, "Since we live in a world of vectors, an algebra or language of vectors is a positive necessity."[138] In the long run, Tait's appeals to the naturalness of quaternions did not result in the replacement of Cartesian methods. Most physicists apparently did not share Tait's opinion about the usefulness of quaternions for representing physical magnitudes. Indeed, quaternions appeared to many physicists as mysterious and unintelligible.

Despite the rhetoric of Hamilton and Tait, the theory of quaternions contributed to the trend toward abstraction in physics. Physicists had to contend with the challenge posed by the interpretation of new imaginary terms. Over and above this question, though, the works of Hamilton and Tait followed the analytic tradition of Lagrange, Monge, and Lacroix in dispensing with graphical representations of phenomena. Tait was critical of the "almost impenetrable meaning" of the uninterpreted analytical formulations that Lagrange and Cauchy had advanced, for example, in the science of hydrokinetics, but he agreed with Lagrange that the results of physics would be better represented *without recourse to graphical constructions*.[139] Tait argued that quaternions, instead of traditional algebraic methods, were so transparently understandable that they made the use of diagrams rather superfluous. In his words, "a *quaternion* equation is quite as suggestively intelligible, to those who understand it, as any geometrical diagram can possibly be. In fact, I might almost say, that it is *more* readily intelligible

---

[137] Tait, "On the Intrinsic Nature of the Quaternion Method" (1894), in Tait, *Scientific Papers*, vol. 2, doc. CXVI, quotation on p. 393; emphasis in the original.

[138] Heaviside, *Electromagnetic Theory*, vol. 1, chap. III: "The Elements of Vectorial Algebra and Analysis," quotation on p. 297.

[139] Tait, "On the Importance of Quaternions in Physics" (1890); in Tait, *Scientific Papers*, vol. 2, doc. XCVII, quotation on p. 308.

than diagrams usually are."[140] Tait cited Louis Poinsot and Joseph Fourier to emphasize that "a mathematical formula, however brief and elegant, is merely a step towards knowledge, and an all but useless one, until we can thoroughly read its meaning."[141] Poinsot had argued that physicists should be on their guard against the delusion that a science is complete once it is formulated analytically. Tait agreed, adding that "Poinsot's remark must be confined to the analytical formulae known to him. For it is certain that one of the chief values of quaternions is precisely this:—that no figure, nor even model, can be more expressive or intelligible than a quaternion equation."[142] But many physicists disagreed.

By the early 1890s the quaternion system was widely known but scarcely used by physicists, after almost forty years in circulation. Meanwhile, in little over a decade, vector analysis gained greater acceptance as a practical tool of working physicists.[143]

Gibbs and Heaviside modified Hamilton's system partly because the concept of quaternions seemed too abstract for the needs of physics. It seemed removed from an easy-to-visualize geometry of space. In an 1893 article titled "Quaternions and the Algebra of Vectors," Gibbs ascribed physicists' lack of acceptance of quaternions to Hamilton's approach, which allegedly had obscured the "geometrical relations of vectors."[144] Heaviside too blamed quaternions as responsible for inducing physicists to perceive vectorial systems as "involving metaphysical considerations of an abstruse nature," and he characterized quaternions as "a positive evil."[145] Whereas Tait had defended quaternions as a means of physical representation, Heaviside argued that the notion "has no physical representative, but is a highly abstract mathematical concept."[146] Heaviside thus banished "the mysterious quaternion" to develop a convenient math, an "anti- or ex-quaternionic"

---

[140] Tait, "On the Rotation of a Rigid Body about a Fixed Point" (1868); reprinted in Tait, *Scientific Papers*, vol. 1, doc. XV, quotation on p. 87; emphasis in the original.
[141] Tait, "On the Importance of Quaternions in Physics" (1890); in Tait, *Scientific Papers*, vol. 2, doc. XCVII, quotation on p. 308.
[142] Ibid.
[143] Crowe, *History of Vector Analysis*, p. 216.
[144] Gibbs, "Quaternions and the Algebra of Vectors" (1893), p. 463; reprinted in *The Scientific Papers of J. Willard Gibbs*, vol. 2, quotation on p. 170.
[145] Heaviside, *Electromagnetic Theory*, vol. 1, chap. III: "The Elements of Vectorial Algebra and Analysis," quotation on pp. 134–135.
[146] Ibid., p. 136.

system.[147] The advocates of vector analysis valued the correspondence of symbols to physical relations more than mathematical elegance.[148] If a quaternion expression did not satisfy any practical needs of physicists, then it did not warrant special attention.

The advocates of quaternions and vectors shared an aspiration to devise an expressive physical mathematics. Vector theorists followed Maxwell in the belief that physicists needed a practical mathematical symbolism to represent physical entities directly and, furthermore, to facilitate thinking and visualization of these entities in a geometric way. What they sought, in a way, was a method that would serve to mathematize operations that were naturally involved in physicists' imagination of physical processes. Hence, Heaviside argued:

> And it is a noteworthy fact that ignorant men have long been in advance of the learned about vectors. Ignorant people, like Faraday, naturally think in vectors. They may know nothing of their formal manipulation, but if they think about vectors, they think of them *as* vectors, that is, directed magnitudes. No ignorant man could or would think about the three components of a vector separately, and disconnected from one another. That is a device of learned mathematicians, to enable them to avoid vectors. The device is often useful, especially for calculating purposes, but for general purposes of reasoning the manipulation of the scalar components instead of the vector itself is entirely wrong.[149]

Here Heaviside conveniently appealed to the example of Michael Faraday, who had successfully conducted ground-breaking investigations in electricity and magnetism practically without recourse to established mathematical methods. Vectors then appeared not as a subsequent reformulation added by Maxwell to his theory of electromagnetism but as the fundamental conceptual tools used by Faraday to secure the discoveries that Maxwell later formulated in explicit mathematical notations. Heaviside did not merely point to the role of vectors in past scientific works; he demonstrated their

---

[147] Ibid., p. 139.
[148] Crowe, *History of Vector Analysis*, p. 217.
[149] Heaviside, *Electromagnetic Theory*, vol. 1, chap. III: "The Elements of Vectorial Algebra and Analysis," quotation on p. 298.

value in currently extending physical theory by polishing and advancing Maxwell's theory by vector means.

Heaviside's work of 1893 quickly gained supporters, including prominent physicists such as George Francis FitzGerald, Oliver Lodge, and August Föppl, who pursued the style of electromagnetic research established by Maxwell. Heaviside's polemics against quaternions, as well as his milder criticisms of Cartesian coordinate geometry, were received relatively well, for they were accompanied with important new physical results. Thus, by the late 1890s, it seemed that perhaps vector theory might eventually replace Cartesian coordinate geometry in physics.

In retrospect, vector theorists' criticisms against quaternions seem excessive. Hamilton's approach was closer to their interests than it might otherwise appear. In particular, even though Gibbs and Heaviside claimed that in Hamilton's approach quaternions were more fundamental than vectors, Hamilton had devoted his entire "First Book" (more than one hundred pages in length) of his *Elements of Quaternions* to a thorough exposition of vectors and their operations, with only *a single* and brief mention of the quaternion, merely as a name for the expression "scalar plus vector."[150] Moreover, Hamilton's presentation of vector concepts and operations was almost entirely geometric, without any excursions into metaphysics. In the second and third books of his *Elements,* Hamilton introduced the quaternion as the quotient of two vectors and as the products and powers of vectors, respectively—geometric conceptions based on the relations of vectors. Furthermore, Hamilton anticipated the application of his methods to the science of electricity.[151] With justice, the great French mathematician Henri Poincaré acknowledged at the end of the nineteenth century that quaternions, having been "so promptly utilized by the English physicists," had not been "a useless fancy" but had "rendered us more apt to penetrate the secrets of nature."[152] Moreover, he explained Maxwell's success

---

[150] Willard Rowan Hamilton, *Elements of Quaternions* (1866); 3rd ed., vol. 1, ed. Charles Jasper Joly (New York: Chelsea, 1969), p. 11.

[151] According to his son, William Edwin Hamilton, writing in 1866; preface to the first edition, *Elements of Quaternions,* 3rd ed., vol. 1, p. vii.

[152] Henri Poincaré, "Analysis and Physics," *The Value of Science,* authorized translation by George Bruce Halsted; reprinted in *Foundations of Science* (New York: Science Press, 1913), p. 282.

in predicting physical symmetries in electrodynamics "twenty years ahead of experiment" on the ground that "Maxwell was accustomed to 'think in vectors.'"[153]

Also, Ernst Mach, in his critical history of mechanics, ventured an appreciation of the new mathematical methods. He praised the essentially vectorial approaches of Grassmann and Hamilton: "These inquirers have developed mathematical conceptions that conform more exactly and directly to our geometrical ideas than do the conceptions of common analytic geometry; and the advantages of the analytical generality and the direct geometrical insight are thus united. . . . every step in the calculation is at the same time the clear expression of every step taken in the thought; whereas in the common method, the latter is forced entirely into the background by the introduction of three arbitrary coordinates."[154] Irrespective of the debates on the adequacy of mathematical methods for physics, the various approaches to quaternions and vectors constituted a significant advance over the traditional methods of analysis.

Physicists claimed that vector procedures were simpler and easier to understand. They argued that vectors served to represent or visualize physical processes better. Hence, as with coordinates and quaternions, physicists emphasized the "naturalness" of their favorite methods. Coordinate algebra continued to be employed in physics as a whole, but vector algebra was gaining popularity. Vector theory proved most valuable in electricity and magnetism, as well as in the general subject of the analysis of motion.

By the 1890s, multiple textbooks on mechanics incorporated vector methods prominently. Introductions to kinematics were increasingly formulated in terms of vector algebra. For example, at the Faculté des Sciences of Paris, Professor Paul Appell began his courses on rational mechanics by introducing vector algebra and the kinematics of points, as "preliminary indispensable notions." Following tradition, Appell maintained that among the mathematical sciences, the first was the science of number, followed then by geometry, which introduces the notion of space, and then by

[153] Ibid., p. 283.
[154] Ernst Mach, *Die Mechanik in ihrer Entwickelung historisch-kritisch dargestellt*, translated by Thomas J. McCormack as *The Science of Mechanics: A Critical and Historical Account of Its Development*, 6th ed., with revisions through the ninth German edition (La Salle, Ill.: Open Court, 1960), pp. 576–577. The last edition of the book during Mach's lifetime was issued in 1912.

kinematics, which introduces the notion of time. Only afterward did the science of mechanics ensue, he said, which dealt with "fictional" causes known as forces, the true causes of physical phenomena being "impossible to discover."[155] In his textbook of 1893, based on his lectures, Appell introduced the notions of force and mass only after dedicating the first two chapters (seventy-seven pages) to the concepts and procedures of vectors and kinematics. Only in chapter 3 did he introduce the principles of mechanics, such as the principle of inertia. Likewise, Xavier Antonmari and C. A. Laisant, in their textbook on mechanics of 1895, began with a section on the "preliminary notions" of vectors, followed by three chapters on kinematics, before proceeding to dynamics and then to statics.[156] Other textbooks also exhibited this order. On the whole, vector methods in the analysis of motion served to integrate a symbolic analysis of geometric figures that served to capture perceptible relations among physical magnitudes. In a textbook of 1905, Appell and James Chappuis commented: "The geometric part concerning *vectors* has been developed by elementary geometric methods which offer the advantage of habituating the student to reason directly on the objects themselves, whereas the abuse of the methods of analytic Geometry destroys the intuition and the spirit of invention."[157]

---

[155] Paul Émile Appell, *Traité de Mécanique Rationelle; Cours de Mécanique de la Faculté de Sciences*, vol. 1: *Statique – Dynamique du Point* (Paris: Gauthier-Villars et Fils, 1893), pp. v, 1.

[156] X. Antonmari and C. A. Laisant, *Questions de Mécanique à l'Usage des Élèves de Mathématiques Spéciales* (Paris: Librarie Nony & Co., 1895).

[157] Paul Émile Appell and James Chappuis, *Leçons de Mécanique Élémentaire à l'usage des Élèves des Classes de Première C et D, conformément aux Programmes du 31 mai 1902* (Paris: Gauthier-Villars, 1905), p. vi.

CHAPTER FIVE

Scientific Definitions
The Concepts of Space and Time

How did physicists reformulate basic concepts of kinematics? Having reviewed how evolving notions of space, time, and motion entered mathematical physics, we will now trace their development in the form of more specific concepts, such as distance, displacement, speed, velocity, duration, and simultaneity. Mathematical refinements led to the reformulation of some of these concepts, including distance, displacement, speed, and velocity. Traditional mathematics continued to be used in the analysis of others, those of duration and simultaneity, but these concepts were redefined critically in terms of measurement procedures.

This chapter illustrates how the growth of new mathematical and conceptual approaches affected the elements of kinematics. To a considerable extent, as charted in the previous chapters, the traditions that developed mathematical and conceptual revisions of the science of motion were independent of one another. Yet the two traditions merged as physicists applied new mathematical approaches to the basic concepts of kinematics. In particular, this chapter analyzes how they applied the distinction between vectors and scalars to the concepts of kinematics. It also traces how certain kinematic concepts continued to develop independently of vector theory. Later we will see how Einstein's relativistic kinematics stemmed from an analysis of the concept of simultaneity in terms of measurement procedures and traditional mathematical methods, alongside a neglect of vector concepts.

The complexion of Newton's mechanics was refined through the work of physicists who developed new mathematical methods to analyze phenomena in a more succinct and powerful manner. This enterprise consisted of at least two parts. First, physicists formulated advanced analytical theorems to efficiently describe complicated physical systems. For example, Joseph-Louis Lagrange, William Rowan Hamilton, and others, developed mathematical methods to deduce the forces and stresses of moving systems from a single formula. Second, and the subject of the present study, physicists developed new systematic methods, mathematical languages, to better represent physical magnitudes. This latter enterprise was crystallized by the emergence of quaternion theory and vector theory. In the hands of physicists such as Peter Guthrie Tait, Josiah Willard Gibbs, and Oliver Heaviside, such methods served to facilitate the progress of specific branches of physics, especially electricity and magnetism.

Despite the fruitfulness of vector methods, though, their application occurred haphazardly, because physicists sought to employ the most efficient calculational tools *and* to be widely understood. For example, as discussed in chapter 4, after the emergence of quaternion theory James Clerk Maxwell, like most physicists, persisted in using Cartesian coordinate equations for the analysis of motion, although he advocated the application of Hamilton's vectorial ideas.[1] For Maxwell, electrodynamics seemed especially susceptible to formulation in Hamilton's vectorial expressions, as indeed Hamilton himself had anticipated by arguing that his system would prove valuable not only in mechanics and optics but also in electricity.[2] Thus, Hamilton's mathematical work served as a stimulus for the *formal* reappraisal of Newton's theory. The proponents of new mathematical systems presented them as new languages that could better express *the same results* as the traditional methods. Yet vector methods did not serve simply to recast traditional concepts in new *form;* they also entailed the revision and refinement of the substantive content of basic concepts of physics, along with the creation of significantly innovative concepts.

While Tait, Gibbs, Heaviside, and other physicists discussed the relative merits of different mathematical methods in the 1890s, several physicists critically analyzed the elements of their discipline. Heinrich Hertz, for example, in addition to his well-known experimental researches in electromagnetism, pursued a theoretical investigation of the logical relationships

---

[1] James Clerk Maxwell, *A Treatise on Electricity and Magnetism* (1873); 2nd ed., vol. 1 (Oxford: Clarendon Press, 1881), p. 9.
[2] William Rowan Hamilton, *Elements of Quaternions* (1866), ed. Charles Jasper Joly, 3rd ed. (repr., New York: Chelsea, 1969), advertisement to the first edition by William Edwin Hamilton, p. iii.

of the concepts of mechanics. He formulated his ideas systematically in his book, *The Principles of Mechanics*. In his introduction, Hertz asked in passing "why is it that people never ask . . . what is the nature of velocity?"[3] He raised this issue after noting that the exact nature of certain physical agencies such as force and electricity remained matters of puzzlement and concern, as they continued to appear mysterious despite the successful investigations of physicists. He surmised that the latter concepts, and not the concept of velocity, continued to attract attention and inquiry because many ideas were routinely attached to the concept of velocity without causing problems, whereas the terms *force* and *electricity* had accumulated "more relations than can be completely reconciled amongst themselves."[4] Physicists encountered what appeared to be "painful contradictions" and consequently labored to clarify ambiguous conceptions. Such difficulties naturally clustered more densely in relatively complex areas of physics, such as electricity and magnetism, yet even a field as apparently simple as the study of motion was not spared critical inspection. While many physicists worked at the frontier of their field, attempting to understand phenomena that remained quite unexplained, a number of theorists became increasingly concerned with the foundations of their discipline.

Concepts of Motion and Distance

Once the geometry of motion was conceived as a science in its own right, it seemed necessary to clearly elucidate the meaning of its concepts. One basic concept in need of clarification was that of velocity. Earlier, in the mid-1700s, d'Alembert had complained in his *Essay on the Elements of Philosophy* that the concept of velocity had not been elucidated "with all the necessary precision" in any mathematical definition.[5] "The velocity of a body that moves uniformly is equal, they say, to the space divided by the time; or, as other mathematicians say, the result of this division is the measure of veloc-

---

[3] Heinrich Hertz, *Die Prinzipien der Mechanik in neuem Zusammenhange dargestellt*, preface by H. von Helmholtz (1894); *The Principles of Mechanics, Presented in a New Form*, trans. D. E. Jones and J. T. Walley (New York: Dover, 1956), p. 7.
[4] Ibid.
[5] Jean d'Alembert, *Essai sur les élémens de Philosophie, ou sur les principes des connaissances humains, avec les éclaircissemens* (1759?); reprinted in *Oeuvres complètes de d'Alembert*, vol. 1, part 1 (Paris: A. Belin, et Bossange père, fils et frères, 1821), p. 317; translations by A. Martínez.

ity. This manner of speaking, applied with rigor, does not convey any clear idea; because one does not know how to divide space by time; one cannot divide a quantity of one nature by another of a different nature; to divide a *place* by an *hour,* it's as if one wanted to know how many times an hour is contained within a place, and one sees well that this question makes no sense."[6] D'Alembert therefore suggested that the concept of velocity should better be reconstrued as the quantitative comparison of the motion of one body relative to the motion of some other body, instead of abstractly comparing motion to sections of pure space and time.[7] But his proposal was only a suggestion, leaving the question of the exact definition of velocity to later writers.

Despite such objections, usual definitions of velocity continued to employ the division of one kind of quantity by another kind. Consider a typical example. Henri Resal defined velocity as

$$\frac{s}{t} = \frac{v}{1},$$

where s is "the arc of the trajectory described or traversed after the time t," such that the "path" or "space traversed" is $s = vt$. He used this notation to represent the uniform motion of a point, whether it be rectilinear or curvilinear.[8] He then discussed "varying" (or nonuniform) motion, where the successive paths traversed are not proportional to the corresponding times (time intervals). He noted that in many applications, such as when one says of "a man, horse, or locomotive" that its velocity is so many kilometers per hour, such rates are only simple approximations—that is, "mean velocities." Resal then used the widespread mathematical concept of function to describe varying motions, such that all such equations have the form

$$s = f(t)$$

for calculating the path traversed, so long as "the paths are reckoned from a common origin." Regarding this general equation, Resal explained that it

[6] Ibid., p. 316.

[7] D'Alembert, *Essai sur les élémens de Philosophie.* See also Jean d'Alembert, *Traité de dynamique,* (1743); 2nd ed. (Paris, 1758; repr., New York: Johnson Reprint, 1968), p. 16.

[8] Henri Resal, *Traité de Cinématique Pure* (1862), p. 2; translations by A. Martínez.

could serve for the geometric representation of motion. Assuming a coordinate system, he let "t represent the abscissa and s the ordinate." The tracing of the curve could be produced by using either the functional relation or a table consisting of the numerical values of s corresponding to those of t, "that which will result notably if the times and the spaces are the results of observation."[9]

Such definitions of velocity intermixed arithmetical algebra, the theory of functions, and analytic geometry. Still, notice that several aspects of motion were not clearly captured in these definitions. For example, ambiguities remained pertaining to the concepts of direction, the distinction between points and intervals, and the distinction between positions and displacements.

Moreover, the algebraic representation of physical relations involved subtleties. The use of a letter, as a name for a physical notion such as a distance, could generate the impression that the *distance* exists as a thing in itself. In contradistinction, one could argue that the name distance stands actually for a *relation* among certain perceptions. This latter path was advocated by Johann Bernhard Stallo, who complained that algebraic symbols seemed to designate objects, whereas numbers better exhibit relations, because they consist of units. He argued that "the use of letters as symbols, i.e., as representatives of numbers, is in itself a serious infirmity of mathematical notation. In the simple formula, for instance, expressive of the velocity of a moving body in terms of space and time ($v = s/t$), the letters have a tendency to suggest to the mathematician that he has before him direct representatives of the things or elements with which he deals, and not merely of their ratios expressible in numbers."[10] Stallo voiced agreement with a claim by Eugen Dühring to the effect that algebraic symbolization is radically deficient inasmuch as it does not display numerical units. Stallo claimed that letters obscure the real relations among natural processes that we aim to represent. He rejected the indiscriminate use of the word "quantity" to refer to both extended magnitudes and abstract numerical aggregates. He claimed that this lack of distinction engendered confusions symptomatic

[9] Ibid., p. 3.

[10] Johann Bernhard Stallo, *The Concepts and Theories of Modern Physics* (New York: D. A. Appleton and Company; London: Scientific Series/Kegan Paul, 1881); 3rd ed. (1888) reissued with an introduction by Percy W. Bridgman (Cambridge, Mass.: Belknap/Harvard University Press, 1960), p. 277.

of the old presumption "that our arbitrary or conventional classifications of natural phenomena are coincident with essential distinctions between them and can be used as a source of inferences respecting their nature and origin." He rejected "the confusion between purely conventional forms of thought and speech and forms or laws of objective existence."[11]

Earlier, efforts to distinguish between facts and conventions in formal physical theories had been made by Hermann von Helmholtz. Owing to his investigations on visual perception, Helmholtz turned to analyze our notions of space. He wanted to distinguish "which propositions of geometry express truths of factual significance and which are merely definitions or a consequence of definitions or of the chosen mode of expression?"[12] Once he became acquainted with the writings of the mathematician Bernhard Riemann, Helmholtz's interest in the subject increased. In an influential lecture of 1854, Riemann had advocated analysis of the "Hypotheses which underlay the Foundations of Geometry," arguing that apparent physical space is only one of many mathematically conceivable kinds of "manifolds."[13] Helmholtz shared the desire to identify the hypothetical concepts involved in traditional geometry. Moreover, because geometry served as a foundation for physical science, Helmholtz wanted to ground geometry itself not on hypotheses but on facts of experience. To this end, he argued that geometric knowledge was based on "the observed fact that the motion of rigid bodies is possible in our space, with the degree of freedom that we know."[14] Thus, he took it as a fact of experience that the size and shape of at least some material bodies do not change when they move. Helmholtz believed that this axiom underlay necessarily the very possibility of applying geometry to the physical world. It seemed that only if material bodies such as rigid

[11] Ibid., p. 276.
[12] Hermann von Helmholtz, "Ueber die tatsächlichen Grundlagen der Geometrie," *Verhandlungen des naturhistorisch-medicinischen Vereins zu Heidelberg* 4 (22 May 1866), 197–202; p. 197.
[13] Bernhard Riemann, "Ueber die Hypothesen, welche der Geometrie zu Grunde liegen," lecture delivered in Göttingen, 1854, *Abhandlungen der Königlichen Gesellschaft der Wissenschaften zu Göttingen* 13 (1867), 132–152; reissued in Riemann, *Gesammelte mathematische Werke und wissenschaftlicher Nachlaß*, ed. Richard Dedekind and Heinrich Weber, 2nd ed. (Leipzig: B. G. Teubner, 1892), pp. 272–287.
[14] Hermann von Helmholtz, "Ueber der Ursprung und die Bedeutung der geometrischen Axiome," lecture delivered in Heidelberg, 1870, in Helmholtz, *Vorträge und Reden*, vol. 2 (Braunschweig: F. Vieweg, 1884), 1–31; p. 19.

rulers keep their form despite their motions is it then possible to carry out physical measurements. Helmholtz tried to show that presuppositions such as are involved in the geometries of Euclid and Riemann could be derived from the axiom of the mobility of rigid bodies.[15]

Helmholtz contended that even the concept of a straight line stems from the physical experiences of bodies actually moving through space. In opposition, some theorists continued to argue that geometry involved axioms that were not derived from experience but were rooted in inborn transcendental intuitions. For example, Albrecht Krause argued instead that a straight line can be defined, irrespective of experience, as a line having only one direction. In reply, Helmholtz objected: "Yet how should one define direction?—surely, only again by using the straight line. We are moving here in a vicious circle."[16] He repeatedly argued that, insofar as the so-called axioms of geometry are regarded as propositions about nature, they should then be susceptible to experimental test. He claimed that actual measurements, involving moving material bodies, should be regarded as a source of knowledge preferable to arguments from tradition or appeals to authorities such as Kant. Instead, Helmholtz voiced agreement with Newton's old but neglected view that geometry is founded on mechanical practice and that geometry is the part of universal mechanics that precisely demonstrates the art of measurement.[17]

A few other theorists also placed notions about the mobility, rigidity, and superposition of bodies at the foundations of geometry. For example, in

[15] For analysis and discussion, see Roberto Torretti, *Philosophy of Geometry from Riemann to Poincaré, Episteme*, vol. 7 (Dordrecht: D. Reidel, 1978), chap. 3. Note also that in 1866 Helmholtz derived conclusions about geometry which he thought were universal, because they applied to the geometries of Euclid and Riemann, but afterward he realized that he had been mistaken, because he had overlooked the geometry of Lobachevshky, which, like Euclid's, is infinite and three-dimensional but in which not all of Helmholtz's derivations applied.

[16] Hermann von Helmholtz, Speech held at the Commemoration-Day celebration of the Friedrich Wilhelm University of Berlin, 3 August 1878; revised, expanded, and translated as "The Facts of Perception," in Hermann von Helmholtz, *Science and Culture: Popular and Philosophical Essays*, trans. and ed. David Cahan (Chicago: University of Chicago Press, 1995), pp. 342–380; appendix I, p. 370. See also, Albrecht Krause, *Kant und Helmholtz: über den Ursprung und die Bedeutung der Raumanschauung und der geometrischen Axiome* (Lahr: Moritz Schauenburg, 1878).

[17] Helmholtz, *Science and Culture*, appendix III, p. 380.

1897 the young philosopher Bertrand Russell formulated a theory of metric geometry based on the so-called axiom of free mobility. He expressed this axiom thus: "Spatial magnitudes can be moved from place to place without distortion; or, as it may be put, Shapes do not in any way depend upon absolute position in space."[18] Russell rejected the idea that geometry should be based on dynamics, because all physical magnitudes, such as time, force, and mass, presupposed the "independent" possibility of measuring spatial magnitudes. Russell conceived of what he called "merely kinematical matter"—that is, matter that has been separated in thought from all forces and dynamical properties. Russell assumed that any such kinematic bodies are rigid *a priori*, because he conceived of no reason, in the absence of all causes, why their shapes should change. He assumed that the axiom of free mobility was the only logically possible axiom for establishing geometric relations among figures. He argued that "two magnitudes, which appear equal in one position, appear equal also when compared in another position. There is no sense, therefore, in supposing the two magnitudes unequal when separated, nor in supposing, consequently, that they have changed their magnitudes in motion."[19] Russell used the axiom of free mobility, alongside the relativity of position, and more, to try to establish the allegedly transcendental *a priori* character of certain aspects of geometry.[20] Likewise, many other mathematicians and philosophers agreed with Kant that certain notions of space are necessary and forever unchangeable, irrespective of anything that we might discover from new physical experiences.

Interest in reformulating the concepts of space grew out of mathematicians' concern for the exact validity of geometric knowledge. They had believed traditionally that geometry is something unique, that its content, methods, and goals exist only in accordance with the ancient tradition established by Euclid. But once non-Euclidean geometries were invented, the idea of the uniqueness of geometry had to be abandoned. Consequently, the elements of geometry fell under critical scrutiny and were variously reformulated. Among them were many that had become essential concepts

---

[18] Bertrand Russell, *An Essay on the Foundations of Geometry* (Cambridge: Cambridge University Press, 1897), p. 150.
[19] Ibid., pp. 153–154.
[20] For analysis, see Torretti, *Philosophy of Geometry from Riemann to Poincaré*, pp. 301–320, 410–411.

in the study of motion, such as coordinate, length, and distance. Because such concepts had been invented originally for the study of figures static in space, it would have been in the interest of physicists to reformulate such concepts in regard to the study of motion. But this was scarcely the case. As professional mathematicians revised such concepts, they mostly made them increasingly more abstract.

Consider now the concept of coordinate. Traditionally, since the early days of analytic geometry, coordinates were construed as points distributed along rectilinear lines. A few mathematicians, such as Leibniz in 1692, made occasional use of curvilinear coordinates, but for the most part, point coordinates were rectilinear. Coordinates were interpreted as the fundamental elements in analytic geometry, just as the point was viewed as the fundamental element in Euclidean synthetic geometry. But this view was replaced in the nineteenth century as revisions in the foundations of geometry led mathematicians to posit a multiplicity of concepts. Mathematicians gradually came to employ a wealth of entities and relations as "coordinates."[21] Such extensions of the meaning of the concept of coordinate were hardly intended to serve to clarify its function in physical science. Moreover, even in analytic geometry, there existed some ambiguity, as coordinates could be interpreted as points or as distances from an axis, that is, as spatial intervals. On the one hand, for example, in Georg Simon Klügel's *Mathematical Dictionary* of 1803, coordinates were defined as "straight lines," that is, as intervals.[22] On the other hand, given the program to arithmetize geometry, coordinates were viewed simply as numbers, a stand forcefully taken by Adrien Marie Legendre.[23]

The concept of distance was likewise extended and made increasingly

[21] See, e.g., Gino Loria, "Perfectionnements, évolution, métamorphoses du concept de 'coordonnées,'" Contributions à l'histoire de la géométrie analytique," *Analele Universitătii din Timisoara. Seria stiinte matematice* **18** (1942), 125–145; **20** (1944), 1–22; **21** (1945), 66–83; *Osiris*, **8** (1948), 218–288.

[22] Georg Simon Klügel, *Mathematisches Wörterbuch; oder, Erklärung der Begriffe, Lehrsätze, Aufgaben und Methoden der Mathematik mit den nöthigen beweisen und literarischen nachrichten Begleitet in alphabetischer ordnung,* 5 vols. (Leipzig: E. B. Schwickert, 1803). Hans Wussing, *Die Genesis des abstrakten Gruppenbegriffes* (1969); translated by Abe Shenitzer with the editorial assistance of Hardy Grant, *The Genesis of the Abstract Group Concept* (Cambridge, Mass.: MIT Press, 1984), p. 28.

[23] Adrien Marie Legendre, *Eléments de Géométrie* (Paris, 1794); *Elements of Geometry,* trans. John Farrar (Cambridge: Hilliard and Metcalf, University Press, 1819).

abstract. Arthur Cayley, for example, was one of the mathematicians who revised the concept of distance. He sought to clarify the relationship between projective and metric geometry. Traditionally, a distance was construed as the magnitude of the spatial separation between two points. Cayley realized that he could make the meaning of this concept more general if he focused on the form of the relation of distance regardless of what mathematical entities, such as points, are deemed to have this relation. Hence, in 1859 Cayley called "a notion of distance" any relation that satisfies the condition

$$\text{Dist. } (P, P') + \text{Dist. } (P'\ P'') = \text{Dist. } (P, P'')$$

for arbitrary positions of three points P, P', P".[24] This expression constituted a generalization of the traditional concept of distance in analytic geometry, as expressed by

$$s = \sqrt{\left(x_2 - x_1\right)^2 + \left(y_2 - y_1\right)^2 + \left(z_2 - z_1\right)^2},$$

and in ordinary differential geometry, as expressed by

$$ds = \sqrt{\left(dx\right)^2 + \left(dy\right)^2 + \left(dz\right)^2}.$$

More influential than Cayley's "notion of distance" was the concept of a "multiply extended magnitude" advanced by Bernhard Riemann. In 1854 Riemann advocated a conception of distance or "linear element" of indefinitely many dimensions, such that

$$ds = \sqrt{\left(dx_1\right)^2 + \left(dx_2\right)^2 + \left(dx_3\right)^2 + ... + \left(dx_n\right)^2},$$

where the value of the square root is always positive, or more simply,

$$ds = \sum \sqrt{\left(dx\right)^2}$$

[24] Wussing, *The Genesis of the Abstract Group Concept*, p. 170.

in Riemann's notation.[25] Riemann further generalized this notion by assuming

$$ds^2 = \sum_{i=1}^{n} \sum_{j=1}^{n} g_{ij} dx_i dx_j,$$

where $g_{ij} = g_{ji}$, which are functions of the coordinates $x_1, x_2, \ldots, x_n$, thus allowing the structure of a "manifold" such as space to vary from point to point.[26] This conception generated much interest in the study of non-Euclidean geometries.

The notion of distance was also analyzed by Bertrand Russell. In his *Principles of Mathematics* of 1903, he commented first that the notion of distance "seldom receives precise definition," even though it "is enormously complex."[27] But instead of proceeding to delineate specific meanings of the term, Russell complained that often "there is more concern for numerical measurement of distance than for its actual definition," and he then proceeded to "endeavor to generalize the notion as much as possible," to use the word distance "to cover a far more general conception than that of distance in space."[28]

Mathematicians thus developed concepts of space beyond the needs of physicists. Hermann Grassmann, Riemann, and others invented algebraic systems of geometry involving more than three spatial dimensions. Remarkably, some of these developments eventually found physical utility in the early 1900s. But in the 1890s, language refinements of immediate practical value in physics arose mainly from vector theory.

Vector mathematics facilitated the reformulation of the concepts of motion to a degree of exactitude that had not been secured with Cartesian

---

[25] Bernhard Riemann, "Ueber die Hypothesen, welche der Geometrie zu Grunde liegen" (1854); translated by William Kingdon Clifford, "On the Hypotheses which Lie at the Base of Geometry," *Nature* **8** (1873); parts I–II in 1 May: pp. 14–17, and part III in 8 May: pp. 36–37.

[26] Detlef Laugwitz, *Bernhard Riemann 1826–1866* (1996); *Bernhard Riemann, 1826–1866: Turning Points in the Conception of Mathematics,* trans. Abe Shenitzer (Boston: Birkhäuser, 1999), p. 236; Morris Kline, *Mathematical Thought from Ancient to Modern Times* (1972), vol. 3; reissued (New York: Oxford University Press, 1990), p. 890.

[27] Bertrand Russell, *The Principles of Mathematics* (1903; rep., New York: W. W. Norton, 1938), pp. 252, 253.

[28] Ibid., pp. 253, 180.

mathematics. Given Hamilton's quaternion approach, physicists tried to analyze the broad subject of mechanics according to the categories of scalars and vectors. All mathematical quantities referring only to magnitude, such as length, volume, and mass, were designated scalars, while all quantities involving direction as well as magnitude, such as velocity, momentum, force, and magnetization, were identified as vectors. That certain physical magnitudes essentially involve directionality in space was already recognized by Newton, as he distinguished forces, for example, by their magnitude, direction, and location. Other physicists followed suit and explicitly isolated questions of directionality whenever it seemed necessary. But the distinction between directed and nondirected magnitudes was not advanced in definitive, systematic mathematical form until the work of Hamilton. In turn, Maxwell stressed its significance by calling it "one of the most important features of Hamilton's method."[29] The concepts of scalars and vectors overlapped the areas of kinematics and dynamics, as both of these fields involved them.

As the distinction between scalars and vectors gained broad currency, it became applied to the concept of velocity. Still, its application often lacked rigor. Consider the work of Maxwell, who had emphasized the value of vectors. The distinction between quantities with or without direction appealed to Maxwell. Overall, he cultivated a profound interest in classification schemes, an interest that led him to advance novel classifications in electricity, magnetism, thermodynamics, mathematics, instruments, research techniques, and more. In the opening pages of his *Treatise*, Maxwell defined length and time as two of "the three fundamental units" of physics (the third being mass). Subsequently, he specified the definitions of "derived units," characterizing the unit of "velocity" as that "in which unit of length is described in unit of time with dimensions $[L/T]$." Only *after* discussion of other derived units, and of the subjects of physical and mathematical continuity, and of periodic and multiple functions, did he treat the subject of vectors and scalars. Although Maxwell specified displacements as vectors, thus tacitly distinguishing between simple scalar *lengths* and *spatially directed changes of position*, he did not distinguish vector and scalar notions of velocity. So instead of stating, say, that "speed" represents the magnitude

---

[29] Maxwell, *Treatise*, 2nd ed., vol. 1, p. 9. See also Oliver Heaviside, *Electromagnetic Theory*, vol. 1 (1893; repr., New York: Chelsea, 1971); p. 97.

of the rate of change of position, Maxwell simply wrote that the "velocity" of a body is an instance of a vector quantity.[30]

This lack of distinction is likewise present in his work of 1876 on the foundations of physics, *Matter and Motion*. There, to describe changes in the location of material particles in space, Maxwell distinguished between displacements and velocities. He called *displacement* any change in the configuration of a system wherein only the geometric change of initial and final positions is considered without reference to time.[31] He called *motion* any such displacements considered in relation to the time it takes a particle to travel from one point to another. "The rate or speed of the motion is called the velocity of the particle, and its magnitude is expressed by saying that it is such a distance in such a time."[32] Thus, Maxwell did not distinguish between speed and velocity, with the former being the scalar measure (nondirected magnitude) of the velocity vector. It is noteworthy, nonetheless, that in the same work he asserted: "It is specially important to understand what is meant by the velocity or rate of motion of a body, because the ideas which are suggested to our minds by considering the motion of a particle are those which Newton made use of in his method of Fluxions, and lie at the foundation of the great extension of exact science which has taken place in modern times."[33]

Following the concepts of the calculus as developed by Newton and Leibniz, physicists commonly defined velocity as the differential of a trajectory s with respect to a time t:

$$v = \frac{ds}{dt},$$

where any variation in the direction of the velocity required further analytical specifications.[34] This concept was extended to quaternion and

---

[30] Maxwell, *Treatise*, 2nd ed., vol. 1, p. 9.

[31] James Clerk Maxwell, *Matter and Motion* (1877; repr., New York: Dover, 1991), p. 18.

[32] Ibid., p. 19.

[33] Ibid., p. 20.

[34] See, e.g., William Thomson and P. G. Tait, *Treatise on Natural Philosophy*, vol. 1, part I; new ed. (Cambridge: Cambridge University Press, 1879), p. 13. August Föppl, *Vorlesungen ueber Technische Mechanik*, vol. 1, *Einfuehrung in die Mechanick* (1899), 2nd ed. (Leipzig: B. G. Teubner, 1900), p. 27. Julius Petersen, *Kinematik*, German edition by R. von Fischer-Benzon (Copehagen: Andr. Fred. Höst & Sohn, 1884), pp. 4, 8.

vector theory simply by treating the trajectory and the velocity as directed quantities, while treating the time as a scalar. Thus, one could write:

$$\mathbf{v} = \frac{d\mathbf{s}}{dt} \quad \text{or} \quad \rho' = \frac{d\rho}{dt},$$

where vectors were designated by bold characters or by Greek letters. In his *Elements of Quaternions*, Hamilton briefly introduced the vector concept of velocity: "In applications to *mechanics*, if t denote the *time*, and if the *term* P of the *variable vector* ρ be considered as a *moving point*, this *derived* vector ρ' may be called the *Vector of Velocity*: because its *length* represents the *amount*, and its *direction* is the *direction* of the velocity."[35] Likewise, Josiah Willard Gibbs in his pamphlet *Elements of Vector Analysis* indicated, "If ρ is the vector drawn from a fixed origin to a moving point at any time t, $d\rho/dt$ will be the vector representing the velocity of the point."[36]

Neither Hamilton nor Gibbs elaborated further the vector concepts of velocity and displacement in their major works. For Hamilton, the details of basic kinematics were not of primary concern. Likewise, Gibbs did not write the sort of elementary treatise on the application of vector algebra to physics that clearly distinguished the basic concepts of distance, displacement, velocity, and speed. His pamphlet was essentially a mathematical work. His ideas, however, were more clearly set forth by Edwin Bidwell Wilson, who prepared Gibbs's lecture notes for publication in 1901 under the title *Vector Analysis: A Text-Book for the Use of Students of Mathematics and Physics*. In the section titled "Kinematics," Wilson defined the "velocity" of a particle as its "rate of change of position" and expressed it in differential form as

$$\mathbf{v} = \operatorname*{Lim}_{\Delta t \doteq 0} \left[\frac{\Delta \mathbf{r}}{\Delta t} \right] = \frac{d\mathbf{r}}{dt},$$

---

[35] Hamilton, *Elements of Quaternions*, ed. Charles Jasper Joly, 2nd ed., (London: Longmans, Green, 1899–1901), vol. 1, p. 99.

[36] Josiah Willard Gibbs, *Elements of Vector Analysis* (New Haven, Conn.: privately printed, 1881); reprinted in *The Scientific Papers of J. Willard Gibbs*, vol. 2 (New York: Dover, 1961), p. 30.

noting, "This is the limit of the increment $\Delta \mathbf{r}$ to the increment Δt."[37] Wilson then distinguished the concepts of velocity and speed:

This velocity is a vector quantity. Its direction is the direction of the tangent of the curve described by the particle. The term *speed* is used frequently to denote merely the scalar value of the velocity. This convention will be followed here. Then

$$v = \frac{ds}{dt},$$

if s be the length of the arc measured from some fixed point of the curve.[38]

Accordingly, Wilson also explicitly distinguished between the concepts of "displacement of translation" and "distance," the former being a vector, the latter a scalar.[39]

In Britain, Oliver Heaviside conveyed the basic distinctions between vectors and scalars in kinematics. In his "Elements of Vectorial Algebra and Analysis" of 1893, Heaviside distinguished a displacement $\mathbf{r}$ from a distance s, even though he also referred to $\mathbf{r}$ as "being a vector distance."[40] He also explicitly distinguished between the "velocity" $\mathbf{v}$ and the "speed" v or ds/dt.[41] Hence, he defined velocity with the differential expression:

$$\frac{d\mathbf{r}}{dt} = \frac{d\mathbf{r}}{ds}\frac{ds}{dt} = v\mathbf{T} = \mathbf{v}.$$

Following Hamilton, Heaviside described the magnitude v as "the tensor" of the vector $\mathbf{v}$. He explained that "the tensor of a vector is its size, or magni-

[37] Edwin B. Wilson, *Vector Analysis: A Text-Book for the Use of Students of Mathematics and Physics,* based on the lectures of J. Willard Gibbs (New Haven, Conn.: Yale University Press, 1901), p. 125.

[38] Ibid., p. 125.

[39] Ibid., pp. 2, 136.

[40] Oliver Heaviside, "Elements of Vectorial Algebra and Analysis," in *Electromagnetic Theory* (1894), vol. 1; (repr., New York: Chelsea, 1971), p. 181.

[41] Ibid., p. 169.

tude apart from direction."[42] Hamilton had introduced the term *tensor* in the same connection, as he designated the direction of a vector *a* as "*the Versor of that Vector, a,*" and proposed to "express the *Length* of the same line *a*, by introducing the *new name* TENSOR, and the *new symbol*, T*a*; which we shall read, as the *Tensor of the Vector a*: and shall *define* it to be, or to denote, *the Number which represents the Length of that line a.*"[43]

Similar clarity of definitions was likewise advanced by other physicists, though inconsistencies and lack of rigor in the use of scalar and vector terminology continued. Oliver Lodge, in *Elementary Mechanics,* published in 1896, wrote that motion "has two primary properties to be studied—Speed and Direction; both of which are sometimes held to be included under the one name, Velocity."[44] Here the distinction between scalar and vector descriptions of motion seemed explicit, because Lodge argued that the velocity of a body could vary because of a change in its speed (rate of change of position) or because of a change of its orientation of movement.[45] Nonetheless, Lodge's text suffers from inconsistencies in his use of terms. For example, the words *speed* and *velocity* are sometimes used as if they were equivalent; velocity is simply defined as "rate of motion," as "distance traveled divided by time," and as "length per time," expressions that omit the notion of direction.[46] Also, although Lodge employed opposite signs for opposite directions, he did not use the term *displacement* but simply stated that velocities are constituted by "distances" or "lengths."

Not everyone adopted novel verbal distinctions. Hermann Grassmann, for example, criticized Hamilton's terminology, noting that "according to Hamilton's procedure one designates simple and long familiar concepts by new, often unsuitable terms, as 'vector' for 'stroke,' 'tensor' for 'length' or 'numerical value,' etc."[47] Still, in German-speaking countries the works

---

[42] Ibid., p. 142.

[43] Hamilton, *Elements of Quaternions,* 2nd ed., vol. 1, pp. 163–164.

[44] Oliver Lodge, *Elementary Mechanics; including Hydrostatics and Pneumatics* (London and Edinburgh: W. & R. Chambers, Ltd., 1896), p. 7.

[45] Ibid., p. 77.

[46] Ibid., p. 8.

[47] Hermann Grassmann, "Der Ort der Hamilton'sche Quaternionen in der Ausdehnungslehre," *Mathematische Annalen* 12 (1877), 375–386; reissued in *A New Branch of Mathematics, The Ausdehnungslehre of 1844 and Other Works,* trans. Lloyd C. Kannenberg (Chicago: Open Court, 1995), p. 525. I translated Grassmann's "Strecke" as "stroke," instead of Kannenberg's "displacement."

of August Föppl and Heinrich Hertz served to propagate the vectorial-kinematic concepts of Hamilton, Gibbs, and Heaviside. Föppl chose to designate vectors by using bold gothic characters, while he designated scalars with Latin and Greek letters or also with absolute-value signs. In his *Introduction to Mechanics*, Föppl designated instantaneous velocity with the character "$\mathfrak{v}$, whereas u accordingly signifies the absolute value of $\mathfrak{v}$."[48] The expression *absolute value* derived partly from the work of the mathematician Karl Weierstrass, who in 1841 introduced the sign $|x|$ to designate nonnegative values of x.[49] This symbolism was later carried over into vector theory to designate the scalar of a vector. Thus, for example, Föppl employed the gothic character $\mathfrak{B}$ to stand for a vector, and he employed $|\mathfrak{B}|$ to signify its scalar magnitude.[50] Föppl defined the velocity vector by

$$\mathfrak{v} = \frac{d\mathfrak{r}}{dt},$$

stating that "its magnitude is equal to ds/dt."[51] Thus, Föppl employed various notations and expressions to demarcate scalar and vector concepts.

The scalar and vector distinctions in the basic concepts of motion also were traced explicitly by Hertz in his *Principles of Mechanics* of 1894. Like Maxwell, Hertz defined displacement as a vector describing the passage of a material point from an initial to a final position irrespective of time. Yet Hertz traced more clearly the distinction between vectors and scalars by referring to the "magnitude of a displacement" from one position to another as the "distance" or "length" between two positions.[52] Hertz noted that this scalar measure was defined independently of its analytical representation and of the choice of coordinates, and he distinguished between coordinate and vectorial methods of representation:

[48] Föppl, *Einfuehrung in die Mechanik*, 2nd ed., p. 123.
[49] Karl Weierstrass, *Mathematische Werke*, vol. 1 (Berlin: Mayer & Müller, 1894), pp. 67, 252; quoted in Florian Cajori, *A History of Mathematical Notations*, vol. 2 (Chicago: Open Court, 1929), pp. 123–124.
[50] August Föppl, *Einführung in die Maxwell'sche Theorie der Elektricität* (1894); 2nd ed.: Max Abraham and August Föppl (Leipzig: B. G. Teubner, 1904), quotation on p. 6.
[51] Abraham and Föppl, *Einführung in die Maxwell'sche Theorie*, 2nd ed. (1904), p. 9.
[52] Hertz, *Die Prinzipien der Mechanik* (1894);*The Principles of Mechanics*, trans. Jones and Walley, quotation on p. 54.

The displacement of a point is completely determined by its initial and final position. It is also completely known when we are given its initial position, its direction, and its magnitude.

The displacement of a system is completely known when we know its initial and final positions. It is also completely known when its initial position, and what are termed its direction and magnitude, are given.[53]

Even though Hertz's distinctions between kinds of physical quantities clearly derived partly from vector concepts, his treatment relied fundamentally on common analytical methods. Only after six chapters on the "analytical representation" of material points and systems did Hertz explicitly introduce the concept of vector, at the beginning of a chapter titled "Kinematics." Accordingly, Hertz defined the concept of velocity as "the instantaneous rate of motion of a system," adding that it "may be regarded as a vector quantity with regard to that system." And he identified the scalar component of a velocity as "the magnitude of the velocity . . . also called the velocity of the system along its path, or, when misunderstanding cannot arise, simply the velocity."[54]

Significantly, vector theorists usually introduced the vector concept of velocity by explicitly extending the concepts and notations of the differential calculus, rather than by reference to algebraic coordinate geometry. This approach agreed with the general tendency to advocate vector methods as an alternative analytical device independent of Cartesian coordinates. For example, by considering the decomposition of a vector into rectangular "scalar components," Föppl admitted that "vector algebra leads back to scalar algebra," but he cautioned that, "as important as this method is, it introduces into vectorial concepts a foreign and arbitrary element."[55] Such criticisms of coordinate geometry emerged from physicists' struggles to earn acceptance for quaternion and vectorial methods. Meanwhile, the definition of velocity in terms of the differential calculus actually involved ambiguities, as a "characteristic incident" from the life of William Thomson, Lord Kelvin, illustrates. Felix Klein recounted how Thomson criticized the pure

---

[53] Hertz, *The Principles of Mechanics*, pp. 52, 54.
[54] Ibid., p. 127.
[55] Abraham and Föppl, *Einführung in die Maxwell'sche Theorie*, 2nd ed., p. 10.

mathematics of Cambridge professors such as Isaac Todhunter: "Entering the lecture-room, he suddenly addressed his students with the question, 'What does dx/dt mean?' He was answered with only strictly logical definitions; all were rejected. 'Forget Todhunter; dx/dt is the velocity!'"[56]

The definition of basic kinematic concepts in terms of differentials involved also subtle ambiguities in the logical interpretation of the calculus, though such ambiguities were imported into vectorial presentations.

Hence, early writers on vector kinematics defined the vector concept of velocity in terms of instantaneous velocities, that is, as the ratio of any *infinitesimally small* displacement and an instant of time. Nevertheless, if physicists instead had defined the velocity vector explicitly in terms of ordinary intervals of space and time along one coordinate axis, they would have defined it as

$$\mathbf{v} = \frac{\mathbf{x}}{t} = \frac{\left(x_2 - x_1\right)}{\left(t_2 - t_1\right)},$$

where x_1 and x_2 are the coordinates that represent the position of the object along the X-axis at the instants of time t_1 and t_2, respectively. As this notation shows, coordinate geometry was suited to convey explicitly the concept of position, a concept that, for the most part, was otherwise lacking in the mathematics of vectors and differential calculus. After all, one of the advantages of both vector analysis and the differential calculus was that these methods enabled physicists to operate directly with magnitudes, irrespective of particular position-references to axes of coordinates. The role of the concept of position in vector theory continued to involve ambiguities even in the twentieth century.[57] At any rate, Hertz was careful enough to distinguish clearly between the concepts of position, displacement, and distance. He designated a displacement with the expression

$$(x'_v - x_v),$$

[56] Felix Klein, *Vorlesungen uber die Entwickelung der Mathematik im 19 Jahrhundert*, Part I (Berlin: J. Springer, 1926); translated by M. Ackerman, in *Lie Groups: History, Frontiers and Applications*, vol. 9 (Brookline, Mass.: Math Sci Press, 1979), p. 223.
[57] For discussion of "the problem of the location of vectors," see Banesh Hoffmann, *About Vectors* (Englewood Cliffs, N.J.: Prentice-Hall, 1966; repr., New York: Dover, 1975), pp. 31–33, 104–105, and also pp. 13–14, 25, 41.

where x_v and x'_v designate the position values of coordinates at the start and end of the displacement, respectively; and given the expression

$$(x'_v - x_v)^2,$$

he designated a distance by the positive square root of this quantity.[58] This extraction of the positive root from the square of a quantity is of course equivalent to the more direct operation of finding the "absolute value" of the quantity.

Despite the elucidation of vectorial concepts in kinematics, ambiguities persisted in their use. The words path, trajectory, position, length, distance, and displacement, were often used synonymously and indiscriminately. Likewise, the distinction between velocity and speed was rarely applied consistently. In Germany, in particular, physicists used the same word, *Geschwindigkeit*, to mean either speed or velocity, and few designated the scalar quantity as "the magnitude of the velocity," as did Hertz and Föppl. Notwithstanding Maxwell's emphasis on the importance of the distinction between vectors and scalars, others usually used this distinction briefly or superficially. This tendency naturally corresponded to their particular interests. For example, Heaviside's main concern was electromagnetic theory; hence, he did not dwell on the details of kinematics. Physicists might dismiss such inconsistencies by regarding them merely as simple questions about the correct application of *language*. However, significant ambiguities lurked in the mathematical application of the basic concepts of kinematics.

Owing to the traditional rules of analytic geometry, the "scalar" quantities of vector theory could be interpreted as involving elements of directionality, notwithstanding the claims of vector theorists. In particular, vector theorists had admitted negative numbers alongside positive numbers as "directionless" scalars, even though they allowed the negative sign to denote changes of direction. For example, Edwin Bidwell Wilson wrote that scalars possess "*magnitude* but no direction," but he allowed that "*positive and negative numbers of ordinary algebra are the typical scalars.*"[59] He asserted, "The negative sign, –, prefixed to a vector *reverses* its direction," but without noting that this statement is entirely inconsistent with his further state-

[58] Hertz, *The Principles of Mechanics*, pp. 52–55.
[59] Wilson, *Vector Analysis*, pp. 1–2.

ment that "scalar multiplication does not alter direction but merely multiplies length."[60] This same ambiguity was present in Gibbs's own pamphlet *Elements of Vector Analysis*.[61] As for Heaviside, he characterized tensors as magnitudes without direction, yet he spoke of resolving a vector into its "scalar components—the Cartesian components—which are the tensor of the three rectangular vector components."[62] But if the scalar tensors were nothing other than Cartesian components, then such tensors, like Cartesian components, would have direction because of their positive or negative values along each coordinate axis. Heaviside also allowed "the – sign" to serve to "reverse direction, or, reverse the tensor without altering direction."[63] Equivalent ambiguities were propagated by other physicists.

The solution to such ambiguities lay in drawing a distinction between scalars that possess positive and negative values and scalars that only may have positive values. This distinction was advanced by Föppl, who identified strictly "positive scalars" such as "mass, volume, and density," by contrast to "positive and negative scalars" such as "potential and quantity of electricity."[64] But even then Föppl did not indicate that the basic scalars of kinematics (i.e., speed, distance, and time) are positive scalars. Other physicists avoided the issue by providing examples of scalars that do not relate to translational motion. For instance, Heaviside and Wilson each listed mass, density, energy, and temperature as scalars (while introducing concepts such as displacement, velocity, acceleration, and force as vectors), even though they recognized scalar concepts of motion.[65] Perhaps the issue was expressed most clearly by Hamilton, in his *Elements of Quaternions*, in a footnote where he argued that scalar magnitudes, such as lengths, are best represented by numbers without signs: "This *number*, which we shall presently call the *tensor* . . . may always be *equated*, in calculation, to a *positive scalar*: although it might perhaps more *properly* be said to be a *signless number*, as being derived solely from comparison of *lengths*, without any reference to *directions*."[66] Ever since mathematicians began to interpret nega-

---

[60] Ibid., pp. 8, 7.

[61] Gibbs, *Elements of Vector Analysis*, pp. 17–18.

[62] Heaviside, "Elements of Vectorial Algebra and Analysis," p. 142.

[63] Ibid., p. 146.

[64] Abraham and Föppl, *Einführung in die Maxwell'sche Theorie*, 2nd ed. (1904), p. 9.

[65] Heaviside, "Elements of Vectorial Algebra and Analysis," p. 13; Wilson, *Vector Analysis*, p. 1.

[66] Hamilton, *Elements of Quaternions*, 2nd ed., vol. 1, p. 111, see also p. 170.

tive numbers as magnitudes oriented in a specific direction, positive numbers likewise came to signify specific directions or orientations in space. Nonetheless, the older tradition that interpreted positive numbers simply as quantities without direction persisted. Thus, the invention of the "absolute value" operation was insufficient to distinguish *numerically* between, say, displacements in a particular (positive) direction and pure quantities such as lengths. By contrast, the devices and notations of vector theory could express the distinction between scalar and directed magnitudes clearly.

On the whole, efforts to improve the mathematical representation of basic kinematic quantities were carried out by physicists, not mathematicians. By modifying the notations and verbal expressions for such basic quantities, vector theorists refined the conceptual framework of physics. Hence, a greater specificity of ideas, and potentially greater expressive clarity, became increasingly available to professional physicists. Meanwhile, mathematicians also were developing relevant concepts in the field of kinematics, especially in relation to the concepts of space. Whereas physicists made the basic concepts more specific and particular, mathematicians aimed to generalize fundamental concepts, making them increasingly abstract.

Vector mathematics facilitated more efficient and succinct descriptions of physical processes than traditional methods of analysis, even though physicists often employed vector concepts and terminology ambiguously. Physicists then were perhaps more interested in formulating physical principles with exactitude than in consistently applying expressions that presumably would be well understood anyway, whether written one way or another.

By 1900 August Föppl aptly characterized the problematic development of physical mathematics. In the foreword to his *Introduction to Mechanics*, Föppl placed special emphasis on the question of the physical adequacy of mathematical methods. In his discussion, he compared the usefulness of the various contemporary mathematical methods:

> It comes especially into consideration that mathematical theories, as they are carried out everywhere nowadays, are cast in a form that is not particularly favorable to their immediate application in mechanics. In mechanics, and, above all, in theoretical physics, one deals mainly with the analysis of directed magnitudes. For such an application, despite all the high esteem we must emphasize for the achievements of contemporary mathematics, the generally introduced articles

of apprenticeship are inadequate. Algebra and Analysis involve only the relations between magnitudes without direction; the one apparent exception, the well-known geometrical interpretation of complex magnitudes, had this aim but in fact damaged more than availed the subject. To be sure, analytic geometry demonstrated how, with the help of the coordinate method, one can also submit directed magnitudes to calculation, but this means of information necessitated a diffuse representation that is little conducive to clear interpretation. Finally, the new synthetic geometry approaches the needs of mechanics, indeed, owing to its avoidance of coordinate systems, and thus was received with spirited delight a generation ago, even by representatives of technical mechanics. On the other hand, most busied themselves so predominantly with pure projective relations that those left for the investigation of directed magnitudes did not suffice.[67]

Like other physicists, Föppl understood that mechanics essentially involved the analysis of directed magnitudes. Accordingly, he emphasized that physics required mathematical methods that would be suited ideally for this purpose. He agreed with physicists such as Tait and Heaviside who argued that coordinate geometry failed to facilitate a clear representation of directed magnitudes. Föppl emphasized his dislike for the coordinate method by stating that modern synthetic geometry had the advantage of avoiding coordinates. The synthetic methods advanced in the nineteenth century had the advantage of treating geometric relations with great generality, as well as incorporating, to some extent, the notion of directed magnitude, but such methods remained insufficiently developed for use in physics. Föppl's reference to the enthusiasm with which such works were received perhaps refers especially to the work of Jakob Steiner, whose synthetic approach was "inordinately admired" in Germany.[68] But such works were hardly put at the service of physical science. Projective geometry, in particular, had been developed in ways far too abstract for easy physical application, especially because it involved concepts such as infinity, imaginary points, and complex planes.

According to Föppl, physicists finally had an appropriate mathematical instrument in the form of vectorial methods:

---

[67] Föppl, *Einfuehrung in die Mechanick*, 2nd ed., p. v; translation by A. Martínez.
[68] Carl B. Boyer, *A History of Mathematics* (1968), 2nd ed., revised by Uta C. Merzbach (New York: John Wiley and Sons, 1989), p. 608.

A healthy start toward the improvement of the mathematical resources used in mechanics, at present, has been made by a few prominent mathematicians for a long while already. For the time being,
they do not extend particularly far and, unfortunately, up to now they
have had but slight influence upon the formulation of the prevalent
mathematical theory. One must hope that the future will bring what
hitherto has been neglected; many signs seem to point to this. In the
meantime, we must try to manage with the mathematical means that
currently have become available to us.

Herewith it should not be said that mechanics, lacking willpower
of its own, must be bound to the usual representation forms of pure
mathematics. Rather, it must be allowed the freedom, if possible, to
aid itself with the means of expression best suited for its aims. As a
matter of fact, in recent treatments of mechanics the concept of directed magnitudes or vectors has gradually moved ever more into the
foreground, though even in such works the vector is still everywhere
analyzed into its components, by calculations. For that reason I have
already resolved, for a long time, in view of the basic mathematical
knowledge that one can assume, to operate with vectors themselves,
wherever possible.[69]

Föppl's arguments can be summarized as follows. The development of
mathematical methods ideally suited to the needs of physics had been neglected. Physicists should not limit themselves to using common mathematical methods but should be free to improve them as necessary. Vector
methods appeared as the best available for the analysis of physical magnitudes, even though they had not yet been developed to their full capacity.
Owing to such observations, Föppl resolved to employ vector methods fully
in his writings. Likewise, other physicists came to rely increasingly on vector mathematics.

At any rate, physicists analyzing motion were troubled by problems that
seemed much more urgent than questions about consistency of terminology. In the analysis of theoretical mechanics, the concepts of neither displacement nor velocity attracted thorough inquiry. Of course, the concepts
of the ether and of absolute space drew widespread attention. Yet the foun-

---

[69] Föppl, *Einfuehrung in die Mechanick*, 2nd ed. (1900), pp. v–vi; translation by
A. Martínez.

dations of kinematics eventually were thoroughly reformulated on the basis of critical revisions of another concept, namely, the concept of time.

Ambiguous Concepts of Time

As indicated earlier, Hamilton's original development of vector concepts stemmed from his analysis of algebra and imaginary numbers based on the notion of time. Nevertheless, physicists later reformulated the kinematic concept of time independently of vector concepts. Despite occasional considerations about "the direction of time," physicists construed all time quantities as scalars, thus remitting the quantitative analysis of time to the traditional methods of algebra. Whereas the concepts of space and motion were extended owing to the new methods developed by mathematicians and vector theorists, the concept of time was revised irrespective of such mathematical developments. Such critical revisions were based on the attempts to define measures of time in terms of the relative motions of bodies.

Alongside the criticisms of the ideas of absolute space and motion, theorists also criticized the traditional conception of absolute time. Just as physicists pondered the question of whether force or motion is more fundamental, they also asked whether motion or time is more fundamental. Ordinarily, physicists assumed that time was the independent parameter, in both a physical and mathematical sense. Thus, in his *Mathematical Principles of Natural Philosophy*, Newton had argued, "Absolute, true, and mathematical time, of itself, and from its own nature, flows equably without relation to anything external."[70] Likewise, Maxwell conceived of time in accordance with Newton's interpretation, that is, as essentially independent of matter and motion.[71] Yet in terms of human experience it could be argued otherwise, that the concept of time was derived from the phenomena of motion.

Physicists attempting to give definite meaning to the abstract concept of time often appealed to kinesthetic ideas. Oliver Lodge, for example, acknowledged, "It is probable that our idea of motion (that is, of free muscular motion) precedes and suggests our idea of time; and that our notion of *equal*

---

[70] Isaac Newton, *Philosophiae Naturalis Principia Mathematica* (1687); translated by Andrew Motte, *Mathematical Principles of Natural Philosophy* (1729), revised by Florian Cajori (Berkeley: University of California Press, 1946).
[71] Maxwell, *Matter and Motion*, pp. 11–12.

intervals of time depends on our recognition of *uniform motion.*"[72] By means of such arguments, physicists understood the equality of time intervals in terms of bodily motions. Thus, Maxwell defined equal intervals of time on the basis of Newton's first law of motion: because bodies are assumed to continue in uniform rectilinear motion if no external forces intervene, then "intervals of time are equal when the relative displacements during those intervals are equal."[73]

In a paper of 1870, Carl Neumann dispensed with Newton's appeal to the idea of absolute time by grounding the scale of time exclusively on "objective time lapses" of freely moving material points. In the 1870s, the philosopher Christoph Sigwart argued that an objective system of time was lacking and thus that it had to be established somehow on the basis of our simultaneous perceptions.[74] In his book of 1881, Johann Bernhard Stallo argued that there exists no absolute time, and hence that "there is no absolute measure of duration."[75] Likewise, Ernst Mach emphatically rejected Newton's notion of absolute time, especially in his *Mechanics.* In his *Analysis of Sensations,* he cautioned that the analysis of time was difficult on various levels: "Much more difficult than the investigation of space-sensation is that of time-sensation."[76] Because the sensations of time seem to be ever present and inseparable from other sensations, Mach argued that, to study time, one has to investigate "the *variations* of time-sensations." Despite such psychological difficulties, Mach argued the analysis of time had to be mainly psychological. He noted that questions of time could hardly be investigated from the physical or physiological sides, partly because "the physiological processes underlying the time-sensation are less known, more radical, and more thoroughly concealed than the processes underlying the other sensations."[77]

In view of arguments such as those advanced by Mach, some physicists increasingly grounded the validity of fundamental principles on sense per-

[72] Lodge, *Elementary Mechanics,* p. 7.

[73] Maxwell, *Matter and Motion,* p. 31.

[74] Christoph Sigwart, *Logik,* 2 vols. (Tübingen: Mohr, 1873 and 1878); *Logic,* trans. H. Dendy (London: Sonnenschein, 1895), vol. 2, pp. 235–238.

[75] Stallo, *The Concepts and Theories of Modern Physics* (1881), p. 221.

[76] Ernst Mach, *Beiträge zur Analyse der Empfindungen* (1886); *Contributions to the Analysis of the Sensations,* trans. C. M. Williams (Chicago: Open Court Publishing, 1897), p. 109.

[77] Mach, *Analysis of the Sensations,* p. 109.

ceptions rather than philosophical intuitions. Physicists realized, for example, that to measure the times of any physical processes requires the coordination of such processes to others, and they increasingly elucidated the specific sorts of operations involved in such comparisons.

Using the concept of an inertial reference frame, Carl Neumann and Ludwig Lange specified a way to conceive of the passage of time. If free bodies travel uniformly relative to a reference frame, then any such body travels equal distances in equal time intervals. Thus, an "inertial time scale" was defined by Lange relative to which the rates of other motions may be determined, be they inertial or accelerated. The concept of an inertial time scale serves then as an ideal standard clock, consisting simply of a free particle moving along a graduated ruler. The idea of an "inertial time scale" presupposed that if a physical process is employed as a clock and does not transpire uniformly, then any measurements of time made by reference to it would not serve for the analysis of phenomena in terms of Newton's mechanics. But even after a particle moving inertially was postulated as a process upon which to ground the measurement of time, Neumann and Lange did not provide any *specific procedures* by which an observer in an inertial system would actually employ such a particle to coordinate the times at which different events transpire. Even as physicists sought clarity of meaning, more ambiguities arose. For if the law of inertia was invoked to compare time measurements, how could they be ascertained in terms of actual physical procedures?

James Thomson argued that the uniformity of motion embodied in the law of inertia lacked a clear meaning because physicists had not established a "principle of chronometry" with which to ascertain uniform measures of motion. This ambiguity added to the already vague concepts of rest and rectilinear motion through space, because Thomson argued that, to specify "that which is called velocity, it is necessary that true chronometry should have been previously attained to, in idea at least, and approximately in fact."[78] Hence, Thomson too asserted that only relative motion could be known or analyzed, and thus he spoke of "quasi-velocity" as opposed to

---

[78] James Thomson, "On the Law of Inertia; the Principle of Chronometry; and the Principle of Absolute Clinural Rest, and of Absolute Rotation," *Royal Society of Edinburgh Proceedings* 12 (1884), 568–578, 571; reprinted in James Thomson, *Collected Papers in Physics and Engineering* (Cambridge: Cambridge University Press, 1912), doc. 57, pp. 379–388.

"true velocity" to refer to a measure of the rate of displacement of a point relative to other points, instead of relative to an absolute space and time.[79] This sentiment was becoming increasingly widespread, as Lodge later emphasized, for example, that when describing or analyzing the motion of an object "we are not thinking of what is called the absolute motion . . . we are thinking of its velocity relative to the earth. Our ignorance of absolute velocities makes this notion of relative velocity not only convenient but essential."[80] This even led him to assert that, "strictly speaking, *motion* appears to be the normal condition of matter at present; . . . and no such thing as *absolute* rest is known."[81]

Yet Thomson's paper of March 1884 delved into issues more subtle than the well-known relativity of motion. While discussing the foundations of kinematics, he indicated that not only was the idea of pinpointing a single location in space at different moments in time ambiguous, but the idea of identifying *the same instant of time at two different locations in space* was ambiguous also. He realized that the problem of determining the simultaneity of distant events was involved in the study of motion. Thus, Thomson argued that the separation of distant objects at a given time could be determined "if, at least, we be content to waive the difficulty as to the imperfection of our means of ascertaining or specifying, or clearly idealising, simultaneity at distant places."[82] He recognized the possibility of a problem in the exact specification of the physical meaning of the concept of simultaneity. He noted that, to determine the distance between objects, people ordinarily used "signals by sound, light, by electricity, by connecting wires or bars, or by various other means." Yet he understood that such methods lacked precision inasmuch as they did not take into account the speed of signal transmission. For example, if a man were to perform an action at the exact moment that he sees someone else perform another action a certain distance away, both actions would *not* be simultaneous, because he actually had to wait for light to travel to his eyes *before* performing his action. This problem did not dissuade Thomson from employing the *idea* of simultaneity, because he admitted that the transmission time for any signal employed to coordinate times "involves an imperfection in human pow-

[79] Thomson, "On the Law of Inertia; the Principle of Chronometry," pp. 571–572.
[80] Lodge, *Elementary Mechanics*, p. 38.
[81] Ibid., p. 5.
[82] Thomson, "On the Law of Inertia; the Principle of Chronometry," p. 569.

ers of ascertaining simultaneity of occurrences at distant places. It seems, however, probably not to involve any difficulty of idealising or imagining the existence of simultaneity."[83] Clearly, Thomson realized that there was something problematic about the exact determination of simultaneity, yet he did not conclude that this ambiguity constituted a fundamental problem for physics.

Meanwhile, similar insights emerged on the other side of the English Channel. In France, Auguste Calinon, a graduate of the École Polytechnique, analyzed similar questions. In his "Critical Study on Mechanics" of 1885, Calinon attempted to revise common interpretations of the concepts of motion, force, mass, and time. He argued, for example, that the notion of absolute motion was metaphysical and mathematically meaningless. In regard to relative motions, he clearly understood that any actual knowledge of velocity depends on a previous measurement of a time. Furthermore, there was a "vicious circularity," because to compare temporal durations one needs an understanding of uniform motion.[84] In principle, a few scientists had long known that there was a vicious circularity in attempts to define uniform motion in terms of time and to define time in terms of uniform motion. Thus, for example, Siméon Denis Poisson, to avoid this logical circularity, had argued that the notion of time did not need to be based on any law of uniform motion.[85] In the 1870s the mathematical economist William Stanley Jevons noted that, "inasmuch as the measure of motion involves time, and the measure of time involves motion, there must be ultimately an assumption."[86] Ludwig Lange, likewise, drew attention to the "methodological circle" in the scientific measurement of motion and of time durations.[87] Calinon added to this understanding by elucidating the importance of determining the same time at distant places—that is, he re-

---

[83] Ibid.

[84] Auguste Calinon, "Étude critique sur la mécanique," *Bulletin de la Société des Sciences de Nancy*, ser. 2, 7 (1885), 87–180 (Paris and Nancy: Berger-Levrault et Cie., Libraires-Éditeurs, 1886), p. 91.

[85] Siméon Denis Poisson, *Traité de Mécanique*, 2nd ed., vol. 1, Imprimeur-Libraire pour les mathématiques, no. 55 (Paris: Bachelier, 1833), p. 205.

[86] William Stanley Jevons, *The Principles of Science* (1874); 2nd ed. (London: Macmillan, 1877), p. 309.

[87] Ludwig Lange, *Die Geschichtliche Entwickelung des Bewegungsbegriffes und Ihr Voraussichtliches Endergebniss. Ein Beitrag zur Historischen Kritik der Mechanischen Principien* (Leipzig: Wilhelm Engelmann, 1886), p. 134.

alized that a knowledge of the simultaneity of distant events was necessary for the measurement of the velocity of a body. He distinguished between the problems of analyzing temporal durations, simultaneity, and succession. The simultaneity of events appeared as a fact of perception, dependent neither on an observer's position or choice of coordinate axes nor on his relative state of motion. The simultaneity and succession of positions of moving distant bodies, to Calinon, consisted of sensations centralized and compared in our brains.[88]

Why did Calinon become interested in the notion of simultaneity? Lacking a definite answer, we may at least note that some of his expressions seem partly reminiscent of the writings of Kant and also that at least one textbook published by the press of the École Polytechnique, in 1861, discussed geometric kinematics by relying heavily on the notion of simultaneity. Its author was Anatole Henri Ernest Lamarle, chief engineer at the School of Bridges and Roads. In that book, Lamarle reformulated the differential and integral calculus on the basis of the kinematics of points, lines, and planes, rejecting the notion of limits. Lamarle construed kinematics as the "*a priori*" geometry of moving figures, and he systematically referred to the simultaneous motions of the constituent points of lines and planes to prove geometric theorems.[89]

At any rate, Calinon became interested in questions concerning the connections between mathematics and experience. At stake were the notions not only of time but also of space. In the 1890s, in his *Study on the Geometric Indeterminacy of the Universe,* he argued that the traditional geometry of Euclid was not the only means of properly representing our visual experiences. Calinon explained that the geometries of Euclid, Lobachevsky, and Riemann all could account for measurements of angles of light rays that coincide on the Earth. Such angles could be determined by actual observations and instruments. However, he pointed out that, in science, the measurement of geometric aspects of light rays other than their angles of coincidence remained "obscure." For example, he discussed the measure-

---

[88] Calinon, "Étude critique sur la mécanique," p. 89.
[89] Ernest Lamarle, *Éxposé Géométrique du Calcul Différentiel et Intégral, précédé de la Cinématique du Point, de la Droit et du Plan, et Fondé tout entiere sur les Notions les plus Élémentaires de la Géométrie Plane* (Paris: Mallet-Bachelier, Imprimeur-Libraire de l'École Impériale Polytechnique, du Bureau des Longitudes, 1861), pp. 6, 15, 18–21, 29, 31, 35, etc.

ment of the form and length of a light ray. On the surface of the Earth such measurements seemed to present no essential difficulties because one could compare the path of the light ray to known material figures. But, he argued, such measurements do not disclose anything about the larger trajectories of light. A light ray could appear rectilinear when referred to our relatively minute terrestrial paths, whereas at an interstellar scale it might actually be curved. He argued that

> if one wants to measure the length of such a ray; assuming rigorously that one can transport all along its trajectory a unit of measure of a fixed length, one does not conceive the possibility of such a measurement but on the condition of referencing exactly the points reached successively by the unit in its transportation along the ray. Thus, in any way that one envisages the question, it goes back to the referencing of points of the light ray in relation to a figure already determined.
>
> Hence, aside from the points that are in the vicinity of a celestial body, and that can be referenced materially by relation to that body, all other points in the Universe cannot be connected to a tactile extension but by light rays, because they are the only lines that we know in space, instead of tactile extensions. In short, the referencing of points of a light ray therefore is not possible but with the aid of light rays, which themselves are not referenced; here lies the vicious circle in which we end.[90]

He realized that it was difficult not only to measure the length of a light ray but "even to conceive that measurement operation and to thus pin down its meaning." Moreover, on the whole, he argued that geometric representations of the universe involve "absolutely irreducible degrees of indetermination." He claimed that the multiple possible geometric representations of experience did not constitute a complication but rather an opportunity for the analysis of experience. Whereas ordinary sensations seemed to be elegantly described with the devices of traditional geometry, perhaps the new geometries would be more suitable for the description of molecular dis-

[90] Auguste Calinon, "Étude sur l'indéterminacion géométrique de l'univers," *Revue Philosophique de la France et de l'Étranger* **36** (1893), 595–607, 603; translation by A. Martínez.

tances and interstellar spaces. He claimed that the free choice of geometric languages was justified only by "reasons of simplicity and commodity."[91]

Calinon developed his ideas further in a booklet titled *A Study of the Various Magnitudes in Mathematics,* published in 1897. In that work he aimed to define succinctly basic magnitudes employed in geometry and mechanics on the basis of a continuous numerical series. Among other topics, he discussed the notion of duration, wanting to disengage it from various "errors which obscure that notion." He noted that durations were commonly assumed to be equal according to the definition, for "when two phenomena of motion are produced at different moments, but in rigorously identical conditions, one says that they are of the same duration." But again he objected that it involves a "vicious circle," because the conditions that are required to be identical must involve "speeds that are functions of duration."[92] Thus, he characterized this common definition as useless. He commented that because the ordinary practical means of measuring time, such as the apparent daily and yearly motions of the heavens, consisted merely of "a simple utilitarian point of view," "it is quite evident that the question of measurable duration remains to be solved entirely, from the scientific point of view."

Accordingly, Calinon proposed a "*Definition of duration.*" He argued that all bodies in motion that have, "at a given moment, simultaneous positions; at another moment will have again simultaneous positions amongst them but posterior to the first." By definition, he required that each of such "successive simultaneities," that is, "successive stages of time," stand in a univocal correspondence with each number in a continuous series.[93] Thus, Calinon envisaged duration as a continuum, such that all physical relations that depend on duration, including trajectories, accelerations, and angles of rotation, would likewise inherit the property of continuity. In this way, he claimed to justify the application of "pure" mathematical analysis to mechanics.

He then designated as "clock" any instrument of which its pointer turns across the portion of an angle proportional to the number x corresponding univocally to a stage in the continuum. He considered as equivalent any two clocks marking simultaneously proportional angles. He consid-

---

[91] Ibid., p. 606.
[92] Auguste Calinon, *Étude sur les diverses grandeurs en mathématiques* (Paris: Gauhier-Villars/Beger-Levrault and Co., 1897), p. 20; translation by A. Martínez.
[93] Ibid., p. 22.

ered many possible clocks as equally legitimate. He then argued that in science one should employ "the clock that is most comfortable for writing the simplest possible formulas expressing the laws of the motions of the bodies in the universe." He expected that if the rate of motion of a clock pointer keeps a constant proportion with the rate of rotation of a planet, then that clock would be particularly useful for describing motions. However, he then demonstrated that the magnitudes of the speeds and accelerations thus measured can vary depending on which clock is used and, moreover, that the direction of an acceleration can also vary accordingly. Without any justification, Calinon argued as though the one clock defined to describe the rotation of one planet as uniform would also describe the rotation of other planets as uniform. Still, he admitted that the use of such a clock was not necessarily imposed but would just constitute a simple and comfortable choice.

Meanwhile, the engineer George Lechalas became engrossed in Kant's analysis of time.[94] Lechalas argued that Kant's notion of reciprocal causal connections served to define the simultaneity of distant events; given Newton's law of gravity, one might say that, for the states of two bodies in which their mutual action equals reaction, the two act simultaneously.[95] An implicit problem, however, was that to know the states of such bodies would involve knowing their velocities and, thus, distant simultaneity.

As for Calinon's work, it influenced at least one prominent individual, the famous French mathematician Henri Poincaré, another graduate of the École Polytechnique. Calinon and Poincaré met in 1886, and soon thereafter Poincaré carefully read Calinon's critical memoir on mechanics. Poincaré sympathized with Calinon's ideas, especially with his effort to revise fundamental concepts that others seemed to take for granted. The works of Calinon and Hertz's book of 1894 motivated Poincaré to refine and publish his own views on the foundations of physics.[96] In 1897 he published a review

---

[94] George Lechalas, *Étude sur l'Espace et le Temps* (Paris: Alcan, 1898), pp. 278–290. See also Bas van Fraassen, *An Introduction to the Philosophy of Time and Space* (New York: Random House, 1970), pp. 54–57.

[95] See also Jules Andrade, *Leçons de méchanique physique* (Paris, 1898), p. 2.

[96] See, e.g., Henri Poincaré, "Les idées de Hertz sur la mécanique," *Revue Générale des Sciences* 8 (30 September 1897), 734–743; reprinted in *Oeuvres de Henri Poincaré*, vol. 7 (Paris: Gauthier-Villars, Libraire du Bureau des Longitudes, de l'École Polytechnique, 1955), pp. 231–250.

of Hertz's book, and in 1898 an article on the foundations of geometry and another on the measurement of time. In the latter article, "The Measure of Time," Poincaré investigated the relationship between time and motion, paying special attention to the concept of simultaneity.

To Poincaré the notion of simultaneity appeared as a long-neglected but important and problematic aspect of the notion of time. The problem of ascertaining the equality of two intervals of time already had been the focus of much attention, resulting in the understanding that such an equality could be established only as a matter of rather arbitrary conventions. Yet the problem of how to reduce to one and the same temporal sequence facts that transpire at different places had remained practically unacknowledged by physicists. The problem consisted of finding the temporal ordering of events or whether or not they happen at the same instant. Poincaré affirmed that although this question had been neglected, the question of how to ascertain the simultaneity of distant events was logically prior to that of finding the equality of time durations.[97] Of course, someone simply could assume the existence of a supreme, universal intelligence who would know the absolute sequence of all events in all places, but this, Poincaré argued, appeared as a crude hypothesis that would fail to facilitate any physical means of discovering such a temporal sequence.[98] He noted that for *some* events an unambiguous temporal sequence could be established, specifically, those events believed to constitute causes of other events. However, Poincaré warned that care should be exercised in identifying temporal causal sequences, because many events are not experienced immediately but only after their effects reach the observer. Thus, even a set of events that appears as a certain sequence of sensations might not take place in that sequence if, say, the impressions take different times to register in the observer's consciousness. Hence, Poincaré asked, "Of two flashes of lightning, the one distant, the other near, can not the first be anterior to the second, even though the sound of the second comes to us before the first?"[99]

[97] Henri Poincaré, "La mesure du temps," *Revue de Metaphysique et Morals* 6 (January 1898), 1–13; reissued in Henri Poincaré, *The Foundations of Science: Science and Hypothesis, The Value of Science, Science and Method,* authorized translation by George Bruce Halsted (New York: Science Press, 1913), p. 228.

[98] Poincaré, *Foundations of Science,* p. 229.

[99] Ibid., p. 231.

Such discussions were not just theoretical; they were practical.[100] Poincaré was not only a mathematician with interests in theoretical physics and the philosophy of science. He also served as an active member and sometime president of the French Bureau of Longitude. Among the labors of this organization were concerted efforts to systematically devise and revise timekeeping conventions, nationally and internationally. Poincaré was thus relatively well acquainted with standard chronometric procedures that, for example, were employed to establish times at distant places. Such procedures were important for map making, navigation, and railroad schedules and, more generally, for disseminating precise measures of time from observatories in Europe out to distant cities and colonies.[101] Telegraphic cables, overland and undersea, served to coordinate "local times" with mother clocks at the heart of the French and British empires.

Discussions about the possible means of exactly determining times and synchrony did not seem to have much practical urgency for physicists in the late 1800s. The one science in which researchers had to rely on procedures for comparing dates of distant events was astronomy, because of the great distances involved in accounting for significant delays in the transmission of light from one point to another. To assign a date to an event at a place far removed from one's location, it was necessary to rely on some means of transmitting information. Thus, an astronomer needs to know not only the distance to a given object or event in space but also the velocity of light to ascertain exactly *when* the event happened. Poincaré argued that astronomers begin their analyses "by *supposing* that light has a constant velocity, and in particular that its velocity is the same in all directions."[102] He suggested that this assumption furnished "a new rule for the investigation of simultaneity," in addition to the various other procedures used by astronomers, navigators, and geographers. Such procedures did not correspond to any general or rigorous rule, but simply served as a convenient means to obtain useful though by no means exact results. Hence, the practical foundations

[100] Peter Galison, "Einstein's Clocks: The Place of Time," *Critical Inquiry* **26** (2000), 355–389; Arthur I. Miller, *Einstein, Picasso: Space, Time, and the Beauty That Causes Havoc* (New York: Basic Books, 2001), chap. 6, pp. 179–213.

[101] Peter L. Galison, *Einstein's Clocks, Poincaré's Maps: Empires of Time* (New York: W. W. Norton, 2003); Alberto A. Martínez, "Material History and Imaginary Clocks: Poincaré, Einstein, and Galison on Simultaneity," *Physics in Perspective* **6** (2004), 224–240.

[102] Poincaré, *Foundations of Science*, p. 232.

for the kinematic study of phenomena involved recourse to measurement procedures dependent on rather arbitrary or variable conventions.

Like Calinon, Poincaré realized that in principle as well as in practice no velocity could be measured "without *measuring* a time."[103] Moreover, this time measure would involve times at spatially separated locations, times that would have to be coordinated by means of signals of known velocity. By such reasoning, a few physicists realized that the practical determination of measures of time and velocity involved a pervasive logical circularity. While in abstract axiomatic form it seemed possible to define and derive fundamental concepts in a deductive sequence, in practice the same concepts appeared to operate simultaneously, because the measurement of the one required a prior procedure for determining the meaning of the other, and vice versa. Here again, as in the analysis of uniform motion, some physicists realized that the measurement of basic physical relations seemed to involve, unavoidably, the introduction of conventions.

The history of physicists' attempts to understand the exact operations involved in the analysis of time illustrates their growing interest in processes of measurement. Physicists did not revise the concept of time in terms of its mathematical representations, as with the concepts of motion and distance. Theorists such as Lange, James Thomson, Calinon, and Poincaré analyzed the notions of duration, simultaneity, and succession in terms of idealized but plausible measurement procedures. The subsequent revision of the foundations of kinematics effected by Albert Einstein in 1905 emerged from this conceptual tradition. Owing to the semblance of arbitrariness in basic mechanics, physicists working in specialized fields were moved to analyze the processes of physical measurement. But physicists' interest in the nature of measurement did not arise only from the theoretical study of motion. It grew mainly out of the necessity for increased precision in physical measurements in engineering and experimental research. Some theoreticians interested in the application of the principles of mechanics to engineering emphasized the importance of measurement. For example, Walter Raleigh Browne argued that, "if we hesitate to accept the well-known apophthegm, 'Science is measurement,' we cannot at least feel any doubt as to its converse, 'No measurement, no science.'"[104]

---

[103] Ibid., p. 234.
[104] Walter R. Browne, *The Foundations of Mechanics*, reprinted from *The Engineer* (London: Charles Griffin and Co., 1882), p. 24.

Table 1. Leading Contributors to Electromagnetic Science and Some of Their
Contributions to the Study of Motion, Listed Chronologically

A. Einstein	1905	"Zur Elektrodynamik bewegter Körper"
H. A. Lorentz	1901	*Zichtbare en onzichtbare bewegingen*
A. Föppl	1899	*Einfuehrung in die Mechanick*
H. Poincaré	1898	"La mesure du temps"
H. Hertz	1894	*Die Prinzipien der Mechanik*
O. Heaviside	1893	"The Elements of Vectorial Algebra and Analysis"
J. C. Maxwell	1877	*Matter and Motion*
C. Neumann	1870	*Ueber die Principien der Galilei-Newton'schen Theorie*
A. M. Ampère	1834	*Essai sur la Philosophie des Sciences*

Not coincidentally, physicists working especially in the field of electricity and magnetism cultivated an interest in the measurement of basic physical quantities. Considering the various contributors to kinematics during the 1800s, we find that several of them were well known for their contributions to electromagnetic theory. We can construct a list of such contributors that reads practically like a who's who in electromagnetic theory (table 1).

Among workers in electromagnetic theory, the interest in the science of motion stemmed from various motivations. One was their attempts to apply electromagnetic theory to bodies in motion, that is, to formulate working theories of electrodynamics and the optics of moving bodies. Also, there was the widespread desire to understand observable phenomena in terms of the motions of invisible entities. A typical example of this approach was the book by Hendrik Antoon Lorentz, *Visible and Invisible Motions,* of 1901. It consisted of basic overviews of many subjects, including the motions of observable bodies, as well as of light, atoms, the ether, molecules, and electrons. Lorentz himself had formulated the most comprehensive theory of the dynamics of electrons. Yet he did not carry out his analyses of motion in terms of systematic distinctions between kinematics and dynamics. In fact, in his book of 1901 he did not use the word kinematics at all. He based much of his presentation on the notion of force. Nonetheless, he did echo some of the concerns of writers on kinematics, such as the emphasis on the importance of observations and measurements. He did not portray the science of motion as an *a priori* subject; instead, he characterized "the so-called mechanics" as the "theory of the phenomena of motion."[105] Like various other writers, he justified the usual "Laws of motion" on both actual experiments and "on experiments which we only make in thoughts."[106]

Another reason why researchers in electricity turned to the elements of kinematics was the need to clearly elucidate precisely how abstract theories were to be applied to measurement and electrical engineering. Thus, George Francis FitzGerald wrote, "Our knowledge of electrical science has advanced *pari passu* with the means of making electrical measurements."[107] FitzGerald, a leading contributor to electromagnetic theory, was perhaps the one who best stimulated the refinement of Maxwell's theory by Hertz, Heaviside, and Lodge.[108] Despite his many contributions, FitzGerald never published them in book form. Still, physicists' interest in measurement may be illustrated by noting that FitzGerald, at the time of his death, left an incomplete abstract on a projected treatise on the measurement of time intervals, lengths, mass, and the like.

The trend toward measurement went counter to the widespread classical tradition that framed scientific knowledge in terms of axiomatic deductive systems. Following the ancient logical styles of Euclid and Archimedes, Newton formulated his account of mechanics in a synthetic manner in which the meaning of empirical propositions was derived from fundamental postulates that purportedly were independent of experience. Physicists who later reformulated the principles of motion in terms of specific measurements sacrificed part of the vaunted generality of the laws of physics, because their validity then depended on operational definitions and conventional stipulations. In this respect, Ernst Mach's critical analy-

---

[105] H. A. Lorentz, *Zichtbare en onzichtbare bewegingen* (1901); German translation: *Sichtbare und unsichtbare Bewegungen; Vorträge auf Einladung des Vorstandes de Departements Leiden der Maatschappij tot nun van't algemeen im Februar und März 1901*, edited from the Dutch original by G. Siebert (Brauschweig: F. Vieweg und sohn, 1902), p. 1.

[106] Ibid., pp. 5, 6.

[107] George Francis FitzGerald, "Observation, Measurement, Experiment: Short Abstract of Methods of Induction: Measurement of Time, Mass, Length, Area, Volume" (1890s, unfinished); published in *The Scientific Writings of the Late George Francis FitzGerald*, ed. Joseph Larmor (Dublin & London: Hodges, Figgis, & Co., and Longmans, Green, 1902), p. 534.

[108] Bruce Hunt, *The Maxwellians* (Ithaca: Cornell University Press, 1991).

sis of the foundations of mechanics, first published in 1883, exercised a profound influence on those who desired to set physics on a more empirical footing. Still, the old axiomatic tradition persisted, as the analysis of Newton's system led theorists to advance revised definitions and laws introduced as statements of intuitive origin and validity. This was the path pursued by Hertz in his work of 1894. Yet Hertz's work deviated substantially in content from the classical presentations of theoretical mechanics. By contrast, other physicists preferred traditional formulations of their science over novel accounts. The most vocal among them was Ludwig Boltzmann.

Like Poincaré, Boltzmann was inspired to publish his views on the foundations of physics by Hertz's work on the principles of mechanics. Boltzmann published the first part of his *Lectures on the Principles of Mechanics* in 1897, and the second part followed in 1904. He also lectured on Hertz's approach in Munich, Leipzig, and the United States. He admired Hertz's work because he agreed that physics should be based on logical concepts and mental pictures of the utmost clarity.[109] However, instead of minimizing the use of certain older concepts, such as Hertz did with the concept of force, Boltzmann gave an essentially traditional presentation. He placed great importance on preserving the traditional approach, while meeting the critical exactitude that Hertz and others had demanded of mechanics. He argued: "I regard it as my life's task to help ensure, by as clear and logically ordered an elaboration as I can give of the results of classical theory, that the great portion of valuable and permanently usable material that in my view is contained in it need not be rediscovered one day."[110] Boltzmann did not deny the value and usefulness of Hertz's approach, despite certain intrinsic

---

[109] See, e.g., Ludwig Boltzmann, "Über die Grundprinzipien und Grundgleichungen der Mechanik" (1899, Clark University 1889–1899, decennial celebration, Worcester, Mass.), in *Populäre Schriften*, Essay 16, pp. 253–307; translated by Paul Foulkes, "On the Fundamental Principles and Equations of Mechanics," in Boltzmann, *Theoretical Physics and Philosophical Problems*, ed. Brian McGuiness (Boston: D. Reidel, 1974), quotation on p. 104.

[110] Ludwig Boltzmann, "Über die Entwickelung der Methoden der theoretischen Physik in neuerer Zeit," Address to the meeting of scientists at Munich (22 September 1899), in *Populäre Schriften*, Essay 14, pp. 198–277; "On the Development of the Methods of Theoretical Physics in Recent Times," in Boltzmann, *Theoretical Physics and Philosophical Problems*, quotation on p. 82.

complications, but he desired to preserve the classical models and concepts as "simple and directly useful pictures" to promote physical understanding, *alongside* novel accounts.[111]

While a number of physicists tried increasingly to ground fundamental concepts on experience and physical operations, Boltzmann argued that clarity in physical theory would be secured sooner by disregarding that approach. In his *Lectures on the Principles of Mechanics,* he argued, "It is precisely the unclarities in the principles of mechanics that seem to me to derive from not starting at once with hypothetical mental pictures but trying to link up with experience from the outset."[112] Boltzmann did not hesitate to employ abstract concepts, such as Cartesian coordinate systems, so long as they served as simple, unambiguous mental pictures upon which to base useful calculations.[113] The concepts of physics need not be directly formulated in empirical terms, because this would reduce the fertility of theories for predicting new phenomena and suggesting new experiments. Hence, neither vague and arbitrary nor completely empirical concepts seemed desirable as the basis of physical theory. To Boltzmann, it made sense that the extremely productive period in which Maxwell's electromagnetic theory had arisen, owing to many "excessively bold" hypotheses, had been followed by a reaction in favor of complete reliance on experience.[114]

By the end of the 1800s, the diversification of new approaches to the foundations of physics was such that the classical approach seemed devoid of vocal supporters. Boltzmann, for one, saw himself as the last active advocate of Newton's mechanics. In an address at a meeting of natural scientists in Munich in 1899, he presented himself "as a reactionary": "I feel like a monument of ancient scientific memories. I would go further and say that I am the only one left who still grasps the old doctrines with unreserved

---

[111] Boltzmann, "On the Fundamental Principles and Equations of Mechanics," p. 113.
[112] Ludwig Boltzmann, *Vorlesungen über die Principe der Mechanik, I. Theil* (Leipzig : Barth, 1897); translated by R. Foulkes as "Lectures on the Principles of Mechanics," in Boltzmann, *Theoretical Physics and Philosophical Problems,* quotation on p. 225.
[113] Boltzmann, "Lectures on the Principles of Mechanics," p. 251.
[114] Ludwig Boltzmann, "Über die Prinzipien der Mechanik I," Inaugural lecture at Leipzig (November, 1900), in *Populäre Schriften,* Essay 17, pp. 309–330; "On the Principles of Mechanics," in Boltzmann, *Theoretical Physics and Philosophical Problems,* quotation on p. 144.

enthusiasm—at any rate the only one who still fights for them as far as he can."[115]

This chapter traces the emergence of various refinements in the basic concepts of kinematics in the 1800s. Physicists used vector theory as a language to distinguish the concepts of displacement and velocity from the concepts of distance and speed. Mathematicians extended the meaning of the terms coordinate, distance, and space, generalizing them and making them increasingly abstract. The concepts of duration, succession, and simultaneity were reformulated in terms of idealized physical relations and measurement procedures. Despite such refinements, conceptual ambiguities remained widespread in kinematics, as physicists used diverse notations, words, and inconsistent expressions to convey basic physical relations.

In the opening pages of his *Principles of Mechanics,* Hertz noted that physicists seemed somewhat embarrassed when presenting the foundations of mechanics.[116] If he was right, then it appears that physicists such as himself at least were sufficiently concerned to invest time and effort to attempt to resolve such difficulties. Their investigations were carried out partly under the conviction that the apparent defects in the foundations of mechanics were "only defects in form."[117] Yet, on close inspection of the theory of motion, physicists eventually encountered problems for which neither their training nor their mathematical knowledge sufficed. While attempting to unify the various branches of physics, theorists were driven to analyze in detail the intricacies of actual and hypothetical measurement procedures. They had to cope with propositions that were hardly meaningful in terms of physical operations, and they had to distinguish between psychological impressions and physical events. They had to deal with inconsistencies in terminology, with logical circularity of definitions, and with intrinsically obscure concepts—all the while applying novel mathematical methods. What physicists found during their investigations of the foundations of mechanics were not simply ambiguities *of form.* What they found were difficulties that were by no means easily resolved.

[115] Boltzmann, "On the Development of the Methods of Theoretical Physics in Recent Times," p. 82.
[116] Hertz, *The Principles of Mechanics,* pp. 6–7.
[117] Ibid., p. 9.

CHAPTER SIX

Discovery and Invention
Conceptual Origins of Einstein's Relativity

A new theory of kinematics appeared in September 1905 in a paper on electrodynamics in the journal *Annalen der Physik*. The author of "On the Electrodynamics of Moving Bodies," Albert Einstein, was not even a university lecturer: "A twenty-six year old patent expert (third class), largely self-taught in physics, who had never seen a theoretical physicist (as he later put it), let alone worked with one, author of several competent but not particularly distinguished papers, Einstein produced four extraordinary works in the year 1905, only one of which (not the relativity paper) seemed obviously related to the earlier papers. These works exerted the most profound influence on the development of physics in the 20th century. How did Einstein do it?"[1] Scarcely any contemporary documents show how he formulated his kinematic theory. Few extant traces exist of his speculations and analyses before his 1905 paper, and none of his rough drafts. Few contemporary letters mention his ideas. Nonetheless, several historians have thoughtfully and carefully labored to reconstruct at least some aspects of Einstein's thoughts, piecemeal, and with considerable success. But despite many varied efforts, "no general consensus has emerged" on the origins of Einstein's theory.[2]

---

[1] John Stachel, "'What Song the Syrens Sang': How Did Einstein Discover Special Relativity?" (1989), *Einstein from 'B' to 'Z'* (Boston: Birkhäuser, 2002), p. 158.

[2] Ibid. Some noteworthy recent accounts are Robert Rynasiewicz, "The Optics and Electrodynamics of 'On the Electrodynamics of Moving Bodies,'" *Annalen der Physik* (Leipzig) **14**, suppl. (2005), 38–57; John D. Norton, "Einstein's Investigations of Galilean Covariant Electrodynamics Prior to 1905," *Archive for History of the Exact Sciences* **59** (2004), 45–105.

What follows is not a history of electrodynamics; rather, it is a biographical account of how Einstein made a new kinematics. Few contemporary documents track his thoughts on that field because, unlike electrodynamics and optics, it was not a main area of his inquiry. Many later accounts and letters, however, bear significant traces pertaining to the roots of his ideas. One must carefully organize the findings, and the task of reconstruction is often like that of a paleontologist, trying to construe a life on the basis of just a few deteriorated bones. It is surprising, nonetheless, how much one can learn from even little bits of evidence.

Yet the excess of speculation that sometimes gets incorporated into the reconstruction is disconcerting. Biographers are often moved by literary aspirations and psychological hypotheses. These tendencies often lead them to concoct plausible and lifelike accounts held together but by the glue of the imagination. Plausible fiction and history, though, are distinct. Conjectures do sometimes lead to previously unknown documentary evidence; but lacking such newfound evidence, historical hypotheses are just speculations. Too often people succumb to the impulse to say much more than the available evidence warrants. Conjectures are often far less imaginative than they seem; what they often belie is a laziness to imagine additional plausible explanations that, in turn, belittle the first as but merely one hypothesis among many.

So reader beware. Accounts of the origins of special relativity have often been thick in conjectures and thin on facts. People tell stories, and fiction sometimes inspires and incites productivity, so all good stories are welcome. It is often fine not to dismiss histories tainted with fiction, but well to ponder the hypotheses there advanced. Yet what follows is not a good story; if you want one, there are plenty at the bookstore.

Adding to the evidence that predates 1905, there exist many retrospective statements pertaining to the roots of special relativity, which were later voiced by Einstein and his peers over decades. Because of his success, he was interviewed by physicists, reporters, biographers, psychologists, and historians. Friends, relatives, and co-workers inquired about his creativity. By systematically piecing together this wealth of evidence, a remarkably detailed picture emerges. Too often, writers have neglected chronology and have overemphasized selected bits of evidence. The following narrative is organized chronologically, to the extent that it seemed possible, and is built from documentary evidence.

Speculations of a College Student

In the late 1880s, Heinrich Hertz demonstrated experimentally that invisible electromagnetic effects traverse space like waves in a medium. His findings fulfilled a main prediction of Maxwell's theory of electromagnetism. The technological possibilities were extraordinary: messages could hence be transmitted with invisible signals without using wires, across considerable distances. Such signals provided further compelling evidence that vacuum space is not empty but is replete with a subtle invisible medium, known then as the ether. Hertz's experiments quickly became widely known among physicists and the public.

Like many others, Albert Einstein was a young lad who became interested in questions about electromagnetic waves in the ether. His father and uncle owned and operated a business that developed and distributed electrical technologies, such as lamps and dynamos. As a boy, Albert had been fascinated by magnetism. In 1895 the sixteen-year-old wrote an essay pondering "the state of the ether in the magnetic field."[3] Having heard about "Hertz's wonderful experiments," Albert proposed to investigate how a magnet affects the structure of the ether. He argued that there should be variations in the speed of an electromagnetic wave, such as a ray of light, passing through a magnetic field. By studying such variations, he hoped, one might measure the deformations of the ether. Such speculations about variations in the motion of light relative to bodies in the ether arose occasionally, even in magazines about science and technology.

Thus, the young Einstein began to cultivate an interest in questions about the propagation of light. As we know, the concept of the relativity of motion had been around for a long time. How did it begin to gain importance for him? In 1895 he was preparing to take entrance examinations to a technical school at Zurich, Switzerland. As reported by his sister, Maja, many years later, one of the sixteen-year-old's books was a textbook by Jules Violle on mechanics.[4] Violle based his treatment of dynamics on

---

[3] Albert Einstein, "Über die Untersuchung des Aetherzustandes im magnetischen Feld" (1895), *The Collected Papers of Albert Einstein*, vol. 1, *The Early Years, 1879–1902*, ed. John Stachel (Princeton: Princeton University Press, 1987), p. 6.
[4] Maja Winteler-Einstein, "Albert Einstein—Beitrag für sein Lebensbild" (1924), in ibid., p. lxiv.

the "principle of relative motions" together with the principle of inertia.[5] The date when Einstein studied Violle, 1895, coincides with his later claim that he pondered the problems of the relative motion of light for ten years before 1905.[6]

Still, we can hardly infer that Einstein's lifelong interest in relative motion originated from Violle, at least not from that textbook alone. Other experiences could also have contributed to his interest. For example, as reported by Max Talmey, a young friend of Einstein's family, already at the age of thirteen Albert had a favorite philosopher.[7] Immanuel Kant had famously advocated, in 1758, what he called "new" concepts of motion and rest, in opposition to Newton's views on absolute motion, by arguing that one should never talk about the motion of any given body without first specifying relative to what other body it is said to move. According to a thirdhand account, the young Albert "especially liked the views of Immanuel Kant and, as he told a friend later, it was through Kant's philosophy that he began to question the truth of generally held beliefs."[8] Reportedly, Albert and Max Talmey discussed "at great length Immanuel Kant's observations concerning the objective and subjective attitude toward knowledge."[9] In any case, we do not know whether at an early age he studied or particularly cared for Kant's views on motion; apparently Einstein left no comments on the matter. The point is that, in addition to Violle's textbook, other sources were at hand.

Alongside ideas on relative motion, ideas on absolute motion also

[5] Jules Violle, *Lehrbuch der Physik,* part 1, *Mechanik.* vol. 1, *Allgemeine Mechanik und Mechanick der festen Körper;* German edition by E. Gumlich et al. (Berlin: Julius Springer, 1892), pp. 90–91.

[6] Einstein, interview of 4 February 1950, R. S. Shankland, "Conversations with Albert Einstein," *American Journal of Physics* **31** (1963) 47–57; p. 48.

[7] Max Talmey, *The Relativity Theory Simplified; And the Formative Period of Its Inventor* (New York: Falcon Press, Darwin Press, 1932), p. 164. That Einstein then studied Kant's *Kritik* was also noted by a fellow student at the Luitpold Gymnasium; see Fritz Genewin to Einstein, 23 October 1924, photocopy in possession of John Stachel, Boston University.

[8] Bela Kornitzer, "Einstein Is My Father," *Ladies' Home Journal* **68** (April 1951), 47, 134, 136, 139, 141, 255–256; p. 255. Kornitzer briefly interviewed Einstein and interviewed his son Hans Albert at length.

[9] Dimitri Marianoff with Palma Wayne, *Einstein: An Intimate Study of a Great Man* (Garden City, N.Y.: Doubleday, Doran, 1944), p. 33. Marianoff was Einstein's stepson-in-law from Einstein's second marriage.

abounded. The young Albert could read even the preface of the *Mathematical Principles of Natural Philosophy* and learn that Newton had composed it mainly to advance the knowledge of true motions rather than of motions merely apparent. In the 1890s most physicists believed that all uniform rectilinear motion is relative, at least in the sense that absolute motion could not be detected by mechanical means; but nevertheless, they also believed that true absolute motion existed, relative to space itself, even if it could not be detected. This belief did not seem to be unverifiable. Rather, owing to electromagnetic theory, physicists thought they had at hand some theoretically plausible ways to establish a kind of seemingly absolute motion. Experiments had demonstrated that light seems to propagate like waves in a medium. Many experiments suggested that this medium, the ether, could not be moved. If the ether were really fixed in space, then perhaps some experiment would detect the motion of bodies relative to it.

Later in life, Einstein recalled that at the age of sixteen he became puzzled by the relationship between light and bodies in motion. When he was a student at a school in Aarau, he conceived a "childish thought-experiment": "When one chases a lightwave at the speed of light, one would have a time-independent wave field. But yet there seemed to exist no such thing!"[10] So "the youth of sixteen pondered: what would happen if a man should try to capture a ray of light?"[11] A psychologist who interviewed him in 1916 noted: "The process started in a way that was not very clear, and is therefore difficult to describe—in a certain state of being puzzled. First came such questions as: What if one were to run after a ray of light? What if one were riding on the beam? If one were to run after a ray of light as it travels, would

---

[10] Albert Einstein, "Erinnerungen-Souvenirs," *Schweizerische Hochschulzeitung* **28** *Sonderheft* (1955), 145–153; reprinted as "Autobiographische Skizze," *Helle Zeit–Dunkle Zeit. In memoriam Albert Einstein*, ed. Carl Seelig (Zurich: Europa Verlag, 1956), p.10; translation by A. Martínez.

[11] Antonina Vallentin, *Le Drame d'Albert Einstein* (Paris: Libraire Plon, 1954), p. 23. A friend of Einstein and his second wife, Vallentin was a well-known writer of biographies. Regarding the quotation, she recounted that "it was with these simple words that Einstein described to me that which he himself regarded as the starting point of his work." Translations by A. Martínez; the French verb *rattraper* can be translated as capture, overtake, catch up. The English edition reads: "The question that fascinated the youth of sixteen was: What would happen if a man should try to imprison a ray of light," Vallentin, *Einstein, a Biography*, translated from the French by Moura Budberg (London: Weidenfeld and Nicolson, 1954), p. 30.

its velocity thereby be decreased? If one were to run fast enough, would it no longer move at all?"[12] "The problem that occupied him," hence, "was the optics of moving bodies, or, more exactly, the emission of light from bodies that move relative to the ether."[13] It seemed paradoxical that, if one could move fast enough, one would see a weird phenomenon: stationary light. He knew no such thing to exist so he expected instead that, "from the standpoint of such an observer," everything should transpire exactly as for an observer at rest.[14] Motion might be relative, but there would be no way to tell, by looking at the behavior of light, whether one is absolutely moving or at rest. Thus, Einstein became curious about "problems in the optics of moving bodies."[15]

Einstein entered the Zurich Polytechnikum in the fall of 1896, at the age of seventeen and a half, six months shy of the required minimum age. Other students, such as Louis Kollros, noted that he was the youngest.[16] Although Albert was lonely and quiet at first, he eventually made a few friends with whom to discuss his interests. Among them was the lively Marcel Grossmann, a diligent student of mathematics. The two would often meet at the Café Metropol to talk.[17] Another new acquaintance was Mileva

---

[12] Max Wertheimer, *Productive Thinking* (New York: Harper & Brothers, 1945), p. 169.

[13] Alexander Moszkowski, *Einstein: Einblicke in seine Gedankenwelt. Gemeinverständliche Betrachtungen über die Relativitätstheorie und ein neues Weltsystem / Entwickelt in Gesprächen mit Einstein* (Hamburg: Hoffmann & Campe; Berlin: F. Fontane, 1921); reissued as *Einstein, the Searcher: His Work Explained from Dialogues with Einstein*, trans. Henry L. Brose (London: Methuen, 1921), reprinted as *Conversations with Einstein* (New York: Horizon Press, 1970), p. 227.

[14] Albert Einstein, "Autobiographisches" (1946), trans. Paul Arthur Schilpp, in *Albert Einstein: Philosopher-Scientist*, ed. Schilpp (Evanston, Ill.: Library of Living Philosophers, 1949).

[15] Albert Einstein, "Wie ich die Relativitätstheorie entdeckte," Lecture delivered at the University of Kyoto, Japan (1922), transcribed into Japanese by Jun Ishiwara, *Einstein Kyôzyu-Kôen-roku* [transcription of Professor Einstein's lecture], first published in the periodical *Kaizo* (1923); also in (Tokyo: Kabushiki Kaisha, 1971), p. 80. Present translation by Fumihide Kanaya and A. Martínez. Also translated by Y. A. Ono, as "How I Created the Theory of Relativity," *Physics Today* **35** (August 1982), 45–47.

[16] Louis Kollros, "Erinnerungen-Souvenirs," *Schweizerische Hochschulzeitung* **28** Sonderheft (1955), 169–173, p. 170; reissued as "Erinnerungen eines Kommilitionen," in Seelig, *Helle Zeit*, pp. 17–31.

[17] Einstein "Erinnerungen" (1955), p. 147; Einstein, "Skizze," p. 11.

Marić, a reserved student likewise entering the Poly in 1896. Albert also met Michele Angelo Besso at a house where various people gathered on Saturday afternoons to play music. Six years older than Albert, he also had broad interests in physics and philosophy. He had graduated from mechanical engineering at the Poly in 1895, and in the fall of 1896 he had started working at a factory of electrical machines in Winterthur.

Michele and Albert discussed various subjects, including popular questions about the existence of atoms and the ether.[18] In this connection, in 1897, Michele recommended to Albert that he read Ernst Mach's critical book on the history and concepts of mechanics.[19] Michele had turned to Mach's writings while he was a student at the Poly, under the suggestion of Aurel Stodola, professor of mechanical engineering.[20] Mach was skeptical of the existence of allegedly invisible entities that could not be detected experimentally. He was particularly critical of the concept of atoms, as well as the notions of absolute space and absolute time. Mach argued that instead of referring motions to absolute space, motions should be established only relative to material systems. Absolute motion had no physical significance, but motions relative to the universe as a whole could perhaps be ascertained by reference to the ether, the stars, or all the bodies in the universe.

Albert was interested in the conceptual underpinnings of physics. Kant had argued that certain organizational categories of the mind are unchangeable, such as the notion of causality. Yet he had also argued that there are other "concepts" that can be constructed from experience. Albert learned that "whatever in knowledge is of empirical origin is never certain."[21] He developed the opinion that laws of physics "owe their origin to the inventive faculty of the human mind. It was this very point that Einstein esteemed so highly in Kant's work. Kant's principal point was that the general laws of science contain not only the result of experience but also an element pro-

[18] Besso to Einstein, 12 October to 8 December 1947, Albert Einstein and Michele Besso, *Correspondance, 1903–1955,* German transcriptions with French translations, notes, and introduction by Pierre Speziali (Paris: Hermann, 1972), p. 386.
[19] Einstein to Seelig, 8 April 1952, Albert Einstein Archives, at the Einstein Papers Project, California Institute of Technology, item 39-018.
[20] Besso to Einstein, 12 October to 8 December 1947, *Correspondance,* p. 386.
[21] Albert Einstein, "Remarks on Bertrand Russell's Theory of Knowledge," in *The Philosophy of Bertrand Russell,* ed. Arthur Schilpp (1944); reprinted in Einstein, *Ideas and Opinions* (New York: Crown Publishers, 1954), p. 22.

vided by human reason."[22] Reportedly, his early interest in Kant's outlook encouraged him "to work out his own original theories."[23]

Thanks to philosophers such as Kant and Arthur Schopenhauer, Einstein could increasingly acknowledge the imperfections of general concepts and understand how they are constructed and rooted in perception.[24] Schopenhauer, for example, argued, "All concepts, all things that are thought, are indeed only abstractions, and consequently partial representations from perception, and have arisen merely through our thinking something away."[25] Likewise, in the evenings, while at the Poly, Einstein began to study the works of Ludwig Boltzmann, who argued, "To construct thought-pictures we constantly need designations for what is common to various groups of phenomena, thought-pictures or intellectual operations; we call such designations concepts."[26]

Once his first academic year at the Poly ended, having studied mainly courses on mathematics, Einstein received good grades. Likewise, Mileva Marić successfully completed her course work, and she then left the Poly to audit courses at Heidelberg. Einstein was disappointed that she left but initiated a lasting correspondence with her. Still, in the fall of 1897, Einstein

[22] Philipp Frank, *Einstein, Sein Leben und seine Zeit* (Munich: P. List, 1949); reprinted, with a foreword by Einstein from 1942 (Brauschweig/Wiesbaden: Friedr. Vieweg & Sohn, 1979); an English translation appeared before the original German text: Frank, *Einstein: His Life and Times* (New York: Knopf, 1947; London: Jonathan Cape, 1948), p. 69. Frank was a colleague and friend who interviewed Einstein.

[23] Kornitzer, "Einstein Is My Father" (1951), p. 255. Kornitzer interviewed Hans Albert Einstein.

[24] See, e.g., Arthur Schopenhauer, *Die Welt als Wille und Vorstellung*, vol. 2 (1844) *The World as Will and Representation*, trans. E. F. J. Payne (New York: Dover, 1966), pp. 15, 66, 141, 148, 179. That the young Einstein read works by Schopenhauer was noted by Rudolph Kayser, Einstein's son-in-law, writing under the pseudonym Anton Reiser, *Albert Einstein: A Biographical Portrait*, with a preface by Albert Einstein (New York: A. & C. Boni, 1930), p. 55.

[25] Schopenhauer, *World as Will and Representation*, vol. 2, p. 378. For commentary on Schopenhauer's notion of concepts and some comparison to Kant's see *The Cambridge Companion to Schopenhauer*, ed. Christopher Janaway (Cambridge: Cambridge University Press, 1999), pp. 77–80, 110, 113–120, 230–231, 243.

[26] Quoted by Arthur I. Miller in *Imagery in Scientific Thought: Creating 20th-Century Physics* (Cambridge, Mass.: Birkhäuser Boston, 1984, and MIT Press, 1986), p. 80. Boltzmann, *Vorlesungen über die Prinzipe der Mechanik* (Leipzig: Barth, 1897); reprinted in Brian McGuiness, ed., *Boltzmann: Theoretical Physics and Philosophical Problems*, trans. R. Foulkes (Boston: Reidel, 1974), pp. 223–254.

continued to spend time with his friend Marcel Grossmann and other class-
mates who also studied mathematics and physics. One classmate, Walter
Leich, recollected their activities in the daily recesses:

> It was our habit to walk back and forth in the long hall off of the lec-
> ture rooms. We carried on lively discussions about subjects we were
> studying and about things in general.
>
> The most animated talker was the alert and capable student Marcel
> Grossmann, the most reserved Einstein. These two young men be-
> came close friends and were much together away from school. . . .
>
> Einstein was a pleasant, gentle, soft-spoken person. In class-room
> tests he acquitted himself as well as but no better than others in the
> class. . . . His spirit of independence asserted itself one day in class
> when the professor mentioned a mild disciplinary measure just taken
> by the school authorities. Einstein for once raised his voice energeti-
> cally to the effect that students at the Polytechnic School should not
> be subjected to strict regulations.[27]

Einstein enjoyed the independence that the Poly allowed. Attendance in
many seminars was not mandatory, so he pursued his interests in physics
independently.

At the time, Einstein was very impressed by "the achievements of mechanics in areas which apparently had nothing to do with mechanics,"
such as "the mechanical theory of light, which conceived of light as the
wave-motion of a quasi-rigid elastic ether."[28] Above all, he was impressed
by the kinetic theory of gases. Yet his faith in mechanics was shaken by
Mach's skeptical attitude toward its foundations; Mach's book "exercised a
profound influence upon me in this regard, while I was a student."[29] One
of the major questions discussed in those days was, Do molecules exist?[30]
Einstein learned that leading scientists such as Mach and Wilhelm Ostwald
had stated boldly that they did not believe in the existence of molecules and

---

[27] Walter Leich to Otto Nathan, 6 March 1957, Einstein Archives, item 60-255.

[28] Einstein, "Autobiographisches" (1946), trans. Schilpp, p. 19.

[29] Ibid., p. 21; Reiser, *Biographical Portrait* (1930), p. 49; Frank, *Life and Times*
(1948), p. 68.

[30] I. Bernard Cohen, "An Interview with Einstein," *Scientific American* **193**, no. 1 (July
1955), 68–73, p. 72.

atoms. His older friend, Michele Besso, also introduced him to Mach's work on the theory of heat, which Einstein enjoyed as well.

Another subject of interest was Maxwell's theory of electricity and magnetism. It was the chief means by which physicists investigated questions about light and the ether. Albert was disappointed that the teachers at the Poly did not offer lessons on Maxwell's theory.[31] So, he proceeded to study it thoroughly on his own, with the help of August Föppl's popular textbook and similar works.[32] Contrary to his early expectations, he learned that physicists believed that the behavior of light should vary depending on the state of motion of the observer. This conviction was based on Augustin Fresnel's wave theory of light, along with Maxwell's theory of electromagnetic fields. In Maxwell's theory, the speed of electromagnetic radiation, such as light in a vacuum, was as a constant. But if an observer moved relative to the ether, then that observer should be able to detect relative variations in the speed of light. Thus, Einstein encountered, "in his second year of college, the problem of light, ether and the Earth's movement. This problem never left him."[33] As the Earth moves through the ether, variations in the speed of light rays should be detectable. So, in 1898, Einstein began to devote much thinking to investigate the question of light and relative motion. Years later he recalled that this problem was essentially "my life for over seven years"— from 1898 until 1905—"and that this was the main thing."[34]

Following Maxwell, most physicists believed that the equations of electromagnetic theory implied that, as bodies move rectilinearly through the luminiferous ether, relative to such bodies there would exist certain asymmetries in the propagation of light—that is, if the ether were absolutely "at rest" in space. Hence, even if perhaps one were never able to identify absolute motions, that is, in space itself, one might be able to identify the motion of bodies relative to the ether. Like many physicists, Albert became increasingly interested in this question, though still a student. It was a subtle issue. Föppl, for instance, appreciated Mach's dislike for unobserv-

[31] Kollros, "Erinnerungen" (1955), p. 170.

[32] August Föppl, *Einführung in die Maxwell'sche Theorie der Elektricität* (Leipzig: B. G. Teubner, 1894). That Einstein read Föppl was noted in Reiser, *Biographical Portrait* (1930), p. 49.

[33] Reiser, *Biographical Portrait* (1930), p. 52.

[34] Einstein, Interview of 24 October 1952, Shankland, "Conversations with Albert Einstein" (1963), p. 56.

able hypothetical entities, and thus he tried to abstain from referring to the ether. In addition to Föppl's book, Einstein also read works by Gustav Kirchhoff, Hermann von Helmholtz, and Hertz, all with "a veritable mania for reading, day and night."[35]

By mid-April of 1898, Mileva Marić was back in Zurich, to enroll again in the Poly for the summer semester, which was just beginning. The nineteen-year-old Albert was glad she returned. At that time, he began to analyze Maxwell's theory intensively by studying Paul Drude's *Physics of the Aether* of 1894.[36] In that book, Drude sought an economical formulation of electromagnetic theory: he avoided any attempts to formulate mechanical models of the ether, avoided the concept of ether velocity, and even suggested ascribing to space itself the properties required to account for electromagnetic phenomena.

Because the Poly gave students leeway to pursue their interests, Albert often skipped the mathematical lectures, preferring instead to spend most of his time in the physical laboratory. He was "fascinated by the direct contact with experience."[37] In the fall semester of 1898, he began courses on electricity and electrical technologies with Professor Heinrich F. Weber. He learned much from Weber, even about the atomistic nature of electrical and thermal phenomena. He enjoyed being at Weber's laboratory, one of the best equipped in Europe. Yet Einstein's dislike of another professor, Jean Pernet, who he saw as incompetent and insane, earned him a reprimand.

On the way home after working in the lab, Einstein and company sometimes talked about ideas in physics. One older classmate, Margarete von Uexküll, who boarded in the same house as Mileva, later recalled Einstein once talking about gravity, mass, and time. According to her, he argued as they walked: "You can well imagine the universe as a sphere filled with any

---

[35] Reiser, *Biographical Portrait* (1930), p. 49. Louis Kollros stated that Einstein read works by Helmholtz, Maxwell, Boltzmann, and Hertz; Kollros to Seelig, 26 February 1952, Eigenössische Technische Hochschule, Zurich Bibliothek, Archives and Special Collections, item Hs 304:740. And in his "Erinnerungen" (1955), p. 170, Kollros stated that at the Polytechnic Einstein read works by Helmholtz, Maxwell, Hertz, Boltzmann, Mach, and Lorentz. Moszkowski recounts that Einstein there read works by Kirchhoff, Helmholtz, Hertz, Boltzmann, and Drude; in Moszkowski, *Conversations*, p. 229.

[36] Paul Drude, *Physik des Aethers auf elektromagnetische Grundlage* (Stuttgart: Ferdinand Enke, 1894). Einstein to Marić, after 16 April 1898, *Collected Papers* 1, p. 213.

[37] Einstein, "Autobiographisches" (1946), p. 15; Einstein, "Erinnerungen" (1955), p. 146; Einstein, "Skizze," p. 10.

mass, the starry world, gases, ether, or however one may call it. Consider now this mass gone, then also there would exist no more time." She listened but felt none the wiser.[38]

After school-hours, Einstein continued his studies of theoretical physics independently: "This extended private study was simply the continuation of earlier habits; and in this Mileva Marić took part."[39] With Mileva, Albert reread some of the works he had already read.[40] To her peers, and even her close friends, Mileva appeared to be too serious and quiet, though good-natured and very intelligent.[41] "She and Einstein," reportedly, "found a common interest in their passion for the study of the great physicists, and they spent a great deal of time together."[42] Thus, Albert had "found a partner in these studies who was working in a similar direction."[43] Although she was four years older than he was, he regarded her, to some extent, as "my 'student.'"[44] Einstein was interested not only in the ideas of scientists but also in aspects of their biographies.[45] According to one writer, one of the books that they read was a biography of Newton:

> Both Albert and Mileva thought that to understand a man's work they ought to know something about the man himself. After they had finished reading a biography of Sir Isaac Newton, the seventeenth-century English mathematician, physicist, and philosopher, Albert asked Mileva, "Have you ever thought that Newton might be wrong in some of his conclusions?" Mileva stared at him as if he had questioned the unquestionable. "Wrong about what?" she asked. "Well, it seems to me that many times Newton doesn't give proofs for what

[38] Margarete J. T. Niewenhuis-von Uexküll, "Albert Einstein nach meiner Erinnerung," manuscript (1953/54?), 14 pages, p. 4, ETH Bibliothek, Hs 304:99; translation by A. Martínez.

[39] Einstein, "Erinnerungen," (1955), p. 146; Einstein, "Skizze," p. 10.

[40] Reiser, *Biographical Portrait* (1930), p. 49.

[41] For example, see the letters from Marić's friend Milana Bota to her mother in 1898, in the collection of Ivana Stefanović, quoted by Michele Zackheim, *Einstein's Daughter: The Search for Lieserl* (New York: Riverhead Books/Penguin Putnam, 1999), pp. 16, 254.

[42] Frank, *Life and Times* (1948), p. 32.

[43] Moszkowski, *Conversations* (1921), p. 229.

[44] Einstein to Marić, 19 December 1901. *Collected Papers* 1, p. 328.

[45] Cohen, "Interview" (1955), p. 71.

he calls laws. He just says, 'This is so.' Couldn't there be forces about which he knew nothing?" "You never simply accept a theorem, a mathematical solution, or a physical law, do you?" asked Mileva. "Always you ask why, as if nothing ever had been said on the subject." Albert grinned but did not try to explain that for him study became a pleasure only when it led to understanding.[46]

Reportedly, Einstein had been convinced by his readings of Kant "that man-found laws of science which stood up to test derived from the inventive element of human reason as well as from experience."[47] Also, Einstein now shared Mach's critical-historical skepticism toward the foundations of mechanics.[48] He appreciated that which could be demonstrated experimentally more than that which was postulated as fundamental assumptions.

The young Einstein believed that there ought to be a way of detecting motion relative to the ether experimentally. In those days, reportedly, "His entire concern was always with the visible processes of physics."[49] He searched in the physics literature but he found no experiment, not "any fact that demonstrates this current of ether, in the physics literature, even in the smallest way."[50] So he decided to pursue the problem: "I wanted to try to demonstrate this ether current against the Earth, in other words, the motion of the Earth. Back then, when I raised this problem in my mind, I did not doubt at all the existence of the ether and the motion of the Earth. Therefore, I sought to examine the difference of amount of heat arising from using two thermoelectric couples, predicting that there should be a difference in the energy between the orientation of the Earth's motion and the opposite direction, by appropriately reflecting with a mirror the light from a single source."[51] He thought that by measuring the heat or pres-

[46] Aylesa Forsee, *Albert Einstein: Theoretical Physicist* (New York: Macmillan; London: Collier-Macmillan, 1963), p. 11. For this biography, Forsee interviewed Hans Albert Einstein. Compare with Vallentin, *Einstein, a Biography,* p. 45.

[47] Peter Michelmore, *Einstein: Profile of the Man* (New York: Dodd, Mead, 1962), p. 29. For this biography, Michelmore interviewed Hans Albert Einstein for two days in 1962.

[48] In 1950 Einstein characterized Mach's *Science of Mechanics* as "truly great," and as "a model for scientific historical writing." Shankland, "Conversations with Albert Einstein," (1963), p. 50.

[49] Marianoff, *Intimate Study* (1944), p. 40.

[50] Kyoto lecture (1922), p. 80; translation by F. Kanaya and A. Martínez.

[51] Ibid.

sure generated by light rays propagating in different directions, one might be able to detect the motion of the Earth through the universal medium. Hence, Einstein entertained the idea of designing such an experiment, but he "did not formulate this experiment with thorough clarity."[52] Meanwhile, practicing physicists had designed and performed many precise experiments with the same aim, but no such effects were found systematically. (At the time, Einstein did not know of such attempts.)[53] Because his professors at the Poly were apparently not interested in carrying out his experiment, or giving him the opportunity to do so, it remained a mere idea.[54]

In any case, basic electromagnetic experiments exhibited seemingly perfectly symmetric effects of relative motion. One particular experiment, an old one, was very well known, even at an elementary level, because it exhibited a dynamic connection between electricity and magnetism. In 1831 Michael Faraday had discovered that the relative motion between a magnet and a conducting wire induces an electric current in the wire, and the effect is the same irrespective of whether one moves the wire or the magnet. For Einstein this was a problem, because Maxwell's theory, as commonly interpreted, provided different explanations for this effect, depending on whether the magnet or the wire moves, whereas the observed effect is identical. In both cases, the same relative motion induces an identical electrical current in the wire. Historians have argued that Einstein may have drawn this case of electromagnetic induction from Föppl's textbook of 1894.[55] Other writers did not describe induction in terms of problematic asymmetries and the violation of an "axiom of kinematics."[56]

During the course of his formal education, Einstein learned about kinematics from a variety of sources. At the Poly he took a course on the mechanics of rigid-body motion in which the professor, Hermann Minkowski, explained: "One speaks of kinematics when one investigates only the geometrical possibilities of motion, where only the concepts of space and time

[52] Ibid.

[53] Reiser, *Biographical Portrait* (1930), p. 52. Einstein described the details of this book as accurate.

[54] Ibid.

[55] Gerald Holton, "Influences on Einstein's Early Work," in *Thematic Origins of Scientific Thought: Kepler to Einstein* (Cambridge, Mass.: Harvard University Press, 1973), pp. 197–217. Miller, *Imagery in Scientific Thought.*

[56] Föppl, *Einführung in die Maxwell'sche Theorie der Elektricität* (1894), p. 311.

are considered."[57] Minkowski, a mathematician, had been an enthusiastic follower of Hertz in his attempt to eliminate the concept of force from the foundations of physics. While studying at the Poly, Einstein began to read Hertz's popular book *Principles of Mechanics* of 1894, in which Hertz complained about the obscurities in the foundations of mechanics and made occasional use of the term kinematics.[58] Hertz even commented: "Why is it that one never asks what is the nature of velocity?" Yet he focused mainly on critically revising the concept of force. Einstein's recurring thoughts involved the more specific question: "What *is* 'the velocity of light'?"[59] He reportedly stopped reading Hertz's book at the point where Hertz replaced all physical forces with purely mathematical connections between particles.[60] Rather than focus on kinematics, Einstein was much more interested in other subjects, especially Maxwell's electromagnetic theory.

Einstein discussed the questions of relative motion in connection with electromagnetic theory with Mileva Marić, who by then was his intimate friend. A good friend later recalled how one day Einstein talked about Marić: "I often work with her. She understands nearly as much mathematics as I myself; for me it is a prize that I've had the opportunity to get to know her. She is not downright beautiful, but she is very clever."[61] In contradistinction, a later friend and colleague of Einstein noted that Marić "was taciturn and reticent" but that "Einstein in his zeal for his studies hardly noticed it."[62]

In the summer of 1899, while studying Hertz's work on the propaga-

---

[57] Hermann Minkowski, "Wintersemester 1898/99. Mechanik"; quoted by Lewis Pyenson, in *The Young Einstein: The Advent of Relativity* (Boston: Adam Hilger, 1985), pp. 24–25.

[58] See, e.g., Heinrich Hertz, *Die Prinzipien der Mechanik* (1894); *The Principles of Mechanics, Presented in a New Form,* trans. D. E. Jones and J. T. Walley (New York: Dover, 1956), p. 127.

[59] According to Wertheimer, the psychologist who interviewed Einstein in 1916; Wertheimer, *Productive Thinking* (1945), p. 169; emphasis in the original.

[60] According to Josef Sauter, statement delivered at the Conference "50 Jahre Relativitätstheorie" in Bern (1955); reprinted in Max Flückiger, *Albert Einstein in Bern: Das Ringen um ein neues Weltbild, Eine dokumentarische Darstellung über den Aufstieg eines Genies* (Bern: Paul Haupt, 1974), p. 154.

[61] Julia Niggli, "Nochmals Albert Einstein. Begegnungen und Briefe," *Aargauer Tagblatt* (Aarau, Switzerland) Friday, 20 June 1952, Year 106, no. 142; translation by A. Martínez.

[62] Frank, *Leben* (1949), p. 39; translation by A. Martínez.

tion of electrical actions, Albert wrote to Mileva that he was increasingly convinced that the theory of electrodynamics of moving bodies did not correspond to reality. He complained that it involved "the name 'ether'" as a medium that was described as being in motion (relative to the Earth), but such motion had no physical meaning, no way of being detected.[63] Perhaps it would be possible to formulate an electrodynamics of moving bodies that did not refer motion to the ether. Einstein hoped to reformulate electrodynamics as the science of "moving electricities and magnetisms" in empty space. He had also studied the papers of Hermann von Helmholtz, who had argued that electricity consisted of free particles. Their existence had been recognized by many physicists, on experimental grounds, by the late 1890s.

By September 1899 Einstein seems to have read a work by Wilhelm Wien that briefly summarized thirteen experiments that had failed to detect the motion of the Earth relative to the ether.[64] Among these experiments were the attempts by Albert Abraham Michelson to detect the "ether wind" by comparing the interference effects of beams of light shot in orthogonal directions and brought back together by mirrors. Michelson's experiment was becoming increasingly notorious, because it had shown, to an unprecedented degree of precision, that the presumed relative motion of the ether through the Earth generated no detectable effect, contrary to expectations. One might begin to suspect that rectilinear motion relative to the ether could hardly be detected. In hindsight, regarding Michelson's result, Einstein once commented: "I guess I just took it for granted that it was true."[65]

Although such experiments had failed, there existed at least slight indications that the ether seemed to be partially affected by the motion of matter. In particular, a famous experiment by Armand Hippolyte Fizeau, published in 1851, had shown that light behaves oddly in moving water. When traversing moving water, the speed of light seemed affected by the water's motion. It was as though the ether itself were dragged along by the moving water, but only very slightly so. If the ether were dragged completely, one would expect that the total speed of light would be the speed of light in water, c/n, *plus* the water's speed, w. Instead, the observations seemed to give a total speed of $c/n + kw$, where $k < 1$. The term k, known as the Fresnel drag

[63] Einstein to Marić, 10? August 1899, *Collected Papers* 1, p. 225.
[64] Einstein to Marić, 28? September 1899, *Collected Papers* 1, p. 233.
[65] Interview of 24 October 1952. Shankland, "Conversations with Albert Einstein" (1963), p. 55.

coefficient, was an important factor in the optics of moving bodies. Also, it constituted an important puzzle in electrodynamics. Because Fizeau's experiment seemed to show that light did not travel at the same rate in stationary as in moving water, it constituted crucial evidence against Hertz's theory of electrodynamics, which assumed that the ether in moving media was fully carried along. Consequently, theorists such as Hendrik Lorentz sought to derive the exact algebraic value of the Fresnel coefficient from a theory of electromagnetism applied to bodies in motion.

At some point Einstein became aware of this puzzle. Fizeau's experiment stood out as a "crucial experiment."[66] It seemed to be "in direct contradiction" to the ether drift experiments, because it seemed to prove that the ether "remains fixed in interstellar space," being scarcely dragged by moving matter.[67] Thus, it seemed that a "decision had to be taken," in favor of either Fizeau's experiment or the ether-drift experiments, yet this seemed "impossible, for both had operated with unsurpassable accuracy. It was impossible to reconcile both views as they were diametrically opposed."[68] Moreover, the slight "drag" suggested by Fizeau's experiment seemed inconsistent with another well-known experimental effect: aberration.[69] As the Earth orbits the sun, the light from any star overhead reaches us at an angle, so any telescope must be slightly tilted such that the light may reach its focal point. Any given telescope, unless full of water, did not exhibit a partial dragging of light. Reportedly, Einstein was especially interested in the experiment of the water-filled telescope, carried out by George Biddell Airy in the early 1870s.[70] Fresnel had devised his partial drag theory to account for such expected effects.

[66] Albert Einstein, "Le principe de relativité et ses conséquences dans la physique moderne," *Archives des sciences physiques et naturelles* **29** (1910), 5–28, especially sec. 2, pp. 7–10; reprinted in *The Collected Papers of Albert Einstein*, vol. 3, *The Swiss Years: Writings, 1909–1911*, ed. Martin J. Klein, A. J. Kox, Jürgen Renn, and Robert Schulman (Princeton: Princeton University Press, 1993), 131–176, sec. 2, pp. 133–136; Albert Einstein, *Über die spezielle und die allgemeine Relativitätstheorie* (Braunschweig: Vieweg, 1917), p. 28.

[67] Moszkowski, *Conversations* (1921), p. 113. Moszkowski interviewed Einstein at length.

[68] Ibid.

[69] Interview of 2 February 1950, and interview of 11 December 1954: Shankland, "Conversations with Albert Einstein" (1963), pp. 52, 57.

[70] Interview of 4 February 1950. R. S. Shankland, "Conversations with Albert Einstein. II," *American Journal of Physics* **41** (1973), 895–901, p. 896.

If theory hardly settled these matters, one might turn to experiment. In September 1899 Albert wrote to Mileva that he "had a good idea for investigating the way in which a body's relative motion affects the velocity of the propagation of light in transparent bodies. I even came up with a theory about it that seems quite plausible to me."[71] Again, he did not carry out any such experiment. Later that month he told Mileva, "I also wrote to Professor Wien in Aachen about my paper on the relative motion of the luminiferous ether against ponderable matter, the paper which the 'boss' handled in such an offhanded fashion."[72] Albert had drafted a paper concerning the ether motion but, apparently, Professor Weber had been dismissive, so he hoped that Wien would find it interesting. But it seems that Wien did not reply. Regardless, Albert found a different kind of success, as Mileva became his girlfriend.

Contrary to standard practice, Einstein himself chose the subject for his diploma paper (and he also chose one for Mileva).[73] Then he spent the spring of 1900 cramming painstakingly for his diploma examinations, relying greatly on class notes lent to him by his friend Marcel. Albert, Mileva, Marcel, Jakob Ehrat, and Louis Kollros took their final examinations. Mileva, unfortunately, did not pass.[74] The exams and the struggle to cram were so exhausting and nerve-wracking that afterward Albert lost most creative interest in physics for about a year.[75] At least he had the freedom and disposition to continue some leisurely readings on his own. On rainy days in July, for example, he read "mainly Kirchhoff's famous investigations of the motion of the rigid body," and he commented to Mileva: "I can't stop marveling at this great work."[76] Nonetheless, one interviewer recounted that "almost a dislike for science and its intellectual technique remained with him after he had finished his course of study. He overcame it only a long time afterward. He approached the broader aspects of thought

---

[71] Einstein to Marić, 10 September 1899, *Collected Papers* 1, p. 230; also in Jürgen Renn and Robert Schulmann, eds., *Albert Einstein and Mileva Marić: The Love Letters*, trans. Shawn Smith (Princeton: Princeton University Press, 1992), p. 14.

[72] Einstein to Marić, 28? September 1899, *Collected Papers* 1, p. 233; *Love Letters*, p. 15.

[73] Louis Kollros to Carl Seelig, 26 February 1952, p. 1, ETH Bibliothek Hs 304:740.

[74] For details see Stachel, "The Young Einstein: Poetry and Truth," *Einstein from 'B' to 'Z,'* pp. 21–38.

[75] Einstein, "Autobiographisches" (1946), trans. Schilpp, p. 8.

[76] Einstein to Marić, 29? July 1900, *Collected Papers* 1, p. 248. Gustav Kirchhoff, *Vorlesungen über mathematische Physik*, vol. 1, *Mechanik* (Leipzig: B. G. Teubner, 1897).

through philosophical studies, chiefly through his readings in Kant and Schopenhauer."[77] Einstein liked to read Kant and Schopenhauer partly "because they made more or less superficial and obscure statements in beautiful language about all sorts of things, statements that often aroused an emotion like beautiful music and gave rise to reveries and meditations on the world."[78]

Looking Misfortune in the Face

Having earned his teaching diploma, Einstein now wanted to be an assistant at the Poly. But he was denied any position there, so he applied to various universities. He was busy applying for job after job, without any success, as he moved between various cities: Melchtal, Milan, Zurich. Away from his beloved Mileva, he visited his family and carried out minor temporary jobs. But then the two spent the winter of 1900 to 1901 together in Zurich. There, he wrote a paper on physics and submitted it to the *Annalen der Physik,* as Mileva proudly informed one of her friends.[79] He also planned to work on his doctoral dissertation.

"Once he regained the taste, after the saturation from the knowledge crammed for his exams, for scientific work, he attacked the problem of the luminous ray, which had haunted him since the age of sixteen."[80] A later friend recounted: "From the time Einstein was fifteen or sixteen years old (so he has often told me) he puzzled over the question: what will happen if a man tries to catch a light ray? For years he thought about this problem."[81]

Because of his travels, his conversations with Mileva were intermittent. Nonetheless, their early discussions on physics seem to have been signifi-

---

[77] Reiser, *Biographical Portrait* (1930), p. 55.

[78] Frank, *Life and Times* (1948), p. 67. Contemporary documents show that Einstein read works by Schopenhauer: Einstein to Grossmann, September 1901, mentions Schopenhauer's *Parerga and Paralipomena;* also Einstein to Marić, 17 December 1901; *Collected Papers* 1, pp. 325, 316.

[79] Mileva Marić to Helene Savić, 20 December 1900, in Milan Popović, *In Albert's Shadow: The Life and Letters of Mileva Marić, Einstein's First Wife,* translations partly by Bosko Milosavljević (Baltimore: Johns Hopkins University Press, 2003), pp. 69–70.

[80] Vallentin, *Le Drame* (1954), p. 36. Vallentin interviewed Einstein; translation by A. Martínez.

[81] Leopold Infeld, *Albert Einstein: His Work and Influence on Our World,* rev. ed. (New York: Charles Scribner's Sons, 1950), p. 41.

cant. In the spring of 1901, in a letter to her, he enthusiastically mentioned: "How happy and proud will I be, when we both together have brought our work on the relative motion victoriously to its end!"[82] His creative interest in the problems of ether and relative motion had reawakened. Back then, his line of inquiry was not in kinematics but in the electromagnetic theory of light applied to moving bodies. He also had occasional discussions with friends, although he had left the academic environment of the Poly. One such friend was Michele Besso. They both spent time in Milan in 1900 and early 1901. Albert admired Michele as "one of the finest and most versatile talents imaginable. He was equally interested in technology, physics and sociology, and was of sound judgment in these as in all phases of life."[83] But despite Michele's erudition, he was inefficient at his job at the Society for the Development of Electrical Enterprises in Italy. Reiterating a story told by Michele's supervisor, Albert wrote to Mileva: "Again, Michele has nothing to do. His manager sends him to the Casale power station to inspect and test the newly installed lines. Our hero decides to leave in the evening, to save valuable time of course, but he unfortunately misses the train. The next day he remembers his assignment too late. On the third day he gets to the train station on time, but to his horror realizes that he has forgotten what he is supposed to do. He immediately writes a card to the office saying that the instructions should be wired!! I don't think this fellow is normal."[84] Despite such absent-mindedness, Albert relished discussing scientific questions with Michele.

In the spring of 1901, the two jointly pondered "the fundamental separation of light-ether and matter," and "the definition of absolute rest," among other things.[85] The luminiferous ether appeared as "an incarnation" of the notion of absolute rest—"all motions of bodily objects could be referred to it, one could in that physical sense speak of 'absolute motion' and also found

---

[82] Einstein to Marić, 27 March 1901, *Collected Papers* 1, p. 282; translation by A. Martínez. Also in *Love Letters*; p. 39. For a discussion of Mileva's role see John Stachel, "Albert Einstein and Mileva Marić: A Collaboration That Failed to Develop," in *Creative Couples in the Sciences*, ed. Helena M. Pycior, Nancy G. Slack, and Pnina G. Abir-Am (New Brunswick, N.J.: Rutgers University Press, 1996); reissued in Stachel, *Einstein from 'B' to 'Z'* (2002), pp. 39–55.

[83] Reiser, *Biographical Portrait* (1930), p. 66.

[84] Einstein to Marić, 27 March 1901, *Collected Papers* 1, pp. 282–283; *Love Letters*, pp. 39–40.

[85] Einstein to Mileva Marić, 4 April 1901, *Collected Papers* 1, p. 285.

mechanics on that notion."[86] They could discuss Mach's comments on Carl Neumann's definition of absolute rest in the hypothesis of the "body Alpha." Such discussions were common, and other sources, soon read by Einstein, also critically discussed Neumann's hypothetical concept.[87] Einstein wanted to formulate a theory characterized by relations among observable quantities, so hypotheses about unobservable bodies seemed unsatisfactory. For most physicists an object's rest or inertial motion could be decided only relative to an arbitrarily chosen body. Depending on where an observer stands, a given object would be seen to move or not. We do not know what Albert and Michele said about such matters. But Albert commented to Mileva that Michele "is a weakling without any glimmers of sound human nature, who can bring himself to no achievement in life or scientific creation, but certainly has an extremely keen mind, the messy workings of which I observe with great delight. . . . He takes great interest in our research, even though he also often misses the big picture by the most petty scruples."[88]

Nonetheless, at some point Einstein and Besso discussed the question of the relative motion of electrical conductors. Decades later Besso reminisced that it was he who highlighted the importance of electromagnetic induction, that the effect obtained from the relative motion between magnet and conductor is the same regardless of which moves, "realizing as an electro-technician that in the framework of Maxwellian theory what appears in the induced part as an electromotive force or as an electric force, depends on whether the inductor of an alternator is at rest or in rotation."[89] Visibly there was no difference in the effects. Einstein found it "intolerable" that physicists gave two distinct theoretical descriptions of the same effect.[90]

[86] Einstein to Solovine, 24 April 1920, Maurice Solovine and Albert Einstein, *Lettres à Maurice Solovine* (Paris: Gauthier-Villars, 1956), p. 18.

[87] For example, Mach's *Mechanics* (1883), and Henri Poincaré, "Relative Motion and Absolute Motion," *Science and Hypothesis;* reissued in Poincaré, *Foundations of Science: Science and Hypothesis, The Value of Science, Science and Method;* authorized translation by George Bruce Halsted (New York: Science Press, 1913), p. 114.

[88] Einstein to Marić, 4 April 1901, *Collected Papers* 1, p. 285; translation by A. Martínez.

[89] Besso to Einstein, 3 August 1952, *Correspondance,* p. 478; translation by A. Martínez.

[90] Albert Einstein, "Grundgedanken und Methoden der Relativitätstheorie, in ihrer Entwicklung dargestellt," manuscript (January 1920), at the Pierpont Morgan Library, New York; a copy is in the Einstein Archives, item 2-070. The consequent edited article is shorter: "A Brief Outline of the Development of the Theory of Relativity," *Nature* **106** (17 February 1921), 782–784.

In April 1901 Besso and Einstein both left Milan. Besso moved, with his wife and son, to Trieste. Albert spent some days with his beloved Mileva; they were eager as usual to see each other. Yet he moved to Winterthur in search of a teaching job. By late May he knew that Mileva was pregnant.[91] Meanwhile his parents bitterly opposed the impending possibility that the two would marry.[92] Mileva also had several arguments with Professor Weber at the Polytechnic.[93] This all caused great stress on her while she had to study to retake the final exams. Unfortunately, she again failed the examination. She then had more difficulties with Weber, and so she left Zurich and decided, embittered, not to return to him.[94]

Despite Einstein's efforts on the problems of relative motion, he still had "no results."[95] By September 1901 he had conceived of yet another experiment for investigating the motion of matter relative to the ether. Einstein wrote to his friend Marcel Grossmann that the experiment would be based on ordinary effects on the interference of light, but he lacked any opportunity to carry it out.[96] Einstein entertained doubts about the full relativity of motion, owing to what he soon identified as a simple mathematical error, but by December 1901 he was thoroughly convinced that all inertial motion is indeed relative, even for electromagnetic processes. Albert wrote to Mileva: "I work eagerly on an electrodynamics of moving bodies, which promises to be a capital treatise. I wrote to you that I doubted the correctness of the ideas about relative motion. My hesitations however were based on a simple calculational error. I now believe in it more than ever."[97] Two days later, he visited Alfred Kleiner, professor of physics at the University of Zurich, and discussed with him the questions of relative motion. Albert wrote to Mileva that Kleiner "advised me to publish

[91] Einstein to Marić, 28? May 1901, *Collected Papers* 1, p. 305.
[92] For example, see Marić to Helene Savić, 20 December 1900, *Life and Letters of Mileva Marić*, p. 70; and also the letter of late 1901, p. 78.
[93] Marić to Helene Savić, (Spring) 1901, *Life and Letters of MilevaMarić*, p. 76.
[94] Marić to Helene Savić, Fall 1901, *Life and Letters of Mileva Marić*, pp. 76–78.
[95] David Reichinstein, *Albert Einstein, sein Lebensbild und seine Weltanschauung* (Prague: Enst Ganz, 1935), p. 23. Reichinstein interviewed Einstein at length for this biography.
[96] Einstein to Grossmann, early September 1901, *Collected Papers* 1, p. 316.
[97] Einstein to Marić, 17 December 1901, *Collected Papers* 1, pp. 325–326; translation by A. Martínez.

my ideas on the electromagnetic theory of light of moving bodies along with the experimental method. He found the method I've proposed to be the simplest and most expedient one imaginable. I was quite happy about the success."[98]

In a manuscript of 1920, Einstein explained that he had been compelled to adopt the relativity of motion, through and through, because of the phenomenon of electromagnetic induction.[99] Moreover, as he became increasingly aware of the failed attempts to detect the relative ether motion, it seemed pointless to carry out his own experiments on that. Indeed, "Einstein was well aware of the null results of a wide range" of such experiments.[100] He recalled, "I was struck by the fact that experiments showed that the velocity of light is constant."[101] Thus, Einstein came to accept, "as a fact," the negative results of all experiments such as Michelson's, "and I came to sense intuitively that it is our mistake to think of motion of the Earth relative to the ether. . . . And from that time on I came to think that, although the Earth goes around the sun, its motion is undetectable by experiments with light."[102] It became just as meaningless to talk about absolute motion, independent of any reference frame, as to talk about absolute rest.

The conviction that there is no unique state of motion, which could be designated as belonging to a special frame of reference, had only gradually solidified in Einstein. Years later he explained that it had not been a "sudden" insight and, hence, that by no means was it a "discovery."[103] He explained, "I was led to it by *steps* arising from *individual* laws of experience." It had taken him several years of reflection, including thought experiments, rough designs for plausible experiments, and some study of actual experiments to arrive at his conviction. Henceforth, it became a fundamental building block for his theoretical endeavors. Years later, Einstein explained to his childhood mentor, Max Talmey, that "the empirically suggested non-existence of any such preferred 'wind-direction' [of ether]

---

[98] Einstein to Marić, 19 December 1901, *Collected Papers* 1, p. 328; *Love Letters*, p. 71.

[99] Einstein, manuscript (1920), Einstein Archives, item 2-070.

[100] Shankland, "Conversations with Albert Einstein. II" (1973), p. 895.

[101] Interviewed in S. J. Woolf, "Einstein's Own Corner of Space," *New York Times*, 18 August 1929, sec. 5 ("Magazine"), 1–2, p. 2.

[102] Kyoto lecture (1922), p. 81; translation by F. Kanaya and A. Martínez.

[103] Moszkowski, *Conversations* (1921), p. 96.

is the main starting point" of what eventually became his special theory of relativity.[104]

Near the end of 1901, Michele Besso gave Einstein a book on the ether, published in 1885. To Einstein the book seemed hopelessly "outdated."[105] Instead, he resolved to study thoroughly the works of Lorentz and Paul Drude on electrodynamics and to ask a friend from the Poly, the earnest Jakob Ehrat, to get him the available literature.[106] Leading theoreticians, such as Lorentz, Drude, Max Abraham, and Henri Poincaré, were already extending electrodynamics much farther than the amateur outsider knew.

Thus, by early 1902, Einstein shifted the direction of his intellectual labors. As Talmey reported, in retrospect, "His special relativity theory was worked out between 1902 and 1905."[107] No longer did Einstein attempt to devise experiments to detect motion relative to the ether. He believed that the motion of all material bodies is relative to other material bodies. He began to intensively study the electrodynamics of Hendrik Antoon Lorentz.

Decades later, Einstein noted in hindsight that before 1905 he had come to know Lorentz's booklet of 1895.[108] Thereby, Einstein learned much about electrodynamics. According to Lorentz, Maxwell's equations were valid exactly in only one reference system, distinguished from all others by its state of rest: the luminiferous ether. Lorentz analyzed electric currents as consisting of true electric masses, independent particles moving about—a conception that Einstein accepted, as he sought a description that would relate free "electricities and magnetisms" in space. He learned about Lorentz's successful attempts to derive the Fresnel drag coefficient by way of a theory that involved the universal stationary ether. To explain the ether-drift experiments, Lorentz relied on a mathematical contrivance: the contraction of moving lengths. He posited that a body moving through the ether contracts along the direction of its motion v, by comparison to an identical body at rest in the ether, by the ratio

$$\sqrt{1 - v^2/c^2} : 1,$$

[104] Einstein to Max Talmey, 6 June 1938.
[105] Einstein to Marić, 28 December 1901, *Collected Papers* 1, p. 330.
[106] Ibid.
[107] Talmey, *Formative Period* (1932), p. 170.
[108] Hendrik A. Lorentz, *Versuch einer Theorie der elektrischen und optischen Erscheinungen in bewegten Körpern* (Leiden: E. J. Brill, 1895); letter from Einstein to C. Seelig, 19 February 1955, Einstein Archives, item 39-070.

where c is the speed of light relative to the ether.[109] Lorentz justified the contraction by hypothesizing that the molecular forces responsible for the cohesion of solid bodies were affected by motion through the ether, such that all bodies contract in the direction of motion, just enough to make such motion undetectable (by eliminating asymmetric path differences among light rays). Perhaps no material bodies were absolutely rigid. Thus, Lorentz tried to explain the many experiments that failed to detect motion relative to the ether. Lorentz also introduced another mathematical contrivance: a term he called *local time*, which served to conveniently make the equations of electromagnetism have the same simple form (very approximately) when referred to moving bodies, as when referred to bodies at rest in the ether. Such artifices, along with the concept of the stationary ether, seemed dispensable for Einstein, who desired to establish a theory that fundamentally involved the symmetries observed in experiments. Nevertheless, Lorentz's elegant works were immensely helpful to Einstein as to many other contemporaries.

Einstein felt that Lorentz had "completely" solved the question of the force of electrons, at least to a first approximation.[110] Hence, at that time Einstein believed that the Maxwell-Lorentz equations were definitely valid and correctly accounted for the facts.[111] Yet he was unsatisfied by Lorentz's account of electrodynamics. The elements of Lorentz's theory did not reflect the symmetry of the phenomena, such as that of electromagnetic induction. If a wire circuit was at rest in the ether, the motion of a magnet in its vicinity was said to produce a real electric field that would set in motion the electrons in the wire. Whereas if instead the magnet was at rest in the ether while the wire was moved, then there would be no electric field, and the same current of electrons in the wire would be said to result from the action of a hypothetical force. Einstein was unsatisfied by this "electromotive force hypothetically introduced by Lorentz."[112] He preferred to think that the symmetric electromagnetic effects of relative motion had but one underlying cause. He reached "the conviction that the electromotive force induced in a conductor moving in a magnetic field is nothing other than

[109] Lorentz actually used the terms p and V, respectively, for the speeds of the body and light.
[110] Kyoto lecture (1922), p. 81; translation by F. Kanaya and A. Martínez.
[111] Ibid.
[112] Einstein, manuscript (1920), Einstein Archives, item 2-070.

an electric field."[113] Einstein thus believed that "the existence of an electric field was therefore a relative one, depending upon the state of motion of the coordinate system being used."[114]

Moreover, Einstein was still puzzled by Fizeau's experiment, so he focused on theoretically examining it. By assuming the existence of the ether, along with Lorentz's force formula, one could account for effects of moving media such as are involved in Fizeau's experiment.[115] But Einstein thought that the formula should apply with respect to an ordinary system of reference rather than to the ether. He "hypothesized" that Lorentz's force formula would work when referred to a moving body instead of to empty space.[116] One could then account for Fizeau's results by thus using the Lorentz force and Maxwell's equations, with "no need of the local time" concept.[117] Still, the equations of electromagnetism did not exhibit complete formal symmetries when referred to systems in relative motion; they assumed complicated forms.

In early 1902 Einstein moved to the old city of Bern, Switzerland, because the father of his friend Marcel had suggested that Einstein be considered for a job there at the federal patent office. By early February he found out that Mileva had given birth to a baby daughter.[118] They hoped that they would soon be reunited. But rather than go to Mileva, Albert tried to secure a job. In the meantime, they continued to exchange affectionate letters.

Although he arrived at beautiful Bern alone, he soon saw acquaintances. He occasionally spent time with Jakob Ehrat, from the Poly. He also spent time with Conrad Habicht, whom he had met at Schaffhausen, and who was working on a doctoral thesis in mathematics. Habicht had hobnobbed with math and philosophy people in Zurich, Munich, Berlin, and Stuttgart.[119] Now he was "very enthusiastic" about Einstein's ideas.[120] Also, Einstein made a new friend: Moritz Solovine, a philosophy student at the University of

---

[113] Message by Einstein, prepared for R. S. Shankland to be read at the centenary celebration of A. Michelson's birth, 19 December 1952, at Case Institute; quoted from *Collected Papers* 2, p. 262.

[114] Einstein, manuscript (1920), Einstein Archives, item 2-070.

[115] Einstein to Besso, 31 October 1916, *Correspondance*, p. 86.

[116] Kyoto lecture (1922), p. 81; translation by F. Kanaya and A. Martínez.

[117] Einstein to Besso, 31 October 1916, *Correspondance*, p. 86.

[118] Einstein to Marić, Tuesday [4 February 1902], *Collected Papers* 1, p. 332.

[119] Franz Paul Habicht to Melania Serbu, 26 October 1943, Einstein Archives, item 39-275, p. 2.

[120] Einstein to Marić, 8? February 1902, *Collected Papers* 1, p. 334.

Bern who sought private lessons in physics. Solovine found in Einstein the kind of friendly teacher he had hoped for: "He presented subjects in a slow and uniform voice, but in a remarkably luminous manner. To render his abstract thoughts more easily intelligible, he sometimes took an example from ordinary experience."[121] The two got along so well that they promptly turned to discussing informally general questions in philosophy and physics.

Einstein was also visited once that spring by Max Talmey, who found him living in considerable poverty.[122] While Einstein was waiting for the job at the patent office to materialize, his family's business had collapsed, so his parents were unable to help him financially. Also, his father's health was failing. Meanwhile, his fiancée and his baby daughter were still far away. Despite such circumstances, Einstein handled himself well, as if following Schopenhauer's advice, to acknowledge that human existence necessarily has its miseries: "If he remembers this, a man will not expect very much from life, but learn to accommodate himself to a world where all is relative and no perfect state exists—always looking misfortune in the face, and if he cannot avoid it, meeting it with courage."[123]

One day Solovine and Einstein decided to jointly read and discuss works in science and philosophy. It seemed to Einstein that most scientists paid no attention to the philosophy of science unfortunately, that it was considered a luxury.[124] Solovine suggested that they begin by reading Karl Pearson's *The Grammar of Science,* and so they did. In connection with Einstein's interests, we may note that Pearson, like Mach, Hertz, Föppl, and others, emphasized relativity, asserting that one "must be careful to notice the relativity of motion; absolute motion, like absolute position, is inconceivable."[125] Furthermore, Pearson noted that, "in order to see our way more clearly through that maze of metaphysics which at present obstructs the entry into physics, we must devote some space to a discussion of the

[121] Solovine and Einstein, *Lettres* (1956), p. vii; translation by A. Martínez.

[122] Talmey, "Einstein as a Boy Recalled by a Friend," *New York Times,* 10 February 1929, p. 145.

[123] Schopenhauer, *Parerga and Paralipomena* (1851); selections reissued in Schopenhauer, *The Wisdom of Life and Counsels and Maxims,* trans. T. Bailey Saunders (Amherst, N.Y.: Prometheus Books, 1995), "Worldly Fortune," p. 90.

[124] Cohen, "Interview" (1955), p. 69.

[125] Karl Pearson, *The Grammar of Science,* part I: *Physical,* Contemporary Science Series (London: Walter Scott, 1892; 3rd ed., London: Adam and Charles Black, 1911), p. 235, and also *Everyman's Library,* no. 939 (London: J. M. Dent & Sons, 1937), p. 177.

elementary notions of kinematics."[126] Einstein was already acquainted with kinematics, from his courses at the Poly, as well as his studies of Hertz and Föppl. Moreover, other sources that touched upon the subject were readily available—for example, Maxwell had written about "Electrokinematics" in his *Treatise on Electricity and Magnetism,* and Thomson and Tait discussed kinematics in their *Treatise on Natural Philosophy.*[127] Both of these influential works were available in German translations in the 1890s. Such works, however, did not highlight the importance of kinematics as Pearson did: as a means for eliminating conceptual obscurities.

Soon, Conrad Habicht joined Einstein and Solovine, and thus they formed a regular discussion group. The three amateurs, none of whom yet held a doctorate, decided to mockingly imitate pretentious scientific societies by giving to their group the bombastic name the Academy Olympia. They held evening meetings, sometimes in one residence, sometimes in another. The three would share a very simple meal, with coffee; Einstein and Habicht sometimes smoked, whereas Solovine detested tobacco in all its forms. Their lively discussions sometimes extended into the late hours of the night; reportedly, "the conversation was often so lively and loud that the other tenants in the house did not find it a matter of spiritual enlightenment, but rather a disturbance of their night's rest."[128]

On 23 June 1902, Einstein finally began work, on a trial basis, as a patent examiner at the Swiss Federal Office of Intellectual Property. The director, Friedrich Haller, was a strict supervisor. He emphasized the need for careful, sharp skepticism toward patent claims, along with the importance of verbal precision. Haller told the patent examiners: "When you take up an application, think that everything the inventor says is wrong. If you don't do that, you follow the inventor's chain of reasoning and thus you become partial. It is a matter of staying critical—vigilant."[129] Haller warned Einstein: "If you do not write clearly, I will throw you out."[130]

---

[126] Pearson, *Grammar,* 3rd ed., p. 221, and *Everyman's* edition, p. 165.

[127] James Clerk Maxwell, *A Treatise on Electricity and Magnetism,* vol. 1 (1873), 2nd ed. (Oxford: Clarendon Press, 1881) p. 326. William Thomson and P. G. Tait, *Treatise on Natural Philosophy,* vol. 1, part I; new edition (Cambridge: Cambridge University Press, 1879), chap. 1: "Kinematics," pp. 1–218.

[128] Reiser, *Biographical Portrait* (1930), p. 67.

[129] Flückiger, *Einstein in Bern,* p. 58.

[130] Franz Paul Habicht to Melania Serbu, 26 October 1943, Einstein Archives, item 39-275, p. 2.

At work, Einstein became acquainted with several other officers. One senior employee was Josef Sauter, who had also studied, years earlier, at the Zurich Polytechnic, and had a doctorate from the University of Zurich. Sauter's dissertation had dealt with magnetism, and he had published it in the *Annalen der Physik* in 1897. He also had recently published there, in 1901, a paper formulating a mechanical system that accounted for Maxwell's equations of electromagnetism. When Sauter found out that Einstein too was interested in Maxwell's theory, he told about his own attempts to formulate a mechanical model. But Einstein replied that he was not interested in mechanical models and remarked "I am a heretic."[131] Later, Sauter discussed with him Hertz's work on the *Principles of Mechanics*, a work that aimed to elucidate mechanical foundations for all of physics. But Einstein repeated "I am a heretic." He was interested in fundamental issues, but he did not seek an account based entirely on mechanics.

Often Solovine visited him at noon, during his lunch break, to carry on their discussions.[132] On days off work, the two would sometimes meet early in the morning and converse all day as they walked around the countryside. Aside from his active interests in physics, Einstein cultivated broader interests. Einstein, Habicht and Solovine met regularly to discuss readings in science, philosophy, and literature. As mentioned before, the first book that Einstein and company read together was Pearson's *Grammar of Science*, which ascribed some importance to kinematics. Moreover, Einstein presumably became increasingly aware of the importance of kinematics from Ampère's own writings. At some point, Einstein and his friends read together "some chapters" of Ampère's *Essay on the Philosophy of the Sciences*.[133] Though Einstein's friends were not quite physicists, they could nonetheless well pursue fundamental subjects such as kinematics. Habicht, the mathematician, was interested in the subject.[134] In mathematics, kinematics was often presented as the pure geometry of motion. Accordingly, some physicists, such as Hertz, and Thomson

---

[131] Sauter (1955), in Flückiger, *Einstein in Bern*, p. 154.

[132] Solovine and Einstein, *Lettres* (1956), p. vii. See also Einstein to Hans Wohlwend, [15 August to 3 October] 1902, *Collected Papers* 5, p. 6.

[133] Solovine and Einstein, *Lettres* (1956), p. viii.

[134] Einstein to Conrad Habicht (18 or 25 May 1905); in *The Collected Papers of Albert Einstein*, vol. 5, *The Swiss Years: Correspondence, 1902–1914*, ed. Martin J. Klein, A. J. Kox, and Robert Schulmann (Princeton: Princeton University Press, 1993), p. 31.

and Tait, had presented kinematics as the *a priori* science of motion, a branch of pure mathematics. By contrast, Ampère's seminal discussion argued that kinematics should be conceived as the science of motion as it appears to observation. The focus on grounding concepts on observations appealed to Einstein, being interested in the visible aspects of physics. He had puzzled over the nature of light by thinking about observers in motion.

Solovine, Einstein, and Habicht also read Mach's *Analysis of Sensations,* among several other works. Einstein admired Mach for his skepticism, "for his general attitude toward the foundations of physics. I see his great merit in that he had disengaged the dogmatism that reigned in the 18th and 19th centuries on the foundations of physics."[135] Einstein also admired Mach for his emphasis on problems concerning the principles of physics. Yet Einstein came to dislike the way Mach tried to connect physics and psychology, as though the objects of physical study were mere sensations, rather than things existing independently of the observer.

Meanwhile, Mileva had been lonely and anguished, living with her parents at Novi Sad.[136] She had suffered from the opposition of Albert's parents, yet she spent time in Bern, in the latter half of 1902, registering as a tenant in rooming houses.[137] She did not bring their baby daughter with her. Albert and Mileva finally married on January 6, 1903, in a small ceremony witnessed by Habicht and Solovine. "That event," Solovine recalled, "did not effect any changes in our meetings. Mileva, intelligent and reserved, listened to us attentively, but never intervened in our discussions."[138] Albert

---

[135] Einstein to Besso, 6 January 1948, *Correspondance*, p. 390; translation by A. Martínez.
[136] For example, see Marić to Helene Savić, late 1901, *Life and Letters of Mileva Marić*, p. 78.
[137] Ann M. Hentschel and Gerd Graßhoff, *Albert Einstein: "Those Happy Bernese Years"* (Bern: Stämpfli, 2005), pp. 112–113.
[138] Solovine and Einstein, *Lettres* (1956), p. xii; translation by A. Martínez. Likewise, in a letter praising Marić to Einstein's biographer Seelig, Solovine yet recalled that "She never took part in our discussions." Solovine to Seelig, 29 April 1952, ETH Bibliothek, Hs 304:1007, p. 1. Contrary to some undocumented claims, there is no evidence that Marić was a member of the discussion group. Likewise, contrary to claims by Arthur I. Miller, in *Einstein, Picasso: Space, Time, and the Beauty That Causes Havoc* (New York: Basic Books, 2001), pp. 76–77, 202, there is no evidence that Lucien Chavan or Michele Besso were members of the discussion group.

and Mileva kept the existence of their daughter a secret. Was she given up for adoption? Some friends of Mileva noticed a change in her attitude, she seemed to brood about a problem; they inquired but she insisted that it was just too personal, "intensely personal."[139] Nevertheless, Albert and Mileva were glad to be together.

Now that Einstein had some peace of mind, he dedicated time to his theoretical investigations. His job at the patent office consumed many hours of each day. It even required him to attend training sessions on Saturdays. Moreover, the majority of the patents that he had to evaluate appeared as "indescribably trivial and lifeless."[140] Nevertheless, study was possible. "He soon discovered that he could find time to devote to his own scientific studies if he did his work in less time. But discretion was necessary, for though authorities may find slow work satisfactory, the saving of time for personal pursuits is officially forbidden. Worried, Einstein saw to it that the small sheets of paper in which he wrote and figured vanished into his desk-drawer as soon as he heard footsteps approaching behind his door. If he had been discovered, he would have been ridiculed as well as harmed. The Director would have laughed at him in addition to being angry."[141]

The discretion seems to have worked. Sauter got the impression that Einstein's daily duties were followed by nightly investigations in physics.[142] In one place or another, he managed to continue his projects. He worked on several subjects, including the foundations of thermodynamics, on which he published a few papers in the *Annalen der Physik*. And he discussed such published papers with Sauter and others. But, above all, Einstein continued to be concerned with the problems of electrodynamics. Reportedly, he explained to his wife his conviction about the relativity of motion: "If I am riding on a train," he said to Mileva, "a raven flying at the exact speed of the train will seem to be making no progress. But a person standing on the ground will say that the raven is flying fast. All motion is relative because it is measured differently by different observers."[143] Examples of this kind

---

[139] Michelmore, *Profile* (1962), p. 42.
[140] Reiser, *Biographical Portrait* (1930), p. 65. See also Marić to Helene Savić, March 1903, *Life and Letters of Mileva Marić*, p. 83.
[141] Reiser, *Biographical Portrait* (1930), p. 66.
[142] Sauter (1955),in Flückiger, *Einstein in Bern*, p. 154.
[143] Forsee, *Theoretical Physicist* (1963), p. 24.

were quite traditional and easy to understand. Yet his main question was "*How can one extend the relativity to the electro-magnetic field?*"[144]

Ruminations of the Hypothetical Sort

Einstein struggled to formulate a simple intuitive theory based on the symmetries of relative motion. To reproduce Lorentz's successful results without using the ether concept, Einstein presumably would have to account for known experimental effects without using the artificial mathematical devices of length contraction and local time. Why would a rigid body contract as it moves against the ether, if no such ether could even be detected? Moreover, why would all bodies contract by the same length though they consist of various substances? Algebraically, length contraction served to make Lorentz's theory yield the known experimental results. Physically, however, this hypothesis seemed artificial. As Einstein recollected, "This hypothesis formally sufficed for the facts of the situation, but in spite of that the mind remained dissatisfied. Should nature have placed us really in an ether gale, and on the other hand exactly so arranged the laws of nature that we can notice nothing of this gale?"[145] "This assumption introduced *ad hoc*"—for Einstein, as for other physicists, the hypothesis of length contraction—"seemed yet but only an artificial expedient, for saving the theory" of Lorentz, for making it agree with experimental results.[146] Such artifices seemed ultimately dispensable.

Einstein knew that not only hypotheses but also the so-called laws of physics might be susceptible to revision. Reportedly, his early readings of Kant had influenced his outlook: "Kant had decided that man-found laws of science which stood up to test derived from the inventive element of human reason as well as from experience. Einstein chewed on this for a long time. If the human mind could invent as well as reason logically, then intuition and imagination were the keys to creation. They could play tricks,

---

[144] Paul Habicht to Serbu, 26 October 1943, Einstein Archives, item 39-275, p. 3, emphasis in the original.
[145] Einstein, manuscript (1920), Einstein Archives, item 2-070; translation by A. Martínez.
[146] Einstein, "Über das Relativitätsprinzip und die aus demselben gezogenen Folgerungen," *Jahrbuch der Radioaktivität und Elektronik* **4** (1907), 411–462, pp. 412–413. Also in *The Collected Papers of Albert Einstein*, vol. 2, *The Swiss Years: Writings, 1900–1909*, ed. John Stachel (Princeton: Princeton University Press, 1989), pp. 433–484, see pp. 434–435.

but they could also lead to the discovery of a truth that had not been known before."[147] To contribute to theoretical physics Einstein turned to the inventive imagination. He had to advance his own hypotheses, a process that could be guided by intuition, by past experiences.

Again, Einstein pondered the mysterious nature of light. If maybe the ether did not exist, then light would not consist of waves in a medium. Presumably, light would consist of a fine granular material, tiny particles spreading out freely across empty space. Since 1899 he had entertained the idea that the agents of electromagnetic actions might move freely in space. Corpuscular conceptions of light had been espoused in the distant past by numerous physicists. For one, he knew that Newton had espoused a corpuscular theory of light, although he did not study Newton's writings in depth.[148] According to classical mechanics, and plain experience, the velocity of a projectile depends on the velocity of its source at the instant of emission. Therefore, if light were a projectile, rather than a wave, its propagation should depend on the motion of its source at the instant of emission. In mechanics, the relativity of motion was assured precisely because the behavior of all bodies projected within an inertial system would all be affected *equally* by its state of motion, such that there is no net difference among their motions. Accordingly, if perhaps light behaved like a projectile, then no experiments involving stationary light sources would reveal any motion of the experimental system. As the ether-drift experiments had all failed, Einstein was reasonably justified in expecting that light obeyed the relativity of motion. So he adopted the hypothesis that the speed of light depends on that of its source.[149]

Relying on this hypothesis, Einstein tried to devise a theory of electrodynamics. Physically, the hypothesis was compatible with experiments in electrodynamics and optics. For example, it agreed with the phenomena of the aberration of starlight in telescopes. This effect can be interpreted according to the relativity of motion together with the hypothetical source-dependence of the speed of light: as the star moves relative to the Earth, the starlight is projected with a velocity that depends on that of the star. The relative

[147] Michelmore, *Profile* (1962), p. 29.
[148] Cohen, "Interview" (1955), p. 71. Cohen notes that Einstein may have learned about Newton's corpuscular conception from Drude's book on light.
[149] Einstein to Paul Ehrenfest, 25 April 1912, *Collected Papers* 5, p. 450. It is unclear whether Einstein entertained this hypothesis before moving to Bern.

velocity of starlight seemed to consist of a simple vectorial sum. Moreover, perhaps Einstein's hypothesis would serve to explain Fizeau's result (and thus the Fresnel coefficient) by somehow accounting for the relative motion between the moving water and the stationary source of light.[150]

The hypothesis seemed to entail some odd novelties, however. One example was that a train of light signals could reverse its sequence: if light were emitted from an accelerating source, then the signals emitted first would move slower than those emitted afterward, so that eventually the latter would overtake the former.[151] This theoretical effect seemed unnatural to Einstein, as no such phenomena was known. Still, the theoretical possibilities were considerable, and his hypothesis seemed promising.

Meanwhile, the meetings of the Academy Olympia continued. For Einstein these meetings were great fun, as they "delighted with a childish joy in all that which was clear and reasonable."[152] According to Solovine, "the foundations and the principles of the sciences" were their constant preoccupation.[153] He also noted, "In the examination of fundamental notions, Einstein employed with predilection the genetic method." That is, that to "clarify" fundamental notions, Einstein appealed to "that which he had been able to observe in children."[154] At a young age, he intensely studied a geometry book that mentioned the "genetic clarification" of a concept, of "specifying in which way it can develop."[155] This sort of expression in formal treatises followed Kant's use of "genetic definition" in his *Logic* of 1800.[156] Yet the "genetic method" identified by Solovine in Einstein, referred to the

---

[150] Albert Einstein, manuscript on the special theory of relativity (1912–1914), in *The Collected Papers of Albert Einstein*. vol. 4, *The Swiss Years: Writings, 1912–1914*, ed. Martin J. Klein, A. J. Kox, Jürgen Renn, and Robert Schulmann (Princeton: Princeton University Press, 1995), doc. 1, pp. 34–35.

[151] Einstein to C. O. Hines, February 1952, Einstein Archives, item 12-251. Interview of 4 February 1950, reported by Shankland, "Conversations with Albert Einstein" (1963), p. 49.

[152] Einstein to Solovine, 3 April 1953, Solovine and Einstein, *Lettres* (1956), p. 125.

[153] Solovine and Einstein, *Lettres* (1956), p. viii.

[154] Ibid.

[155] Theodor Spieker, *Lehrbuch der ebenen Geometrie mit Übungs-Aufgaben für höhere Lehranstalten*, 19th rev. ed. (Potsdam: Aug. Stein, 1890), p. 3.

[156] The expression was used earlier, for example by Christian Wolff, in his *Philosophia Rationalis, Sive Logica* (1728; 3rd ed. 1740), secs. 195, 686, 1126; he equated genetic and "real" definitions.

development of ideas in children, so it was more in the realm of developmental psychology.

At the time, the genetic method aimed to investigate the formation of ideas and habits, to replace static metaphysical presuppositions with scientific statements. Owing to Darwin's explanation of the variations of species, notions of evolution were being increasingly applied to the sciences. In the 1890s James Mark Baldwin pioneered this approach in psychology by performing systematic experiments on children, advocating "the genetic method." Einstein actually read a few texts by authors who voiced interest in the growth of concepts in children. In his *Analysis of Sensations*, Mach argued that "the aim of all scientific research" is the "adaptation of thoughts to facts" and that that process can be essentially observed in children.[157] Mach sought to understand the gradual development of abstract concepts and intuitions; and for that, he pondered how people develop such notions, especially at early ages. He even gave examples from what he had been able to observe in his own child. Furthermore, Hermann von Helmholtz argued that a person's knowledge of axioms of geometry and principles of mechanics stems not from transcendental sources but from ordinary, everyday experiences.[158] Throughout the years, Einstein maintained a playful interest in how children learn.[159]

On the evening of his twenty-fourth birthday, on 14 May 1903, Einstein lectured to his friends at length about "the Galilean law of inertia." As usual, it seemed that "when he boarded a problem he would forget completely

[157] Ernst Mach, *Beiträge zur Analyse der Empfindungen* (1886); *Contributions to the Analysis of Sensations*, trans. C. M. Williams (Chicago: Open Court, 1897), p. 156.

[158] Hermann von Helmholtz, "Über den Ursprung und die Bedeutung der geometrischen Axiome"; "Origin and Significance of Geometrical Axioms," Lecture delivered in the Docenten Verein in Heidelberg in 1870, translated and reissued in Hermann von Helmholtz, *Science and Culture: Popular and Philosophical Essays*, ed. David Cahan (Chicago: University of Chicago Press, 1995), pp. 226–248; see pp. 228–229, 245. Helmholtz, "The Facts in Perception," Speech delivered at the Commemoration Celebration of the Friedrich Wilhelm University of Berlin, 3 August 1878; also in *Science and Culture*, pp. 342–380; see pp. 354–358.

[159] Einstein to Marić, Tuesday [4 February 1902], *Collected Papers* 1, p. 332. Jean Piaget, *Le développement de la notion de temps chez l'enfant* (Paris: Presses Universitaires de France, 1946), p. v. In his *Einstein, Picasso* (2001), p. 187, Miller claimed in passing that "Solovine recalled Einstein's curiosity about how children develop their knowledge of time," but that was a mistake; no such recollection is documented.

the Earth and its miseries and joys."[160] That day he was so absorbed in his account of inertia that he entirely ate, without even noticing what it was, all the expensive caviar that his friends had bought for him as a gift. That same year, as a gesture of appreciation, Solovine presented Einstein with a humorously flattering "Dedication," as "Member of the Academy Olympia." It included a drawing of Einstein's head as a sculpture, and it read:

EXPERT IN THE NOBLE ARTS. VERSED IN ALL LITERARY FORMS—LEADING THE AGE TOWARD LEARNING, A MAN PERFECTLY AND CLEARLY ERUDITE, IMBUED WITH EXQUISITE, SUBTLE, AND ELEGANT KNOWLEDGE, STEEPED IN THE REVOLUTIONARY SCIENCE OF THE COSMOS, BURSTING WITH KNOWL-EDGE OF NATURAL THINGS, A MAN OF THE GREATEST PEACE OF MIND AND MARVELOUS FAMILY VIRTUE, NEVER SHRINKING FROM CIVIC DUTIES, THE MOST POWERFUL GUIDE TO THOSE FABULOUS, RECEPTIVE MOLECULES, INFALLIBLE HIGH PRIEST OF THE CHURCH OF THE POOR IN SPIRIT.[161]

Above the picture of Einstein's bust, Solovine drew a line of hanging sausage links.

One day, at a bookstore, Habicht noticed and bought a recent publication that seemed promising: *Science and Hypothesis,* by the mathematician Henri Poincaré, first published in 1902.[162] Habicht and company read this book at their academy, and according to Solovine, it "engrossed us and held us spellbound for weeks."[163] Poincaré lucidly explained the important functions of hypotheses in mathematical and experimental physics. He even discussed the functions of hypotheses in mathematics.

Traditionally, most mathematicians had believed that the axioms of geometry were necessary and self-evident truths. Following the invention of non-Euclidean geometries, however, increasingly more mathematicians admitted that some of the axioms of Euclid's geometry were based on physical hypotheses, generalizations from experience. This outlook had been fostered thanks to the work of Bernhard Riemann, "About the Hypotheses Which Lie at the Foundations of Geometry," which was also read

[160] Solovine and Einstein, *Lettres* (1956), p. xi; translation by A. Martínez.
[161] Solovine, "Dedication, Einstein as Member of the Olympia Academy" (1903), *Collected Papers* 5, pp. 7–8.
[162] Carl Seelig, *Albert Einstein und die Schweiz* (Zurich: Europa Verlag, 1952), p. 63.
[163] Solovine and Einstein, *Lettres* (1956), p. viii.

by Habicht, Einstein, and Solovine. Poincaré reviewed such developments and argued that the axioms of Euclidean geometry are *"conventions"* freely chosen in accord with ordinary empirical knowledge.[164] He argued that Kant had been wrong in construing the axioms of geometry as *a priori* judgments. Different axioms could be adopted, as in the geometries of Riemann and others. Furthermore, Poincaré explained that some definitions in traditional geometry were accepted merely as assumptions. For example, the definition that figures are equal if they can be superposed exactly presupposes the axiom that a rigid body can be moved without altering its shape when it returns to rest. But to know whether a figure changes when moved, we would need a way to establish the equality of figures. Thus, Poincaré noted that this definition involves a vicious circle, and the usual ideas about the motion of rigid figures were not self-evident truths.[165]

Science and Hypothesis was also the only work among those read jointly by Solovine, Einstein, and Habicht that discussed electrodynamics and Lorentz's theory. After analyzing fundamental mathematical and empirical conditions under which physicists labor, Poincaré illustrated the development of hypotheses in the history of optics and electromagnetism. He reviewed the ideas of Fresnel, Ampère, Maxwell, Helmholtz, Lorentz, and others. Above all, Poincaré argued that, to date, Lorentz had formulated the most satisfactory and serviceable theory of electrodynamics. He praised Lorentz's theory for its simplicity in accounting for the optics of moving bodies, especially for Fizeau's experiment.[166] But he complained that Lorentz's seductive theory violated Newton's law of action and reaction, inasmuch as electrons could allegedly act on the ether without there being any immediate and observable reaction. Moreover, Lorentz's mathematical theory was not rigorously compatible with the principle of relativity. Poincaré disliked that Lorentz had introduced convenient constructs precisely whenever experiments did not agree with theoretical predictions; he preferred that a theory should rigorously agree with phenomena from within, rather than by the addition of auxiliary fixtures.

---

[164] Poincaré, *Science and Hypothesis,* in *Foundations of Science,* p. 65. For a lucid discussion, see Scott Walter, "Hypothesis and Convention in Poincaré's Defense of Galilei Spacetime," in *The Significance of the Hypothetical in the Natural Sciences,* ed. Michael Heidelberger and Gregor Schiemann (Berlin: De Gruyter, forthcoming).
[165] Poincaré, *Science and Hypothesis,* in *Foundations of Science,* p. 61.
[166] Ibid., pp. 149–150, 196.

Poincaré dedicated an entire chapter to discussing questions about relative and absolute motion. He placed great importance on "the principle
of relative motion." The "contrary hypothesis," that phenomena, when referred to systems in inertial motion rather than stationary, might be described by different laws, seemed "singularly repugnant to the mind."[167] He
expected that all experiments would disclose only relative motions among
material bodies.[168] He alluded to the history of old hypothetical subtle fluids,
of heat, of electricity, which were eventually abandoned, and accordingly he
asked: "And does our ether really exist?" (In 1888 he had anticipated that
"no doubt, some day the ether will be thrown aside as useless.")[169] Yet he
admitted the ether as still a very convenient hypothesis. He concluded *Science and Hypothesis* by noting that even if some experiment were to succeed
in detecting what seemed to be absolute motion, that still, for the followers
of Lorentz, "the principle of relativity is safe" because any such seemingly
absolute velocity would be understood as really just a velocity relative to the
ether (rather than to absolute space).[170] Thus, one of the benefits of the ether
hypothesis was that it served to safeguard the principle of relativity.

Aside from acknowledging that the ether hypothesis could be preserved,
Poincaré had asserted some of the ideas that Einstein now entertained.
Some of the other ideas Einstein was considering had likewise been voiced
by a few of the writers with whose writings he was familiar. For example,
Wilhelm Ostwald had suggested, in a work Einstein read, that the ether
hypothesis could well be rejected.[171] Likewise, while discussing "the Hypothetical Method," John Stuart Mill gave skeptical reasons against "the
prevailing hypothesis of a luminiferous ether" in his *System of Logic*, a work
Einstein read with Solovine and Habicht.[172] Furthermore, Mill remarked on

---

[167] Ibid., p. 107.

[168] Ibid., p. 147.

[169] Henri Poincaré, *Théorie mathématique de la lumière*, Sorbonne lectures: 1887–1888,
ed. J. Blondin (Paris: 1889), foreword.

[170] Poincaré, *Science and Hypothesis*, in *Foundations of Science*, p. 197.

[171] Wilhelm Ostwald, *Lehrbuch der allgemeinen Chemie*, 2nd rev. ed., vol. 2, part 1,
Chemische Energie (Leipzig: Wilhelm Engelmann, 1893), pp. 1014–1016. See Einstein
to Ostwald, 19 March 1901, and Einstein to Marić, 10 April 1901, *Collected Papers* 1,
pp. 278, 286.

[172] John Stuart Mill, *A System of Logic, Ratiocinative and Inductive: Being a Connected
View of the Principles of Evidence and the Methods of Scientific Investigation*, 8th ed., vol.
2, (London: Longmans, Green, Reader, and Dyer, 1872), pp. 12, 20.

several phenomena that seemed to be more amenable to explanation on the hypothesis that light propagates as projected corpuscles.[173]

Although Einstein entertained appealing concepts, he still failed to find a solution to the problems of relative motion in electromagnetic theory. In particular, his hypothesis about light seemed to involve theoretical and mathematical difficulties. One difficulty was exactly how to explain Fizeau's experiment, and Einstein found no straightforward way to do so on the assumption that the speed of light depends on that of its source.[174] Again, he could not escape the old conflict between the negative results of the ether-drift experiments and the apparent slight dragging of light in Fizeau's experiment: "This contradiction remains, even if we assume a different hypothesis, not involving the ether, for Fizeau's experiment."[175] Another difficulty was how to explain the formation of a shadow by a screen moving relative to the light source.[176] Einstein conjectured that one would have to introduce "the ugly assumption" that the kind of light emitted from a resonating source depends on the velocity of the light that first impinges on it.[177] Also, he expected that light approaching a mirror perpendicularly at a speed of $c + v$ would have to be reflected at a speed of $c - v$ instead of keeping its same speed.[178] The ensuing mathematical difficulties seemed formidable. He increasingly disliked the idea of characterizing light by various speeds, and it seemed easier to follow tradition by defining light only by frequency and intensity.[179] Einstein also surmised that the speed of light passing through a film could change in such a way that interference effects "would give rise to entirely unbelievable phenomena."[180] Furthermore, he apparently imagined that light of different colors would travel at different speeds; but it then "seemed absurd" that yet all the light emitted from any stationary body, of whatever color, has the same speed.[181] To him this became the "strongest argument," *a priori*, against the hypoth-

[173] Ibid., p. 23.

[174] Einstein, manuscript (1912–1914), *Collected Papers* 4, pp. 34–35; and *Relativitätstheorie* (1917), p. 28.

[175] Moszkowski, *Conversations* (1921), p. 113.

[176] Einstein to Mario Viscardini, April 1922, Einstein Archives, item 25-301.

[177] Ibid.

[178] Einstein to Ehrenfest, June 1912, in *Collected Papers* 5, doc. 409, p. 485.

[179] Ibid.

[180] Einstein to C.O. Hines, February 1952, Einstein Archives, item 12-251.

[181] Ibid.

esis that the speed of light depends on that of its source. This argument also had an empirical component, at least within the range of observational precision available—namely, that if light of different colors propagates at different speeds, one would expect that when a star is eclipsed and then uncovered, then, for the first moment after the star is visible, starlight of all colors would not reach us simultaneously.[182] But no such effect was known.

Still, such arguments were mainly theoretical, not experimental, so Einstein struggled to develop this imaginable way of establishing full relative motion. It seemed that if he were to succeed, then the theory of the optics of moving bodies would become increasingly complicated. But even if such difficulties could be overcome, there remained a greater difficulty. Einstein did not find any way to formulate a mathematical theory of electromagnetism compatible with his hypothesis. He sought differential equations that would allow that in every direction electromagnetic waves may propagate with different speeds, but finally "gave up this approach because he could think of no form of differential equation which could have solutions representing waves whose velocity depended on the motion of the source."[183] Having reached this mathematical impasse, Einstein turned to look for another "way out."[184]

Returning to Lorentz's theory, Einstein weighed what seemed more plausible: that the speed of light depends on its source or it does not.[185] Lorentz's theory accounted very well for the optics of moving bodies, whereas Einstein's alternative scheme entailed complications. He therefore decided to abandon his hypothesis and instead adopt again the common "hypothesis of the independence of the velocity of light" from the motion of its source, owing to "its simplicity and easy practicability."[186] Even though Einstein conceived of radiation as free structures in empty space, he could

[182] Einstein, *Über die spezielle und die allgemeine Relativitätstheorie* (1917); *Relativity: The Special and the General Theory,* authorized translation by Robert W. Lawson (New York: P. Smith/H. Holt, 1931); reprinted (New York: Crown Publishers, 1961), sec. VII, p. 21.

[183] Interview of 4 February 1950. Shankland, "Conversations with Albert Einstein" (1963), p. 49.

[184] Draft of a letter from Einstein to Albert P. Rippenbeim, 1952, Einstein Archives, item 20-046.

[185] Einstein to Ehrenfest, before 20 June 1912, *Collected Papers* 5, doc. 409, p. 485.

[186] Ibid.

not ascribe to them the commonsense notion that their speed depends on the state of the source at the moment of emission. So now he supposed that even if light were made of particles, they would behave like waves in the sense that all light signals would have the same speed. This was a counterintuitive assumption. Einstein had begun to accept that "a theory fundamentally different from that of Lorentz, which would be based on simple and intuitive assumptions and would accomplish the same ends, could not be formulated."[187] He still rejected the stationary ether, but he borrowed Lorentz's fundamental hypothesis that "every light ray in a vacuum always propagates (at least with respect to a certain coordinate system K) with the same definite constant velocity."[188]

The constancy of the speed of light was a physical hypothesis and a convenient way of organizing empirical data. Again, Einstein appreciated the function of concepts in physics; some common concepts were posited by the imagination for the purpose of systematically organizing and analyzing experiences. But some concepts could turn out to be unnecessary, if only there were a more economical way of accounting for phenomena. Einstein surmised, in particular, that the concept of motion relative to an undetectable ether was superfluous. It had even engendered additional concepts that seemed artificial, such as length contraction. Lorentz had tried to justify the latter with hypotheses about intermolecular forces. The concept of "local time" also seemed artificial, lacking explanation. Nevertheless, such concepts exerted a suggestive power.

The concept of local time was significant because it simplified the equations of electrodynamics and also because it constituted a rare example of innovation in the notion of physical time. Poincaré came to construe it as the "most ingenious" idea in Lorentz's electrodynamics.[189] Guided by

---

[187] Einstein, "Die Relativitäts-Theorie," *Naturforschende Gesellschaft in Zürich. Vierteljahrsschrift* **56** (1911), 1–14; *Collected Papers* 3, doc. 17, pp. 425–439.

[188] Albert Einstein, "Die Relativitätstheorie," *Die Kultur der Gegenwart. Ihre Entwicklung und ihre Ziele*, ed. Paul Hinneberg, part 3, sec. 3, vol. 1, *Physik*, ed. Emil Warburg (Leipzig: Teubner, 1915), pp. 703–713, quotation on p. 706. Reprinted in *Collected Papers* 4, p. 539. See also, "Zum Relativitäts-Problem," *Scientia* **15** (1914), 337–48; especially pp. 339–340; reprinted in *Collected Papers* 4 (see pp. 611–612).

[189] Henri Poincaré, "L'état actuel et l'avenir de la Physique mathématique," lecture delivered at the International Congress of Arts and Science in Saint Louis, Missouri, in 1904, in *La Valeur de la Science* (1905); translated and reprinted as "The Present Crisis of Mathematical Physics," in *Foundations of Science*, p. 306.

the results of experiments, Lorentz had contrived a useful mathematical construct, an artificial kind of time. Here indeed was a "concept."[190] Traditionally, several philosophers believed that true time was not a concept in this sense. For Kant, especially, time was not a "concept" and could not be made to depend on others. He claimed that the "axioms" or "principles" of time "cannot be drawn from experience, for this would yield neither strict universality nor apodictic certainty."[191] Time was known intuitively as an *a priori* truth underlying the very possibility of all perceptions. Kant argued emphatically, "Time is not an empirical concept," it is not derived from experience. Schopenhauer had also asserted similar claims.[192] Einstein was familiar with such ideas. From age thirteen he had pored over Kant's *Critique of Pure Reason*. At seventeen, while his friends often attended parties to drink beer, he would instead get "intoxicated on Kant's *Critique*," saying that "beer makes one dumb and lazy."[193] He had further studied Kant's philosophy at the Zurich Polytechnikum in 1897, in a course that included a discussion of Kant's ideas on space and time.[194] Philosophers and metaphysicians thus analyzed "true time." By contrast, Lorentz's local time was a concept that, precisely, claimed no universality or certainty.

In Bern, studying fundamental issues in philosophy and science, Einstein, Solovine, and Habicht read several works that critically discussed notions of time. They discussed Mach's book on mechanics (which Einstein had read earlier), where Mach denounced the notion of absolute time as an idle metaphysical fiction. Likewise, Pearson, in his *Grammar of Science*, had argued that "time is a relative order of sense-impressions . . . there is no such thing as *absolute* time," and that one must not "project the absolute time of conception into a reality of perception."[195]

---

[190] Einstein, "Relativitätsprinzip und die aus demselben gezogenen Folgerungen" (1907), p. 413; *Collected Papers* 2, p. 435.

[191] Immanuel Kant, *Critique of Pure Reason*, trans. and ed. Paul Guyer and Allen W. Wood, Cambridge Edition of the Works of Immanuel Kant (Cambridge: University Press, 1998), p. 162.

[192] Schopenhauer, *World as Will and Representation*, e.g., p. 180.

[193] Hans Byland, "Aus Einsteins Jugendtagen. Ein Gedenkblatt," *Neue Bündner Zeitung*, 7 February 1928.

[194] Stachel, *Collected Papers* 1, p. 364.

[195] Pearson, *Grammar of Science*, pp. 159–162, ellipsis added.

Incidentally, Einstein, Habicht, and Solovine also read the *Ethics* of Baruch Spinoza (a work for which Einstein felt great affinity subsequently) in which Spinoza attempted to prove, among many other things, the proposition: "We can have only a very inadequate knowledge of the duration of things external to us."[196] Moreover, they discussed *Science and Hypothesis*, the book that "engrossed us and held us spellbound for weeks."[197] There Poincaré claimed: "There is no absolute time; to say two durations are equal is an assertion that by itself has no meaning and which can acquire a meaning only by convention. Not only have we no direct intuition of the equality of two durations, but we have not even that of the simultaneity of two events produced in different theaters; this is what I have explained in an article titled *The Measure of Time*."[198] Among the other works they jointly read, a few touched upon the narrow concept of simultaneity in relation to other aspects of time. In his *System of Logic*, John Stuart Mill discussed the character of propositions concerning "Order in Time" in terms of the notions of "Sequence" (succession) and "Co-existence" (simultaneity).[199] In his "Numbering and Measuring," Hermann von Helmholtz discussed in passing the question of "Time measurement," stressing that the equality of durations is ascertained on the basis of whether physical processes begin and end "simultaneously."[200] Also, in his *Grammar of Science*, Karl Pearson argued that a "mode of discrimination" of abstract science must concern the analysis of sequence in time, because "the immediate groups of sense-impressions are not really simultaneous, and most things perceived in space are 'constructs' involv-

---

[196] Baruch Spinoza, *The Ethics* (1677); *The Ethics and Selected Letters*, ed. Seymour Feldman, trans. Samuel Shirley (Indianapolis: Hackett, 1982), p. 85.

[197] Solovine and Einstein, *Lettres* (1956), p. viii.

[198] Henri Poincaré, *La Science et l'Hypothèse* (Paris: Flammarion, 1902), p. 111; translation by A. Martínez. Reissued in *Foundations of Science*, p. 92.

[199] Mill, *A System of Logic*, 8th ed., e.g., book I, chaps. V–VI, pp. 65–70.

[200] Hermann von Helmholtz, "Zählen und Messen," *Philosophische Aufsätze Eduard Zeller zu seinem fünfzijärigen Doktorjubiläum gewidmet* (Leipzig: Fues' Verlag, 1887), pp. 17–52; reprinted in Helmholtz, *Wissenschaftliche Abhandlungen*, vol. 3, pp. 356–391; translated by Malcolm F. Lowe, "Numbering and Measuring from an Epistemological Viewpoint," in Helmholtz, *Epistemological Writings*, ed. Robert S. Cohen and Yehuda Elkana, Boston Studies in the Philosophy of Science, vol. 37 (Boston: D. Reidel, 1977), p. 93.

ing stored sense-impresses."[201] Moreover, Einstein, in particular, was in a position to be especially aware of the neglected but significant notion of simultaneity, owing to his early interest in Kant, who had analyzed it at length.[202]

In 1770 Kant criticized Leibniz and other philosophers for their view, which "at once shows itself erroneous by involving a vicious circle in the definition of time, and also by entirely neglecting simultaneity, a most important consequence of time."[203] Kant also discussed the concept of simultaneity in his work of 1781, *Critique of Pure Reason*. From Kant, Einstein could learn to subdivide the notion of time into the notions of duration, succession, and simultaneity and to appreciate the fundamental role these notions play in human perception.[204] Kant denied the idea that time has any absolute reality, at least in the sense that time could not be said to be "objectively real" if abstracted from the sensibility of intuition and applied to "things in general."[205] Time had "objective validity" only in respect to appearances.[206] Moreover, Kant argued that the temporal ordering of events in space must be constructed according to ideas about physical interactions.[207] Again, time was not a concept; it could not depend on experience. Kant argued that "the proposition that different times cannot be simultaneous cannot be derived from a general concept."[208] Einstein could become further acquainted with such ideas through the works of Schopenhauer. Following Kant, Schopenhauer denied the absolute existence of time, emphasized its

---

[201] Pearson, *The Grammar of Science* (1892), p. 456.

[202] E.g., Kant discusses a "Principle of Simultaneity" in the second edition of his *Kritik der Reinen Vernuft; Critique of Pure Reason*, Cambridge Edition, pp. 316–319.

[203] Immanuel Kant, *De Mundi Sensibilis atque Intelligibilis Forma et Principiis*, Dissertatio Pro Loco, 1770: *Kant's Inaugural Dissertation and Early Writings on Space*, trans. John Handyside (Chicago: Open Court, 1929), p. 57. There is no specific evidence that the young Einstein read Kant's *Inaugural Dissertation*, but Max Talmey reported that the *Critique of Pure Reason* was but one among several philosophical writings by Kant that were read by the young Einstein, in Talmey, "Einstein as a Boy Recalled by a Friend," p. 145.

[204] Kant, *Critique of Pure Reason*, Cambridge Edition, e.g., pp. 296, 301, 319.

[205] Kant, *Inaugural Dissertation and Early Writings on Space*, pp. 56–57. See also Kant, *Critique of Pure Reason*, Cambridge Edition, pp. 164–165.

[206] Kant, *Critique of Pure Reason*, Cambridge Edition, pp. 164, 282.

[207] Ibid., p. 310.

[208] Ibid., p. 162.

subjective origin, and made significant use of the notion of simultaneity.[209] Still, Schopenhauer also commented on variations in our perceptions of time. In his *Parerga and Paralipomena,* which Einstein read, Schopenhauer noted that a month seems longer when people travel than when they stay at home and, also, that time seems to go faster as people get older.[210] He also commented on the problem of being right when everyone else is wrong: "He who can see truly in the midst of general infatuation is like a man whose watch keeps good time, when all clocks in the town in which he lives are wrong. He alone knows the right time; but what use is it to him? for everyone goes by the clocks which speak false, not even excepting those who know that his watch is the only one that is right."[211]

We do not know whether any such varied statements provoked any interest in Einstein, but at least such were the comments about time that were voiced by some of the authors he read.

Furthermore, Einstein, Habicht, and Solovine could have pursued in full Poincaré's discussion of the measurement of duration and simultaneity by seeking his article of 1898. But even if they did not obtain the original publication, it was reissued in Poincaré's latest anthology, *The Value of Science,* published in 1905, which they read sometime that year.[212] Moreover, and earlier still, if Einstein or either of his friends read *Science and Hypothesis* in its German translation of 1904, they could read an editorial note on Poincaré's arguments on simultaneity, including a translation of the relevant passage from "The Measure of Time."[213]

In his article, Poincaré analyzed fundamental problems about the measurement of time. He asserted the need for specifying the exact procedures used to determine the duration of any physical process, as well as the si-

[209] See, *e.g.,* Schopenhauer, *Die Welt als Wille und Vorstellung,* vol. 2 (1844); *World as Will and Representation,* trans. Payne, pp. 8, 33, 35, 38, 48–51, 301, 314, 493. For example, he argued (pp. 48–50) that "Different times are not simultaneous but successive," "We know the laws of time *a priori,*" and that "every part of time is everywhere, i.e., in the whole of space simultaneously."
[210] Schopenhauer, *Parerga und Paralipomena* (1851); *The Wisdom of Life and Counsels and Maxims,* trans. Saunders (1995), "The Ages of Life," pp. 108–109.
[211] Schopenhauer, *Wisdom of Life,* "Position, or a Man's Place in the Estimation of Others," p. 61.
[212] Maurice Solovine to Carl Seelig, 14 April 1952, ETH Bibliothek, Hs 304:1006, p. 2.
[213] Ibid., p. 308.

multaneity of events. He stressed that "we have not a direct intuition of simultaneity, nor of the equality of two durations." Decades earlier, a few theorists, such as Siméon Denis Poisson, Auguste Calinon, and Ludwig Lange, had noted that there exists a logical circularity between the concepts of time and velocity. Following Calinon, Poincaré emphasized the logical circularity in any definition of velocity in terms of time, and vice versa. The essential difficulty is that to measure the speed of a signal traveling in a single direction, one needs to know the instants when the signal passes the end points of a given distance, and these times must be obtained from *synchronized* clocks. The measurement of any one-way speed presupposes knowing the simultaneity of distant events. But the synchrony of clocks at distant places can be set or verified only by means of signals of known speed. Years later Einstein noted, "It would thus appear as though we were moving in a logical circle."[214]

Poincaré argued that the only way then to establish velocities and simultaneity was by means of some convention:

> When an astronomer tells me that some stellar phenomenon . . . happened [at a certain time] . . . I seek his meaning, and to that end I shall ask him first how he knows it, that is, how he has measured the velocity of light.
>
> He has begun by *supposing* that light has a constant velocity, and in particular that its velocity is the same in all directions. That is a postulate without which no measurement of this velocity could be attempted. This postulate could never be verified directly by experiment.[215]

Terrestrial and astronomical measurements presuppose the equality of the speed of light in different directions. This was a useful convention. Poincaré noted that the assumed constancy of the speed of light "furnishes us with a new rule for the investigation of simultaneity." One could call distant events "simultaneous," by convention, by coordinating distant clocks using light signals assumed to travel at the same speed.

Poincaré's concern with practical measures of time found a useful theoretical application in electrodynamics. He soon realized that Lorentz's con-

---

[214] Einstein, *Relativity*, p. 23.

[215] Henri Poincaré, "La mesure du temps," *Revue de Métaphysique et de Morale* 6 (January 1898), 1–13; reissued in *Foundations of Science*, p. 232; ellipses added.

cept of "local time" could be explained in terms of the synchronization of clocks. In a paper presented in December 1900, Poincaré pointed out that the fictitious times that Lorentz had ascribed to moving reference systems, namely,

$$\text{local time} = t - \frac{vx}{c^2},$$

where t is the true time, would actually be obtained by observers moving together but separated by a distance x, who synchronized their watches by exchanging light signals.[216] The procedure can be illustrated as follows.[217] Suppose that two observers stand on opposite ends of a platform moving at a velocity v with respect to the ether. One sends a flash of light to the other, across the length of the platform x. Because the speed of light is taken to be independent of the speed of its source, the signal travels at speed $c - v$ relative to the platform, so that it takes a time $x/(c - v)$ to reach the "receding" observer. In turn, that observer sends a signal back, which travels at a relative speed of $c + v$, so that it reaches the first observer after a delay of $x/(c + v)$. The total lapse for the light signal to go from one clock to the other and back is just the sum of the two:

$$\frac{x}{c-v} + \frac{x}{c+v} = \frac{2xc}{c^2 - v^2}.$$

Not knowing their motion relative to the ether, the observers would assume that the light takes *half of the total lapse* to go from a clock at one end of the platform to another at the opposite end. The discrepancy between that inference and the *real* travel time would be

$$\frac{1}{2}\left(\frac{2xc}{c^2 - v^2}\right) - \left(\frac{x}{c-v}\right) = \frac{-vx}{c^2 - v^2}.$$

---

[216] Johannes Boscha, ed., *Recueil des travaux offerts par les auteurs à H. A. Lorentz à l'occasion du 25ème anniversaire de son doctorat le 11 décembre 1900* (The Hague: Martinus Nijoff, 1900). Each moving system would have infinitely many local times for each true time instant t in the ether system, each local time corresponding to each value of x.

[217] The following example is essentially the same provided by Olivier Darrigol in *Electrodynamics from Ampère to Einstein* (New York: Oxford University Press, 2000), p. 359.

Because a moving clock would not be synchronized with a clock at rest in the ether, the time of any event given as t for the stationary clock would instead be

$$t - \frac{vx}{c^2 - v^2}$$

for the moving clock. The square of the speed of light was known to be far, far greater than, say, the orbital velocity of the Earth; thus, one might well disregard the term v^2. Hence, Poincaré could identify Lorentz's local time with the apparent time of moving clocks thus synchronized.[218] Besides, Lorentz's algebraic expression for the local time was just an approximate value. (Moreover, an additional correction could be made by taking into account the contraction of moving lengths, but the approximate apparent time remains the same.) The important point is that Poincaré found a physical meaning for Lorentz's seemingly arbitrary concept.

In his 1900 paper, Poincaré formulated his argument succinctly in words, but any physicist could easily carry out a calculation like the preceding one. If Einstein read such lines, he would readily grasp that Lorentz's local time could be construed in terms of plausible physical procedures. Poincaré's article of 1900 appeared in a celebratory volume dedicated to Lorentz, which, virtually, "was read by every expert in the field" of electrodynamics.[219] (By early 1906, at the latest, Einstein was familiar with it, because in a paper written in May he cited Poincaré's paper.)[220] Moreover, Einstein and his friends could later encounter the same argument in yet another one of Poincaré's papers in his popular anthology of 1905.[221] There, too, Poincaré explained that moving observers can adjust their clocks with light signals

---

[218] In approximation theory, one may write $t - vx/c^2$ as the first order approximation of $t - kvx/c^2$ where $k = 1/(1 - v^2/c^2)$ by neglecting terms of order higher than v/c.

[219] Olivier Darrigol, "The Electrodynamic Origins of Relativity," *Historical Studies in the Physical Sciences* **26**, no. 2 (1996), 241–312, 303.

[220] Einstein, "Das Prinzip von der Erhaltung der Schwerpunktsbewegung und die Trägheit der Energie," *Annalen der Physik* **20** (1906), 627–633; reprinted in *Collected Papers* 2, pp. 360–366.

[221] Poincaré, "L'état actuel et l'avenir de la Physique mathématique," *La Valeur de la Science*.

such that they obtain Lorentz's local time. But he noted, "It matters little, since we have no way of perceiving it."[222]

Poincaré complained that in Lorentz's theory, even the novel concept of local time was not enough for a systematic accounting of phenomena but that "complementary hypotheses" were also needed.[223] In particular, he complained that physicists had to hypothesize the existence of length contraction and of forces that precisely compensate the effects of inertial motion on gravity, elasticity, and electromagnetism, all to explain why motion relative to the ether could not be detected. This accumulation of artificial hypotheses suggested that Lorentz's theory was not final, but that aspects of it would have to be changed. In Einstein's words, "This sort of way of accommodating experiments with negative outcomes through *ad hoc* contrived theoretical hypotheses is very unsatisfactory."[224]

In July 1903 Habicht had completed his dissertation in mathematics at the University of Bern. Afterward, he moved back to Schaffhausen, where his family lived. Einstein was annoyed to have their academy strained by distance, so he often sent letters to Habicht, trying to encourage his visits. He denounced him as a "miserable sloth."[225] Habicht's occasional visits were not enough: "Do come again, you miserable creature, you most pitiable of all the membra academica who ever inhabited this world! I would scold you even more, but space doesn't allow it."[226] In the meantime, Einstein attended some meetings of the Society of Natural Inquiry in Bern. He gave a talk on 5 December, titled "Theory of Electromagnetic Waves."[227]

In early March 1904, Solovine too left Bern, moving to Strasbourg, and soon to Lyon, to visit the local university. So Einstein was separated from his main conversation partners, aside from Mileva, who was then preg-

---

[222] Poincaré, "The Present Crisis of Mathematical Physics" (1904), in *Foundations of Science*, quotation from p. 309. If one could somehow detect a difference between the local time and time in the ether frame, then the motion of bodies relative to the ether would be measurable.

[223] Ibid., p. 307.

[224] Einstein, "Die Relativitätstheorie," *Die Kultur der Gegenwart*, ed. Hinneberg, part 3, sec. 3, vol. 1, *Physik*, ed. Warburg, pp. 703–713, quotation on p. 707. Reprinted in *Collected Papers* 4, p. 540.

[225] Einstein to Habicht, 30 November 1903, *Collected Papers* 5, p. 24.

[226] Einstein to Habicht, 20 February 1904, *Collected Papers* 5, p. 25.

[227] Flückiger, *Einstein in Bern*, p. 71. Einstein's membership began on 2 May 1903.

nant. Apparently, scientific discussions with her were lacking. A later friend and colleague of Einstein recounted, "When he wanted to tell her, as a fellow specialist, his ideas, which overflowed from him, her reaction was so scant and faint, that often he just did not know whether she was interested or not."[228] Still, Einstein continued his correspondence with friends. His willingness to simplify physical theories is illustrated in a letter to Marcel Grossmann: "You treat geometry without the parallel axiom, and I treat the atomistic theory of heat without the kinetic hypothesis."[229]

On 14 May 1904, Mileva gave birth to a son, whom they named Hans Albert. Meanwhile, Einstein had earlier encouraged his good friend Besso to apply for a job at the patent office. Besso had been working in Trieste, as a freelance engineer, but was willing to take up a more regular employment. Luckily for both of them, Besso was granted the job as a patent examiner and began work in the summer of 1904. Living now in the same city, Einstein and Besso renewed their discussions. After work, the two would often converse about many subjects, including physics, mathematics, and philosophy.[230] Twenty years later, Einstein expressed his appreciation for Besso's "brilliant capabilities and his extraordinary knowledge in technological and purely scientific matters."[231] He was also a good listener. "He was the only one possessed of enough patience to listen long hours to Albert."[232]

At the federal patent office, examiners had to carefully inspect applications to judge their merits and claims to novelty. The description of new devices had to be compared to other devices listed in patent records. The proposals submitted for examination were very diverse, especially owing to the rapid growth of electrical technologies. There were proposals for lamps, motors, engine components, batteries, and many other devices.[233] Many

---

[228] Frank, *Leben* (1949), p. 44; translation by A. Martínez; Frank, *Life and Times* (1948), pp. 34–35.

[229] Einstein to Marcel Grossman, 6? April 1904, *Collected Papers* 5, p. 25.

[230] Carl Seelig, *Albert Einstein, a Documentary Biography,* trans. Mervyn Savill (London: Staples Press Limited, 1956), p. 71.

[231] Einstein to Heinrich Zangger, 21 December 1926; translation by Heinrich A. Medicus, "The Friendship among Three Singular Men: Einstein and His Swiss Friends Besso and Zangger," *Isis* 85 (1994), 456–478.

[232] Marianoff, *Intimate Study* (1944), p. 44.

[233] Some writers have conjectured that patents for timing technologies influenced Einstein creatively toward his special theory. See Peter Galison, "Einstein's Clocks: The Place of Time," *Critical Inquiry* 26 (2000), 355–389; and Miller, *Einstein, Picasso* (2001),

contrived gadgets seemed worthless, but on occasion, some appealed to Einstein. Years later he recalled having liked the design for "an improved slide rule."[234] Sometimes, during slack moments at the office, Einstein and Besso discussed "sensible subjects," such as physics.[235] On some days after work, "they went to a café patronized by university professors and assistants for talk and coffee. It was always coffee because Einstein did not drink wine and he regarded beer as poison. Tobacco, on the other hand, was part of his diet. He was addicted to cigars—the longest, fattest, blackest ones he could find. He smoked one after the other so that, even as a young man, his teeth were stained brown and his throat was raw."[236]

Einstein sometimes met applicants for patents, such as Julius Friedrich Ries, a graduate student of physiology who designed a blood lancet. Ries reported the following chat from 1904:

This example came to me as a kind of thought-flash while reading [Ostwald's] "Naturphilosophie." I explained the matter to Einstein in the following way: Suppose that I am sitting in a railway carriage and it travels at the speed of light along a wall of wooden boards. I have a big pot of paint and a paintbrush. With these I can then make a brushstroke on the wall at particular intervals of time, e.g., every second. Thus, I would divide the wall into sections of 300,000 km each. Now suppose also that the train travels from infinity to infinity. But then it always stays "equally" distant from both end points; so, according to mathematical theory, the train does not move from its location, so here the concepts of space and time must collapse.—I was astonished by how seriously Einstein took this example and how often he returned to it.[237]

---

pp. 179–213; 206–207, 210, 247. However, no documentary evidence shows that Einstein or any of his co-workers ever identified any such influence. For a thorough documentary account of Einstein's time at the patent office, see Karl Wolfgang Graf, "Die Vorprüfung und Erteilung von Erfindungspatenten beim eidgenössische Amt für geistiges Eigentum 1888–1910" (Ph.D. dissertation, History Department, Section on History of Science and Technology, University of Stuttgart, 2001).

[234] Michelmore, *Profile* (1962), p. 43.

[235] Einstein to Besso, 14 December 1946, *Correspondance*, p. 382; Michelmore, *Profile* (1962), p. 43.

[236] Michelmore, *Profile* (1962), p. 43.

[237] Julius Friedrich Ries, in Flückiger, *Einstein in Bern* (1974), pp. 169, 171; translation by A. Martínez.

Back home, Albert and Mileva occasionally invited friends, such as Michele and his wife Anna, for an evening at their apartment. To entertain, Albert sometimes played the violin. Often Mileva "would start the guests off on a subject that she knew interested Albert and then sit back and listen. He argued with gusto and laughed uproariously at the least excuse."[238] Still, Einstein missed the "academic" meetings with Habicht and Solovine.

Despair on the Path to Insanity

Meanwhile, Einstein continued to struggle with electrodynamics. The problem of relative motion "was always with him."[239] He still had not solved it. To some extent, however, "the arguments and building blocks were being prepared over a period of years."[240] "His method was to work from basic truths and try every conceivable deduction, no matter how ridiculous it seemed."[241] He later characterized his labors as "desperate efforts."[242]

Like Poincaré, he now believed that Lorentz's theory included many valuable contributions, but that some parts of it needed to be replaced or revised. Hence, he began to try to somehow modify Lorentz's electrodynamics.[243] The concept of the ether, he thought, hindered the possibility of symmetrically describing electromagnetic effects attributed to bodies in relative motion. He disliked that the ether concept had entailed various other unobservable hypotheses, such as the electromotive force, the contraction of rigid bodies, and the concept of local time. He expected that the equations of electromagnetism should apply in their basic form when referred to bodies in motion. The laws of electromagnetism, like the laws of mechanics, should have the same form when applied to bodies "at rest" as when applied to bodies in uniform rectilinear motion. The example of electromagnetic induction had convinced him that the relativity of motion had to be rigorously valid in electrodynamics. Various experiments in the optics of moving bodies seemed to confirm this conjecture. He believed that fundamental physics should describe relations among empirically mean-

---

[238] Michelmore, *Profile* (1962), p. 43.

[239] Interview of 4 February 1950. Shankland, "Conversations with Einstein" (1963), p. 48.

[240] Einstein to Seelig, 11 March 1952, Einstein Archives, item 39-013.

[241] Michelmore, *Profile* (1962), p. 44.

[242] Vallentin, *Le Drame* (1954), p. 36.

[243] Kyoto lecture (1922), p. 82; translation by F. Kanaya and A. Martínez.

ingful concepts, rather than, say, relations among unobservable metaphysical structures, such as "absolute" space and motion.

He had adopted the widespread hypothesis that the speed of light is independent of the motion of its source at the instant of emission. This hypothesis was at the heart of Lorentz's electrodynamics; why would all light move at the same speed if not because it all consisted of waves in a medium? But Einstein wanted to formulate a theory independently of any such medium. According to one writer, Einstein and Besso allegedly exchanged the following words:

"Do you suppose ether simply does not exist?" he asked Besso. "Newton proved its existence," Besso protested. "Did he?" asked Einstein. "Couldn't it have been just a fabrication to account for something for which he and other scientists had no explanation?" Besso chuckled. "Sometimes I wonder about you, Albert." He tapped his head significantly. Einstein grinned and dropped the subject, but in his own mind he had already abandoned the concept of ether.[244]

Light was very well known to exhibit effects that were understood only in terms of the idea that light consists of transverse waves in a medium. The propagation of light from the stars to the Earth suggested that outer space, the apparent voids between the heavenly bodies, were not empty. Also, Newton had considered the existence of an ether at least in order to explain the transmission of forces across the spaces between heavenly bodies.[245] Even Mach was willing to admit the existence of the electromagnetic ether. By contrast, Besso's younger friend disliked this ether concept, though he lacked alternative explanations for key optical phenomena.

Einstein was not the only physicist trying to reformulate electrodynamics without this ether concept. In particular, the well-known theorist Emil Cohn labored to revise the Maxwell-Lorentz equations to describe electromagnetic processes only with respect to observable bodies.[246] Following Mach, Cohn avoided speculations about microphysical structures and processes. He

---

[244] Forsee, *Theoretical Physicist* (1963), p. 22.
[245] Cohen, "Interview" (1955), p. 71.
[246] Emil Cohn, "Zur Elektrodynamik bewegter Systeme," Akademie der Wissenschaften zu Berlin, mathematische-physikalische Klasse, *Sitzungsberichte* (1904), 1294–1303, 1404–1416.

even avoided the concept of empty space. He rejected Lorentz's hypothesis of length contraction because it involved arbitrary assumptions about molecular forces. Following Poincaré, he interpreted Lorentz's local time as the time given by clocks synchronized by light signals on a moving system. He argued that Lorentz's theory required ideal measuring rods and clocks in order to distinguish Lorentz's distorted quantities from true coordinate relations. Cohn required that light in any homogeneous medium would propagate at the same speed in all directions. His theory also involved the tacit assumption that the speed of light depends on the speed of its source. Cohn did not believe that electromagnetic theory should be reduced to mechanical principles, such as the principle of relative motions. Instead, he allowed that a kind of absolute motion could be defined by referring velocities of bodies (like the Earth) to the stars.

Another physicist who already had rejected the ether concept was Alfred Bucherer. He argued that only relative motions among material bodies could produce variations in observable phenomena of light, electricity, and magnetism. In 1903 Paul Nordmeyer, under Bucherer's direction, had demonstrated experimentally that the intensity of light rays remained constant irrespective of their directions. Bucherer denounced the concept of the stationary ether as an artificial "scaffolding" that had not led to observed effects and should therefore be eliminated.[247] He further rejected Lorentz's hypothesis of contraction as very artificial. Bucherer argued that Lorentz's equations had to be somehow reinterpreted such that all velocities involved would be only relative. (At any rate, we do not know whether Albert Einstein, for one, was acquainted with works such as those of Cohn and Bucherer in 1904.)[248]

Even without an ether, the problems of electrodynamics did not seem easy to solve. In any case, the speed of light should still be susceptible to apparent variations. If there was no ether, then light could move at a constant velocity only relative to something else, such as relative to space itself. Here one returns to the unverifiable concept of absolute space. Otherwise, light might have a constant velocity relative to a given frame of reference, such as a material medium, a given body, or the stars. But then light rays traveling

---

[247] Alfred Bucherer, *Mathematische Einführung in die Elektronentheorie* (Leipzig: Teubner, 1904), p. 131.

[248] In 1907 Einstein mentioned knowing one of Cohn's papers. Einstein to Johannes Stark, 25 September 1907, *Collected Papers* 5, p. 74.

with the same speed in opposite directions relative to that one system could yet have different velocities when relative to other systems. Moreover, *even if* light was independent of the motion of its source, its relative speed should still depend on the motion of reference systems. For example, a single ray of light observed from two reference systems moving in opposite directions could seem to travel with different speeds relative to each system.

The optical experiments that continued to be most influential to Einstein were the well-known effects of the aberration of starlight in telescopes and Fizeau's results on the speed of light in moving water.[249] Einstein's investigations continued to be guided by "the desire to make physical theory fit observed fact as well as possible."[250] He believed that the equations of electromagnetic phenomena, as developed by Maxwell, Hertz, and Lorentz, should apply equally relative to moving or to stationary bodies. Yet these equations seemed to involve intrinsically the constancy of the speed of light. Einstein explained in 1922 that "the fact that these equations work also in the moving ordinary system leads to the invariance of the speed of light. However, it doesn't agree with the laws about the addition of velocities, which we knew in force physics. Why do these two things conflict with each other? I felt like I was hitting a dead end of extreme difficulty."[251] Likewise, he later reiterated that, in his efforts to attain an effective approach to electrodynamics, "the difficulty that had to be overcome was in the constancy of the velocity of light in vacuum which I first thought I would have to give up."[252]

The concept of the invariance of the speed of light contradicted the basic law that all velocities vary according to a simple addition rule. This rule was practically a mathematical statement of the relativity principle. In hindsight, Einstein explained, "The constancy of light propagation seems to conflict with the classical principle of relativity, according to which the velocity of a ray of light assumes different values in the moving system according to the direction of the ray."[253] For example, if a person walks at a speed w inside a railway carriage (reference system) traveling uniformly at a velocity v rela-

[249] Interview of 4 February 1950. Shankland, "Conversations with Albert Einstein" (1963), p. 48.

[250] Albert Einstein, "On the Theory of Relativity," (1921), *Mein Weltbild* (Amsterdam: Querido Verlag, 1934); reprinted in *Ideas and Opinions* (1954), p. 246.

[251] Kyoto lecture (1922), p. 82; translation by F. Kanaya and A. Martínez.

[252] Einstein, manuscript (1920), Einstein Archives, item 2-070.

[253] Einstein, quoted in Moszkowski, *Conversations* (1921), p. 213.

tive to the ground (a second reference system), then this person moves relative to the ground with a velocity of $v + w$. Because the behavior of all bodies on the train (or "moving reference frame") is affected in the same way,
that is, by the addition of a velocity v, then they all behave in the same way
with respect to one another, so that "the laws of physics" hold in the same
way relative to the train and to the embankment. Because the invariance of
light velocity contradicted the rule of addition of velocities, the relativity of
motion and the constant speed of light were "apparently irreconcilable."[254]

Struggling to solve the problem, to modify parts of Lorentz's theory, Einstein felt that there was scarcely any secure theoretical foundation on which
to build. Owing to the works of Max Planck and others, Einstein believed that
neither mechanics nor electrodynamics could be exactly valid.[255] He had also
lost his early faith in the exact validity of Maxwell's equations. Einstein had
found that Maxwell's theory gave incorrect predictions for the fluctuations
of radiation pressure and the motions of tiny bodies exposed to radiation:
"In my view, one could not get around ascribing to radiation an atomistic
structure which, of course, does not fit into the frame of Maxwell's theory."[256]
Thus, he "had already found that Maxwell's theory did not account for the
microstructure of radiation and could not therefore have general validity."[257]
Accordingly, optics could hardly provide a secure foundation. The corpuscular behavior of light and electromagnetic radiation in some experimental areas did not harmonize with the well-known wavelike effects of light, such as
interference and polarization. In sum, "Einstein saw that his efforts began
to endanger the supposedly impregnable foundations of his science."[258]

Failing to devise an electrodynamics on the basis of constructive efforts,
that is, of other theories and hypotheses about the constitution of matter
and radiation, Einstein then hoped to find some principle that would at least

[254] Albert Einstein, "Zur Elektrodynamik bewegter Körper," *Annalen der Physik* **17**
(Leipzig, 1905), 891–921; reprinted in *Collected Papers* 2, pp. 276–306.
[255] Einstein's "Autobiographisches" (1946), p. 53, states "mechanics nor thermodynamics," but he corrected it to read "mechanics nor electrodynamics" in a list of errata,
and the correction was included in the German edition of 1955; see Seiya Abiko, "On
Einstein's Distrust of the Electromagnetic Theory: The Origin of the Light-Velocity
Postulate," *Historical Studies in the Physical Sciences* 33 (2003), 193–215, pp. 204–205.
[256] Einstein to Max von Laue, 1952, quoted by Holton, in *Thematic Origins*, pp. 201–202.
[257] Albert Einstein to Carl Seelig, quoted by Max Born, in *Physics of My Generation* (Oxford: Pergamon Press, 1956), p. 194; see also, *Collected Papers* 5, p. 309.
[258] Reiser, *Biographical Portrait* (1930), p. 68.

outline the general form of the theory: "By and by I despaired of the possibility of discovering the true laws by means of constructive efforts based on known facts. The longer I tried, the more I came to the conviction that only the discovery of a universal formal principle could lead us to assured results. The example I saw before me was thermodynamics. The general principle there was given in the theorem: the laws of nature are such that it is impossible to construct a perpetuum mobile (of the first and second kind). How, then, could such a universal principle be found?"[259]

In 1916 Einstein told an interviewer that, as his original ideas began to arise, "I was plagued by all sorts of nervous conflicts; I went around confused for weeks," as if in a "stage of stupefaction."[260] In 1932 he recalled that, "after seven years of searching in vain, the desire to discover and overcome this generated a state of mental tension in me."[261] From his comments, one early biographer gathered that Einstein's mind "ruminated" over his relativity thoughts even "while he sat alone undisturbed in the work room of the patent office."[262] "To Besso," reportedly, "were poured out all the disappointments and anguish of frustrated moments, of the travail of research. He would cry in despair to him: 'It is no use to try farther; I must give it up.' Then Besso, with an understanding so great and a patience so infinite, would calm the troubled thought."[263]

Besso was encouraging. Regarding Einstein's ideas, he liked to say "If they are roses, they will bloom."[264] Another biographer, who in 1962 interviewed Einstein's son, Hans Albert, recounted: "He worked on and on through 1904. One puzzle led to another. His brain was on fire. His body was exhausted. He could not eat. He could not sleep. *The speed of light never varies.* When people spoke to him, he did not hear. He wandered about in a daze. At times he wondered if this was the way to insanity."[265]

Einstein tried to divide his time between his investigations, his work, and his family. But his theoretical thoughts overran him. He would take his son in "a baby carriage around the streets of Bern. When he felt that he had

[259] Einstein, "Autobiographisches" (1946), trans. Schilpp, p. 53.

[260] Einstein, in Moszkowski, *Conversations* (1921), p. 4; Moszkowski, *Einblicke*, p. 18.

[261] Einstein to Erika Oppenheimer, 13 September 1932, quoted in *Collected Papers* 2, pp. 261–262.

[262] Reichinstein, *Sein Lebensbild* (1935), p. 23; translation by A. Martínez.

[263] Marianoff, *Intimate Study* (1944), p. 45.

[264] Frank, *Leben* (1949), p. 90.

[265] Michelmore, *Profile* (1962), p. 44; italics in the original.

fulfilled his fatherly duties, he would hurry home to pore over problems, which, like an overwhelming passion, possessed him, never letting him rest."[266] He had become obsessed with the problem, it drove him to despair: "Now he experienced the full tragedy of the researcher: his toil was not successful, yet he could not leave it, he could not stop his ruminations; Einstein feared for his health. And only with an intense exertion of the will did he succeed in freeing himself from these thoughts, in compelling himself to think about that no more."[267] Reportedly, all of his efforts "to obtain a theory consistent with the experimental facts had failed."[268] Einstein forced himself to stop working on the problem.

The Eagle and the Sparrow

In October 1904 Solovine moved back to Bern. And the "abominable creature," Habicht, responded to Einstein's *"enraged"* scolding by occasionally spending time in Bern.[269] Thus, the trio could resume meetings of their academy. It was a rewarding distraction for Einstein from his obsessive thoughts about electrodynamics.

In early 1905 Solovine, Einstein, and Habicht read Hume's *Treatise of Human Nature,* "in a very good German edition."[270] "What Einstein liked most about Hume was the unsurpassable clarity of his presentation and his avoidance of any ambiguities intended to give an impression of profundity."[271] Einstein later reminisced that above all other subjects, they busied themselves with discussions of Hume.[272] Likewise, Solovine later recalled that, in particular, "we discussed for weeks David Hume's singularly incisive critique of the notions of substance and causality."[273] Regarding the notion of causality, Hume argued that although certain happenings seem to regularly follow one another, we can never know whether they are really

---

[266] Reiser, *Biographical Portrait* (1930), pp. 67–68.

[267] Reichinstein, *Sein Lebensbild* (1935), p. 23; translation by A. Martínez.

[268] Interview of 4 February 1950, Shankland, "Conversations with Albert Einstein" (1963), p. 48.

[269] Einstein to Habicht, two postcards, both dated 6 August 1904, *Collected Papers* 5, pp. 28–29.

[270] Einstein to Besso, 6 March 1952, *Correspondance*, p. 464.

[271] Frank, *Life and Times* (1948), p. 67.

[272] Ibid.; Einstein to Besso, 6 March 1952, *Correspondance*, p. 464.

[273] Solovine and Einstein, *Lettres* (1956), p. viii; translation by A. Martínez.

necessarily connected.[274] Events appear joined in a temporal sequence, as we experience them, but we do not know if one event is necessarily the cause of its successor. Thus, the fundamental notion of causality is a human conception; it is not justified even by our scientific knowledge. For Einstein, this reading of Hume was "of considerable influence upon my development—alongside Poincaré and Mach."[275] As with Kant, Einstein felt that he could "perfectly understand" Hume.[276] In hindsight, he wrote that "Hume's clear message seemed crushing": human sensations are the only source of knowledge, they are linked only by expectations and habit, and they cannot lead to real laws of nature.[277] "All concepts," Einstein became convinced, "even those which are closest to experience, are from the point of view of logic freely chosen conventions, just as is the case with the concept of causality, with which this problematic concerned itself in the first instance."[278] Poincaré had also drawn attention to the extent to which certain fundamental concepts in science are conventions. Given Hume's formulation, Einstein increasingly overcame a teaching of Kant: that certain fundamental notions are indispensable, as they are not concepts, they are not based on experience.[279] Thus, Einstein felt "a special kinship" with Hume.[280]

Kant had written his *Critique of Pure Reason* partly as a reply to Hume's *Treatise*. Einstein happened to read these works in reverse, so in his case, it was Hume's earlier work that undermined Kant's later arguments. Hume showed a "more sound instinct" than Kant.[281] Einstein readily showed how impressed he was by Hume. In early March 1905 Solovine proposed that instead of holding an academic meeting they could attend an upcoming violin concerto by a famous Bohemian string quartet. Despite Einstein's love of music, he replied that they should not, that it was much preferable to

---

[274] Moszkowski, *Conversations* (1921), p. 161.

[275] Einstein to Besso, 6 March 1952, *Correspondance*, p. 464; translation by A. Martínez.

[276] Einstein to Ernst Fraenkl, no date [1930s], Albert Einstein, *Oeuvres choisies 4, Correspondances françaises*, ed. Michel Biezunski (Israel: Éditions du Seuil, 1989), p. 105.

[277] Einstein, "Remarks on Bertrand Russell's Theory of Knowledge," in *The Philosophy of Bertrand Russell*, ed. Schilpp (1944); reprinted in *Ideas and Opinions* (1954), p. 22.

[278] Einstein, "Autobiographisches" (1946), trans. Schilpp, p. 13.

[279] Ibid.

[280] Reiser, *Biographical Portrait* (1930), p. 55.

[281] Einstein to Max Born, [after 28 June 1918], *The Collected Papers of Albert Einstein*, vol. 8, part B, *The Berlin Years: Correspondence, 1918*, ed. Robert Schulmann, A. J. Kox, Michel Janssen, and József Illy (Princeton: Princeton University Press, 1998), p. 818.

read Hume.[282] On Friday 17 March, when the trio was to hold its meeting at Solovine's apartment, Solovine suddenly could not resist but attend the concert on his own.[283] He left the door open so that Einstein and Habicht could meet at his place nevertheless. Einstein and the "mischievous" Habicht were outraged.[284] They decided to smoke as many cigars as they could in his apartment, with closed windows, in order to saturate it with the smell that Solovine despised. Then they smeared the ashes and cigar butts on Solovine's bed, table, chairs, forks, teacups, teakettle, and more. When Solovine returned he could not stand it. The next day Einstein chastised him for not fulfilling his academic duty: "You miserable subordinate, how! you have the audacity to neglect an academic meeting to go hear violins? Barbarian, Boetian, if ever you allow yourself again such an outrage, you will be excluded and shamefully chased from the Academy!" That night their discussion of Hume extended late until an hour before sunrise.

Einstein also continued to work on physics, while he avoided the problems of electrodynamics. He avoided the puzzles about the speed of light and the motion of reference frames. Nonetheless, he still thought about electrons and light. According to one writer (who at best may have conveyed a third-person account), Einstein and Besso had the following discussion:

> "One thing baffles me," said Besso one night when they had been theorizing on the action of electrons—the smallest quantity of electric charge known to exist independently. "A metal plate exposed to a beam of light gives forth a shower of negatively charged particles. Right?" "*Ja*," said Einstein, nodding amiably. "Why?" asked Besso. "*Ach*, that I cannot explain."
>
> For days Einstein pondered the question put to him by Besso. Finally he told his friend. "I think I have an explanation for your electrons, but first you'll have to accept the idea of light as composed of individual particles of energy."[285]

[282] Solovine and Einstein, *Lettres* (1956), p. x.

[283] "Freitag, 17 März, abends punkt 8 Uhr: Konzert gegeben von berühmten Böhmischen Streichquartett," *Der Bund, Eidgenössisches Zentralblatt* 56 Jahrgang, Nr. 119, 11 March 1905, p. 4.

[284] Einstein to Habicht, [late Summer] 1905, *Collected Papers* 5, p. 33.

[285] Forsee, *Theoretical Physicist* (1963), pp. 16–17. See also Karl K. Darrow, "The Quantum Theory," *Scientific American* 186 (March 1952), 47–54.

Einstein argued that processes of the emission and absorption of radiation can be best understood if radiation is assumed to have a corpuscular nature, that is, if light is made up of particles. Einstein told Besso of his "heuristic viewpoint," and Besso "immediately recognized it as a discovery of the utmost importance and of the greatest consequence."[286] Reportedly, Besso said, "It may take time to absorb that idea," that light consists of particles.[287] It became their main subject of discussion.[288] Hence, in March 1905 Einstein completed a paper about the production and transformation of light.[289] Still, Einstein could not explain how or why, if indeed light consists of particles, one yet observes experimentally properties such as interference, which seem to suggest wave motions in a medium. Moreover, he still espoused the traditional hypothesis that the speed of light, like the speed of waves, is independent of the speed of its source. Virtually all physicists believed that light consisted of waves in the ether. Thus, the irreverent idea of light quanta seemed "very revolutionary."[290]

Einstein also managed to complete works that analyzed matter in terms of corpuscular conceptions. At the end of April, he finished a paper on determining the size of molecules, which he then submitted as a doctoral dissertation to the University of Zurich. He dedicated this work to his friend Marcel Grossmann, in appreciation for his help at the Polytechnic and toward getting the job at the patent office. Very soon Einstein also submitted yet another paper to the *Annalen der Physik*, explaining the motions of small particles suspended in stationary liquids as resulting from microscopic collisions with molecules. Following the kinetic-molecular theory of heat, Einstein made predictions about the motion of small particles that, in turn, could serve to provide experimental evidence of the theory of molecules and atoms. Einstein's departure from the problems of electrodynamics was hence very productive.

Nevertheless, early in the spring of 1905, Einstein still had not solved the problems of light and relative motion. According to a biographer (who

---

[286] Seelig, *Documentary Biography* (1956), p. 71.
[287] Forsee, *Theoretical Physicist* (1963), p. 17.
[288] Seelig, *Documentary Biography* (1956), p. 71.
[289] Einstein's paper on the photon theory of light was the result of five years of pondering to explain the quantum of action introduced by Max Planck in 1900. Interview of 24 October 1952; Shankland, "Conversations with Albert Einstein" (1963), p. 56.
[290] Einstein to Habicht, May 1905, *Collected Papers* 5, p. 31.

once interviewed Einstein's son), "Frustration drove him out to wander the farm lands around Bern. He took time off from the office. Mileva helped him solve certain mathematical problems, but nobody could assist with the creative work, the flow of fresh ideas."[291] He had been trying to reformulate Lorentz's electrodynamics since mid-1904: "Expecting that I had to somehow modify Lorentz's theory, I had spent almost a year thinking in vain, and I couldn't help but think that this puzzle is not easily solved."[292] After "ten years of reflection,"[293] including seven years of focused efforts, Einstein had not solved his puzzles on the optics of moving bodies and electrodynamics.

An account of those days was published in 1930 by Einstein's stepson-in-law through his second marriage, Rudolf Kayser, writing under a pseudonym. Einstein wrote a brief note in the preface to the completed biography: "The author of this book is one who knows me rather intimately" and that "I found the facts of the book duly accurate." Kayser described Einstein's conceptual struggles: "As a man possessed, he was carried away by these most difficult problems of theoretical physics, and talking about them helped more than silence. Going home from his duties he discussed his ideas and investigations with his friend Besso." Years later, Einstein characterized Besso: "He is brilliant in discussions. There he aims to bridge conflicts and discrepancies. This and his talent to penetrate into the thoughts of others make him a fertile critic."[294]

In early May 1905, Einstein lived at 49 Kramgasse, at the heart of old downtown, east of the patent office building (at the corner of Genfergasse and Speichergasse). Besso sometimes walked Einstein home after work. Kayser recounts how Einstein informed Besso of his attempts to solve his theoretical problem: "If he felt that he was nearing a solution of his problems, he would tell his friend—his own eyes glowing—that the success of his efforts was at hand. But the next day he would merely inform him sadly and gently that all his attempts of the past were wrong. Through many long years of hope and disappointment Albert carried on his attempts."[295] Ein-

[291] Michelmore, *Profile* (1962), p. 45.

[292] Kyoto lecture (1922), p. 82; translation by F. Kanaya and A. Martínez.

[293] Einstein, "Autobiographisches" (1946), trans. Schilpp, p. 53.

[294] Einstein to the Patent Office, early 1927; *Correspondance*, p. 546; translation by A. Martínez.

[295] Reiser, *Biographical Portrait* (1930).

stein had fought many battles without success. Finally, one day he decided to take the entire problem to Besso, to discuss it at length. Perhaps Besso's sharp attention to "petty" things would be helpful.[296] In 1922, in a lecture delivered in Kyoto, Japan, Einstein recounted: "But by chance, a friend of mine in Bern helped me out. It was a beautiful day when I visited him, saying, 'Lately, I have one problem that I cannot figure out in any way. Today I've brought you that war.'"[297] Reportedly, Einstein "admitted that he had run up against a complete mental block."[298] So he shared many aspects of the problem: "I tried various discussions with him."[299] Years later, Besso used the following analogy: "Einstein, the eagle, took me, the sparrow, under his wings into the heights, and then there I fluttered yet a little farther above."[300] Einstein recalled: "By that, suddenly I came up with a clear idea."[301] During so many years, Einstein had pursued and "abandoned many fruitless attempts, 'until at last it came to me that time was suspect!'"[302]

In 1916 a psychologist studying scientific creativity, Max Wertheimer, interviewed Einstein at length regarding the thoughts that originally led to the solution of his conundrums on relative motion. Apparently, Einstein realized that the phenomena of motion in question (the phenomena of electrodynamics and light) necessarily involve a tacit measurement of time. Reportedly, he asked himself: "Do I see clearly, the relation, the inner connection between the two, between the measurement of time and that of movement? Is it clear to me how the measurement of time works in such a situation?"[303] We can infer more specifically that the concept of simultaneity emerged as having crucial importance. Wertheimer reported "characteristic remarks" of Einstein—to wit, "the discovery that the crucial point, the solution, lay in the concept of time, more particularly in that of simultaneity. . . . The very moment he saw the gap, and realized the relevance of simultaneity, he knew this to be the crucial point for the

---

[296] Einstein to Marić, 4 April 1901, *Collected Papers* 1, p. 285; *Love Letters,* p. 41.

[297] Kyoto lecture (1922), p. 82; translation by F. Kanaya and A. Martínez.

[298] Michelmore, *Profile* (1962), p. 45.

[299] Kyoto lecture (1922); translation by F. Kanaya and A. Martínez, p. 82.

[300] Besso, quoted in Paul Winteler to Seelig, 6 March 1952, p. 5, ETH Bibliothek Hs 304:1068; translation by A. Martínez.

[301] Kyoto lecture (1922), p. 82; translation by F. Kanaya and A. Martínez.

[302] Einstein, 4 February 1950, Shankland, "Conversations with Albert Einstein" (1963), p. 48.

[303] Wertheimer, *Productive Thinking* (1945), p. 174.

solution."[304] Presumably, "the gap" was between traditional basic concepts and physical experience.

Did Einstein find the solution to his problem right then in the company of Besso? Apparently not. Further evidence suggests that what he found right then was not quite the solution but something about the essence of the problem, the "key" to its solution: "When he held the key, with which he was to open the closed door, in his hand, he despaired, and said to his friend 'I'm going to give it up.'"[305] Another biographer echoed: "It's no use," he told Besso. "I'm on the wrong track."[306] Apparently Einstein then left Besso, perhaps frustrated, but having found at least the crux of the problem, something about the relationship between the kinematic concepts of time and motion, as we will see.

Back home, Einstein likely analyzed the problem further that evening. "A notion acquires its right to existence uniquely by its clear and univocal enchainment with events, with respect to physical experiences," he believed.[307] It was precisely for that reason that already he had long rejected the notion of "absolute velocity," that is, because its "univocal connection with experiences showed itself to be impossible."[308] Likewise, how was the notion of simultaneity connected with experience? As reported by Max Wertheimer, Einstein then "said to himself":

If two events occur in one place, I understand clearly what simultaneity means. For example, I see these two balls hit the identical goal at the same time. But . . . am I really clear about what simultaneity

---

[304] Ibid., p. 183.

[305] Reiser, *Biographical Portrait* (1930), p. 68. Michelmore echoed that he told Besso "I've decided to give it up—the whole theory." *Profile* (1962), p. 45. Michelmore's account is indirect and derivative; it is based partly on two days of interviews with Hans Albert Einstein in 1962 and mainly on many earlier accounts, including Reiser's book. Accordingly, Michelmore's suffers from some mistakes; one peculiarity is that he seems to construe Reiser's metaphorical statement about Einstein holding "the key" literally, as though Einstein "reached the steps to his apartment house, he paused briefly and said . . ." As Michelmore noted, p. vii, Hans Albert Einstein did not check his notes or edit the book.

[306] Forsee, *Theoretical Physicist* (1963), p. 28.

[307] Einstein to Solovine, 24 April 1920, Solovine and Einstein, *Lettres* (1956), p. 20; translation by A. Martínez.

[308] Ibid.

means when it refers to events in two different places? What does it mean to say that this event occurred in my room at the same time as another event in some distant place? Surely I can use the concept of simultaneity for different places in the same way as for one and the same place—but can I? Is it as clear to me in the former as it is in the latter case? . . . It is not![309]

According to Wertheimer, Einstein proceeded to analyze the question along the following lines. He asked himself how one can determine whether two lightning bolts strike distant places at the same time.[310] By conceiving of a feasible physical procedure, one would be able to ascribe a practical meaning, a definition, for the word *simultaneity.*

How could the notion of the simultaneity of distant events be connected with experience? How was the notion of a rigid body connected with experience? How was the geometric shape of a moving body connected with experience? Perhaps Einstein realized that the latter questions depend on the former: to determine the length of a body, we need to know the *simultaneous* location of its extremities. One would need to somehow ascertain the simultaneity of distant events in order to in any way measure a moving length. Moreover, measurements of velocities depend on measurements of lengths and times. Whether or not the speed of light varies relative to a moving system depends precisely on how one measures that speed. Later Einstein recalled: "Only after groping for years did I notice that the difficulty rests on the arbitrariness of the fundamental kinematic concepts."[311] Finally there was some clarity, but no solution.

At some point that night, Einstein went to sleep. Incidentally, he once explained that he routinely stopped thinking about problems at bedtime: "I break off when I want, and banish all difficulties when the time to sleep arrives. A thoughtful dreamwork, as in the case of artists, whereby poets and composers weave the day into the night, is foreign to me."[312] But maybe that one night was different. As one biographer claimed, "That night when he

---

[309] Einstein, quoted in Wertheimer, *Productive Thinking* (1945), p. 174. Ellipses in the original.

[310] Einstein, *Relativitätstheorie* (1917), p. 28.

[311] Einstein, manuscript (1920), Einstein Archives, item 2-070.

[312] Remark from 1916, quoted in Moszkowski, *Einblicke* (1921), p. 18; translation by A. Martínez.

went to bed, he lay awake tossing and turning."[313] Another biographer, his other son-in-law, Dimitri Marianoff, recounted what Einstein once replied when asked how he had arrived at his theory: "He said that one night he had gone to bed with a discouragement of such black depths that no argument would pierce it. 'When one's thought falls into despair, nothing serves him any longer, not his hours of work, not his past successes—nothing. All reassurance is gone. It is finished, I told myself, it is useless. There are no results. I must give it up.'"[314] But the pieces of the puzzle were falling into place. "In vision," according to Marianoff, Einstein's thoughts proceeded as the "underlying unity of size, structure, distance, time, space slowly fell piece by piece, like a monolithic picture puzzle, into place."

If thinking during sleep hours was unusual, thinking in the morning was another matter. Jakob Ehrat, his friend from the Polytechnic, reported what Einstein once told him: "One morning on awakening, as I was sitting up in bed, the breakthrough thought came to me." He referred to the notion that two events that are simultaneous for one observer are not necessarily simultaneous for another.[315] Years later, another friend too recalled that "Einstein said his basic discovery came on waking up one morning, when he suddenly saw the idea. This had been going around and around at the back of his mind for years, and suddenly it wanted to thrust itself forward into his conscious mind."[316]

Einstein realized that if one were to determine the simultaneity of distant events by employing a definite measurement procedure, then that procedure might yield different results when applied in systems in relative motion. For example, a person aboard a moving train could perceive two lightning bolts in a different sequence from what is observed by someone on the ground.[317] This would serve to formulate a new concept of simultaneity, a relative concept. Einstein could reject the traditional notion of "absolute simultaneity" because it seemed to be impossible to connect it univocally with experience.[318] Einstein believed that in physics there should

---

[313] Forsee, *Theoretical Physicist* (1963), p. 28.

[314] Marianoff, *Intimate Study* (1944), p. 68.

[315] Jakob Ehrat to Carl Seelig, 20 April 1952, Einstein Archives, item 71-212; translation by A. Martínez.

[316] Banesh Hoffmann, tape-recorded interview by Denis Brian, 29 October 1982; quoted in Denis Brian, *Einstein, a Life* (New York: John Wiley & Sons, 1996), p. 61.

[317] Wertheimer, *Productive Thinking* (1945), p. 176; Einstein, *Relativitätstheorie* (1917).

[318] Einstein to Solovine, 24 April 1920, Solovine and Einstein, *Lettres* (1956), p. 20.

be "no notion of which its use is *a priori* necessary or justified."[319] Even the fundamental notion of absolute time was susceptible of being modified.

Einstein believed that physical concepts should acquire meaning by their connections to definite empirical relations. Kinematic concepts would be meaningful insofar as they related to experience. Notions such as distance, speed, and acceleration should all be defined by the relations among definite material things, such as rigid rods and synchronized clocks. Length could be defined by measuring sticks. Instants of time could be defined merely as the indications of a pointer on a clockface. If moving observers disagreed over the simultaneity of distant events, then they would disagree accordingly about the lengths of moving bodies. They would also disagree over other related measurements. But such disagreements could be treated systematically and consistently. In this way, now "length contraction" would not be an artificial and independent hypothesis at all; it emerged as a necessary consequence of specific measurement procedures. Moreover, observers would disagree over the interval of time between two events. Therefore, they could obtain *the same measurement* for the speed of a single ray of light, even while the observers move in opposite directions. The measured speed of light could thus be independent of the motion of reference systems.

As suggested by several sources, including his childhood mentor and friend, Max Talmey, Einstein's conceptual breakthrough was thus the idea of the relativity of simultaneity: "The discovery that the simultaneity of distant events is relative was the origin of the relativity theory."[320] If we accept this proposition then we may conjecture that Ehrat's quotation indeed referred to the very morning after Einstein had his productive conversation with Besso. Then that next day, as he recalled in 1922, Einstein promptly informed his friend: "So the next day I came back to him again and approached him, without any greetings: 'Thank you. I have completely solved my problem.' My solution was about the very concept of time. Thus, time is not absolutely defined, and there is an inseparable relationship between time and signal speed. That previous extraordinary difficulty was solved completely by that, for the first time."[321]

If we continue to trust that all the recollections here quoted describe correctly the sequence of events, it appears that that same "next day" when

---

[319] Ibid.
[320] Talmey, *Formative Period* (1932), p. 3.
[321] Kyoto lecture (1922), p. 80; translation by F. Kanaya and A. Martínez.

Einstein saw Besso again was a workday at the patent office, as Kayser's account continues: "The next day it was in the greatest excitement that he took up his duties at the office. He could apply himself to the dull routine only with effort. Feverishly he whispered to his friend that now at last he was on the right track. He made the revolutionary discovery that the traditional conception of the absolute character of simultaneity was a mere prejudice, and that the velocity of light was independent of the motion of coördinate systems."[322] Perhaps Einstein "whispered" his findings because he did not want to upset his supervisor at work, or perhaps he was not ready to share his ideas just like other young innovators. We do not know. Other colleagues such as Dr. Sauter did not enjoy the intimate friendship that Einstein shared with Besso. What matters is that Einstein suddenly and finally found a solution. In 1924 he recalled: "After seven years of reflection in vain (1898–1905), the solution came to me suddenly with the thought that our concepts and laws of space and time can only claim validity insofar as they stand in a clear relation to our experiences; and that experience could well lead to alteration of these concepts and laws. By a revision of the concept of simultaneity into a more malleable form, I thus arrived at the special theory of relativity."[323]

Finally, Einstein moved decisively forward—he took "the step," as he called it years later.[324] Yet regarding his progress toward his new conception,

---

[322] Reiser, *Biographical Portrait* (1930), p. 68. Peter Galison has emphasized that Einstein told Besso about simultaneity (allegedly) while standing on a hill, pointing at clock towers; in Galison, *Einstein's Clocks, Poincaré's Maps: Empires of Time* (New York: W. W. Norton, 2003), pp. 253, 254, 255, 261. But that was a mistake, for there is no evidence of any such incident; see Alberto A. Martínez, "Material History and Imaginary Clocks: Poincaré, Einstein and Galison on Simultaneity," *Physics in Perspective* 6 (2004), 224–240, pp. 228–231. Likewise there is no evidence that Einstein "spoke often with friends from the Federal Postal and Telegraph Administration about issues in wireless telegraphy and synchronization of clocks," as claimed by Miller in *Einstein, Picasso* (2001), p. 247.

[323] Einstein, discographic recording of 6 February 1924, transcribed in Friedrich Herneck, "Zwei Tondokumente Einsteins zur Relativitätstheorie," *Forschungen und Fortschritte* 40 (1966), 133–135; translation from Stachel, *Collected Papers* 2, p. 264.

[324] According to Abraham Pais, "When I talked to him [Einstein] about those times of transition, he expressed himself in a curiously impersonal way. He would refer to the birth of special relativity as 'den Schritt,' the step." See Abraham Pais, *"Subtle is the Lord . . .": The Science and the Life of Albert Einstein* (New York: Oxford University Press, 1982), p. 163.

Einstein once explained that his thoughts had not moved *directly*, step by step, to the solution, but by a tortuous path. He commented, "It is only at the last that order seems at all possible in a problem."[325]

In 1915 Einstein commented that "a great influence" on his labors was effected by readings of "Mach, and ever much more Hume, whose treatise on [human] understanding I studied with fervor and admiration shortly before figuring out the theory of relativity. It is very much possible that without these philosophical studies I would not have arrived at the solution."[326] Late in life, Einstein reiterated the influential roles of Hume and Mach. He wrote that all efforts to resolve the problematic relationship between electromagnetism and mechanics had to contend with the great difficulty that "the axiom of the absolute character of time, viz., of simultaneity, unrecognizedly was anchored in the unconscious. Clearly to recognize this axiom and its arbitrary character really implies already the solution of the problem. The type of critical reasoning which was required for the discovery of this central point was decisively furthered, in my case, especially by the reading of David Hume's and Ernst Mach's philosophical writings."[327] Accordingly, when someone once asked Einstein how he had come to formulate the theory of relativity, Einstein replied: "By refusing to accept an axiom."[328]

He promptly shared his new ideas with Besso. According to a third-person account, they discussed the subject of simultaneity:

> "Did it ever occur to you that many events that seem simultaneous are not truly so?" Einstein asked Besso one evening as they sat munching Swiss chocolate. "This morning as I left the house," said Besso, "a factory whistle emitted a blast. Are you going to tell me that the whistle blast and my departure from the house were not simultaneous?" "The sound of the whistle came from some distance away," Einstein pointed out. "So it took an interval of time—although an extremely

---

[325] Interview of 4 February 1950, Shankland, "Conversations with Albert Einstein" (1963), p. 48.

[326] Einstein to Moritz Schlick, 14 December 1915, *Collected Papers* 8A, p. 220; translation by A. Martínez.

[327] Einstein, "Autobiographisches" (1946), trans. Schilpp, p. 53.

[328] Marianoff, *Intimate Study* (1944), p. 148. The exchange reportedly transpired between Einstein and a professor of the California Institute of Technology.

short one, I'll admit—for the sound to reach your ears. Therefore the whistle blast and your exit were not *exactly* simultaneous." Besso rubbed his chin thoughtfully. "What you're really saying is that the idea of simultaneity depends on velocity, time, and the position of the observer of the event." "Precisely!" exclaimed Einstein, pleased that Besso grasped the idea so quickly.[329]

Traditionally, physicists had assumed that in principle all observers should be able to agree on the time of any given event. Thus, for two reference frames in relative motion, the time of an event according to observers on each frame would be the same: $t' = t$. Likewise, such observers should agree on the duration of any physical process: $(t'_2 - t'_1) = (t_2 - t_1)$. These notions (or "transformation equations") seemed so natural, and necessarily true, that they were nearly everywhere employed tacitly. Problems in electrodynamics had brought them to some physicists' attention. Einstein hence realized that if observers in relative motion disagree about the simultaneity of distant events, they could accordingly disagree about the precise time of events and about durations. So he abandoned the traditional conception of time, dismissing it as arbitrary.

In its place, Einstein posited a concept of time that could be established by physical procedures. Like Poincaré, he knew that Lorentz's local time would actually be the measures of time given by moving clocks synchronized by the exchange of light signals. Einstein decided that this procedure for establishing time could be used systematically to establish the time in all nonaccelerated material systems. It occurred to him that Lorentz's artificial time could be defined as *the* physical time, pure and simple. It alone seemed meaningful:

Surprisingly, it became evident that to overcome the aforementioned difficulties, it was only necessary to make the concept of time sufficiently precise. It required only thc knowledge that the auxiliary quantity introduced by Lorentz, which he called "local time," can be defined simply as the "time." Holding firmly to this definition of time, the fundamental equations of Lorentz's theory then agree with the principle of relativity if only the above-mentioned transformation

[329] Forsee, *Theoretical Physicist* (1963), pp. 23–24.

equations are replaced with such as correspond to the new concept of time.[330]

Einstein had concluded that the idea of a universal time was just an unjustified assumption. By abandoning that concept, he then could apply Lorentz's equations in any nonaccelerated systems. If t describes the time of an event as observed from one system, then the equation that expressed the time of the same event with respect to another system was

$$t' = \frac{t - xv/c^2}{\sqrt{1 - v^2/c^2}}.$$

Here, x is the position of the event in the first system, and v is the relative speed between the two systems. Despite its seeming complexity, this equation gave systematic and consistent results. Accordingly, several other transformation equations would serve to relate measurements of position, length, velocities, and so forth between reference systems.

Maxwell's equations did not change form when reformulated with the exact transformations that stemmed from Lorentz's works. Hence, Maxwell's equations were valid in all inertial systems; they constituted "laws" of physics. By extending the validity of these transformations to kinematics, and thus to all branches of physics that depend on kinematics, Einstein found the formal principle analogous to that which underlay thermodynamics.[331] The theory of heat and energy was based on the principle of the nonexistence of a *perpetuum mobile*. That principle restricted the possible laws of physics. Likewise, Einstein realized that the laws of mechanics, like the laws of electricity and magnetism, should be invariant with respect to Lorentzian transformations. Yet the laws of mechanics were not invariant with respect to these transformations, so Einstein proceeded to modify such laws.

When transforming the length of a rigid rod stationed in one system into its determination with respect to a system moving relative to the first, the latter length would be shorter than the former. Moreover, the effect would be reciprocal: observers in one system would judge that the meter sticks on

---

[330] Einstein, "Relativitätsprinzip und die aus demselben gezogenen Folgerungen" (1907), p. 413; *Collected Papers* 2, p. 435.
[331] Einstein, "Autobiographisches" (1946), trans. Schilpp, p. 57.

the other system were contracted, whereas observers on the latter would judge that instead it is the meter sticks in the former system that are contracted. Thus, Einstein found that the "hypothesis" of length contraction, which had earlier been introduced as an artificial and independent assumption, now emerged "as a compelling consequence" of the new approach.[332]

Likewise, the traditional "law" of the addition of velocities could be replaced with a proposition based on the relations between measured distances and times. It was not *a priori* necessary that the new "addition" rule would be the same as the traditional one, and Einstein found that indeed it was not. The new rule would account for why certain optical experiments did not seem intelligible in terms of the traditional "law." Thus, Einstein proceeded to change the classical laws of kinematics and dynamics.

Alongside the critical outlook that enabled Einstein to reject traditional concepts, he also took creative liberties. On the one hand, he had found that the fundamental categories that Kant had construed as being unchangeable could actually be modified. On the other hand, he had found a way in which he could contribute innovations into the so-called laws of physics. "This had come from the inventive element of human reason that Kant had written about. Einstein had proved his creativity. He was ecstatic."[333]

On Sunday, 14 May 1905, Hans Albert Einstein had his first birthday. At around that time, the Einstein-Marić family moved from the heart of downtown Bern to a residence just a few blocks away from Michele and Anna Besso. Einstein and Besso began daily to walk back from the patent office to their homes together, and they continued to discuss science. In long conversations, Besso played "the role of a critical disbeliever," and he "defended Newton's recognized concepts of space and time," which he justified by appealing to Mach's arguments.[334] Despite Mach's rejection of the notions of absolute space and time, one might yet uphold a relative conception of space and time that agreed with Newton's mechanics. But Einstein upheld his new and radical concepts with conviction. For Einstein, their "conversations on the way home had an incomparable charm."[335]

[332] Einstein, "Relativitätsprinzip und die aus demselben gezogenen Folgerungen" (1907), p. 413; *Collected Papers* 2, p. 435.

[333] Michelmore, *Profile* (1962), pp. 45–46.

[334] Seelig, *Documentary Biography* (1956), p. 71.

[335] Einstein to Vero Besso and Bice Rusconi, 21 March 1955, *Correspondance*, p. 538.

Decades later, a writer (who interviewed Hans Albert) narrated a conversation that allegedly transpired between Einstein, Besso, and Marić. Reportedly, Einstein argued:

> "Nothing except light is ever the same for two consecutive seconds," he told Mileva and Besso one night when they sat talking in the apartment. "It always journeys at the same speed." "In space, yes; but how about light from that?" Mileva asked, glancing toward a lamp. "Whether it comes from a candle, a lamp, an electric-light bulb, or a star, it is unalterable." "Suppose," asked Besso, "that you could put light aboard a rocket. Wouldn't its velocity be increased?" Einstein drummed his fingers on the arm of the wooden rocker in which he sat. "It might be expected that the velocity of the light traveling from the source in the rocket would be 186,000 miles per second, plus the velocity of the body on which it traveled. But nothing can exceed the speed of light. I hope to prove that it remains independent of the system in which it travels."[336]

Still in May, there were plenty of news to tell another good friend. Einstein wrote to Habicht to tell him about his latest papers and more. As usual, he bantered:

> So what are you up to, you frozen whale, you smoked, dried canned piece of soul, or whatever else I would like to hurl at your head, filled as I am with 70% anger and 30% pity! You have only the latter 30% to thank for my not having sent you a can full of minced onions and garlic after you so cravenly did not show up on Easter. But why have you not sent me your dissertation? Don't you know that I am one of the 1½ fellows who would read it with interest and pleasure, you wretched man? I promise you four papers in return. . . . The fourth paper is currently a rough draft and is an electrodynamics of moving bodies which employs a modification of the theory of space and time; the purely kinematic part of this paper will surely interest you.[337]

---

[336] Forsee, *Theoretical Physicist* (1963), pp. 25–26. Forsee interviewed Hans Albert, who might have heard of the conversation, if it is even roughly true, from his parents or from Besso, whom he knew well.

[337] Einstein to Habicht, Thursday, presumably 18 May 1905, *Collected Papers* 5, p. 31.

Einstein outlined his paper: the "Kinematic Part" would precede the "Electrodynamic Part," to solve problems in the latter. He devised a new kinematics based on the realization that Lorentz's notion of local time could be rigorously applied to physics in general.

Einstein was impressed by the agreement between his kinematic approach and Lorentz's equations. Aspects of Lorentz's theory, such as length contraction and the electromotive force, which previously seemed to be independent hypotheses, now emerged as consequences of Einstein's approach. Such effects stemmed from the relativity of motion and of simultaneity. His account corresponded better with the symmetries evidenced by electrodynamic experiments, because it did not pick out any one reference frame as special, as being truly at rest. Thus, the concept of a universal stationary ether seemed superfluous. "Only the introduction of a light ether as the carrier of electric and magnetic forces" did not harmonize with his conception, where instead "electromagnetic fields appear here not as the state of any matter but as independently existing things," as he had anticipated.[338] Yet there remained a fundamental aspect of Lorentz's ether conception, namely, the constancy of the speed of light. Following Lorentz, Einstein had assumed that light speed is constant, at least relative to a given reference frame, and that it is independent of the speed of its source. In 1912 Einstein explained that "it is impossible to base a theory of the transformation laws of space and time on the principle of relativity alone. As we know, this is connected with the relativity of the concepts of 'simultaneity' and 'shape of moving bodies.' To fill this gap, I introduced the principle of the constancy of the velocity of light, which I borrowed from H. A. Lorentz's theory of the stationary luminiferous ether."[339] Accordingly, in a long manuscript of 1912–1914, Einstein emphasized the connection between the light principle and Lorentz's theory. He remarked that "this principle is demanded by Lorentz's theory," and later he even called it "the principle of the constancy of the velocity of light (Lorentz's theory)."[340]

---

[338] Einstein, "Relativitätsprinzip und die aus demselben gezogenen Folgerungen" (1907), p. 413, *Collected Papers* 2, p. 435.

[339] Albert Einstein, "Relativität und Gravitation. Erwiderung auf eine Bemerkung von M. Abraham," *Annalen der Physik* **39** (1912), 1061; reprinted in *Collected Papers* 4, p. 183.

[340] Einstein, manuscript (1912–1914), published in *Collected Papers* 4, pp. 4, 29, 33, 36.

Einstein found that he could formulate his kinematics of clocks, rigid bodies, and electromagnetic processes on the basis of two key assumptions. Allegedly, according to a third-person account, "As he continued to pursue the subject of simultaneity, Einstein began to use the word 'relativity' more and more."[341] Rather than having the principle of relativity be valid as a by-product of several compensating effects, as in Lorentz's theory, he chose to raise the status of the principle to a postulate. That is, he chose to posit it as a fundamental assumption. Likewise, he postulated the constancy of the speed of light in an arbitrary inertial system. From those two postulates, along with physical hypotheses about the behavior of clocks, rods, and light, Einstein proceeded to derive the equations of his kinematics.

Thus, Einstein chose to formulate his new approach as a structured series of mathematical propositions deduced from fundamental axioms. In so doing, he followed an ancient tradition: the axiomatic presentation of mathematical knowledge, as carried out throughout the centuries by mathematicians and scientists such as Euclid, Newton, and, more recently, Hertz. This approach is distinguished by its logical elegance and pedagogical usefulness, yet it often seems to suggest, incorrectly, that the axioms on which a theory is based constituted also the original sources of the theory. In the case of Einstein's kinematics, accordingly, one should not confuse the axioms with its roots. In a series of interviews, the psychologist Max Wertheimer documented Einstein's recollections in this regard, noting: "The axioms were not the beginning but the outcome of what was going on. . . . The axioms were only a matter of later formulation—after the real thing, the main discovery had happened." Underscoring that point, Wertheimer added: "I wish to report some characteristic remarks of Einstein himself. Before the discovery that the crucial point, the solution, lay in the concept of time, more particularly in that of simultaneity, axioms played no role in the thought process—of this Einstein is sure. (The very moment he saw the gap, and realized the relevance of simultaneity, he knew this to be the crucial point for the solution.) But even afterward, in the final five weeks, it was not the axioms that came first. 'No really productive man thinks in such a paper fashion,' said Einstein."[342] The ideas that Einstein analyzed since the mid-1890s included the concepts of relativity and of the

---

[341] Forsee, *Theoretical Physicist* (1963), p. 24.
[342] Wertheimer, *Productive Thinking* (1945), p. 183.

velocity of light; but only in 1905, having found a solution in his analysis of basic kinematic concepts, did the two "principles" emerge as a foundation for his novel theory.

Once Einstein had formulated his general sketch of his paper on kinematics and electrodynamics, it took him about five weeks to complete the article. One biographer echoed: "The transfer of the broad concept of the theory to its logical mathematical progression on paper took five weeks of sapping work."[343] At the end of his paper, Einstein included the following acknowledgment: "In conclusion I wish to say that in working at the problem here dealt with I have had the loyal assistance of my friend and colleague M. Besso, and that I am indebted to him for several valuable suggestions."[344]

Einstein worked himself to exhaustion. He recalled later that he "suffered a collapse.... The upheaval was violent."[345] A biographer, who interviewed his son Hans Albert, recounted that when he finished his paper, "Benumbed, groggy, and terribly tired, he went to bed."[346] Likewise, another third-person account recounted: "When it was over, Einstein's body buckled and he went to bed for two weeks. Mileva checked the article again and again, then mailed it. 'It's a very beautiful piece of work,' she told her husband."[347]

"He was ill for fifteen days" then.[348] "For two weeks," reportedly, "Einstein, worn out by his exertions, ate very little and slept almost not at all. The slightest effort tired him."[349] Back at the office, Einstein told his senior colleague, Josef Sauter, about his latest work. Having finished his paper, he told Sauter that he was overjoyed to find, after all, that his results were perfectly compatible with the equations that Lorentz had ascertained. Decades later, Sauter recalled that Einstein said: "I cannot express the intensity of my joy."[350]

---

[343] Michelmore, *Profile* (1962), p. 46.

[344] Einstein, "Zur Elektrodynamik bewegter Körper," p. 921.

[345] Vallentin, *Le Drame* (1954), p. 38.

[346] Forsee, *Theoretical Physicist* (1963), p. 29.

[347] Michelmore, *Profile* (1962), p. 46.

[348] Vallentin, *Le Drame* (1954), p. 38.

[349] Forsee, *Theoretical Physicist* (1963), p. 29. See also Marianoff, *Intimate Study* (1944), p. 47.

[350] Sauter (1955), in Fluckiger, *Einstein in Bern*, p. 158.

Solovine remarked that Einstein's work benefited from his good life at home in the company of Mileva: "I am convinced that her influence was beneficial, in the environment and affection and in allowing him to work peacefully. I cannot forget that the memoir On the Electrodynamics of Moving Bodies, which established his reputation, was published in 1905, where a perfect harmony reigned between them."[351]

Creative Logical Sins

Einstein's contribution to kinematics arose suddenly in the context of his prolonged labors to solve problems in other fields of physics. In 1922, at a reception in Paris, someone asked him about special relativity: "Did it come to you all at once, or did it take you a long time?" To which he replied: "It came to me all at once, after having thought about it for my whole life."[352] Accordingly, Einstein once explained to his biographer, Carl Seelig, that there transpired "five to six weeks between the conception of the idea for the special theory and the completion of the relevant publication. But it would hardly be correct to consider this as a birthday, because earlier the arguments and building blocks were being prepared over a period of years, although without bringing about the fundamental decision."[353]

Was the formulation of the special theory of relativity a discovery or an invention? Nowadays, many writers call it a "discovery."[354] But throughout his life, Einstein emphasized the importance of invention, when characterizing his theoretical contributions. Alexander Moszkowski was very interested in scientific discovery but was shocked to hear Einstein argue: "The use of the word 'discovery' in itself is to be deprecated. For discovery is equivalent to becoming aware of a thing which is already formed. . . . Discovery is really not a creative act."[355] By contrast, the formulation of the special theory of relativity was a creative act. But that is not to say that Einstein

---

[351] Maurice Solovine to Carl Seelig, 29 April 1952, ETH Bibliothek, Hs 304:1007, p. 3.

[352] This exchange was recalled by Einstein's cousin, Paul Koch (Caesar Koch's son), in a tape-recorded interview with John Stachel.

[353] Einstein to Carl Seelig, 11 March 1952, Einstein Archives, item 39-013.

[354] E.g., John D. Norton, "Einstein's Investigations of Galilean Covariant Electrodynamics Prior to 1905," *Archive for History of the Exact Sciences* 59 (2004), 45–105, pp. 48, 74, 75; and likewise, Miller, *Einstein, Picasso* (2001), pp. 7, 173, 177, 246.

[355] Einstein, quoted in Moszkowski, *Conversations* (1921), pp. 94–95.

invented concepts such as the principle of relativity, local time, the constancy of the speed of light, or the contraction of lengths. Such concepts had already been conceived and developed by physicists such as Lorentz and Poincaré. It was thanks to the labors of Lorentz, above all, that Einstein and many other physicists were able to pursue electrodynamics along definite lines. Above all other physicists, Einstein loved Lorentz.[356] Accordingly, late in life he complained: "People do not realize how great was the influence of Lorentz on the development of physics. We cannot imagine how it would have gone had not Lorentz made so many great contributions."[357]

Thus, Einstein's contribution was scarcely the inception of the basic concepts in electrodynamics. Instead, what Einstein invented was a new way of arranging such concepts in relation to one another. In so doing, he changed the meaning of such concepts. He elaborated: "Invention occurs here as a constructive act. This does not, therefore, constitute what is essentially original in the matter, but the creation of a method of thought to arrive at a logically coherent system."[358]

Reportedly, in conversation, Einstein once commented that there existed "a definite connection between the knowledge that he gained at the Patent Office and the theoretical results that appeared at the same time as products of his intensive thought."[359] At the patent office, as in his father's business and in the laboratories of the Zurich Polytechnikum, Einstein maintained contact with innovative technologies. Haller, the director of the patent office "taught them [the patent clerks] to think sharply and logically, and to select every word in its most exact sense."[360] He encouraged Einstein to cultivate critical scrutiny and a precise clarity of expression; as Einstein recalled, "This man [Haller] taught me how to express myself correctly; he was stricter than my father."[361] And at the patent office, Einstein could discreetly discuss his theoretical investigations with Besso.

Besso himself tried to ascertain the origin of Einstein's insights. Rather

[356] See, for example, Cohen, "Interview" (1955), p. 72.

[357] Interview of 11 December 1954. Shankland, "Conversations with Albert Einstein" (1963), p. 57.

[358] Moszkowski, *Conversations* (1921), p. 96. See also Albert Einstein, "H. A. Lorentz, Creator and Personality," *Mein Weltbild* (1934); reprinted in *Ideas and Opinions* (1954), pp. 73–76.

[359] Moszkowski, *Conversations* (1921), p. 229; Moszkowski, *Einblicke*, p. 227.

[360] Reiser, *Biographical Portrait*, p. 65.

[361] As quoted in 1955 by Sauter, in Flückiger, *Einstein in Bern* (1974), p. 154.

than drawing any connection with modern technologies, Besso believed that the distinctive roots of Einstein's creativity lay in a critical philosophical outlook. In 1947, at the age of seventy-four, Besso wrote to Einstein to ask: "Is it true" that his having introduced the young Einstein to Mach's writings exerted such an influence in the developing student that it "turned his attention to observable things—even maybe indirectly, to 'clocks and measuring rods?'"[362]

Einstein soon replied, acknowledging that Besso had introduced him to Mach's writings while he was a student at Zurich and that Mach's influence on his intellectual development in general had been great. But he added, more specifically, that "frankly" he could "not clearly see" any direct influence of Mach on his work of 1905. By contrast, Einstein claimed to have been influenced directly by Hume: "As far as I am aware of, the immediate influence of D. Hume upon me was greater. I read this together with Conrad Habicht and Solovine in Bern."[363] Einstein reminisced that in their evening discussions, the reading of Hume's *Treatise* had been a major influence on him "—alongside Poincaré and Mach."[364]

Einstein's approach to physics was consonant with Hume's arguments that exact scientific laws cannot be induced from experience and that the truth of generalizations from experience is often unjustifiably asserted owing to merely habitual expectations. The universality of time appeared as just such an unjustified concept. In hindsight, Einstein commented that "Hume saw clearly that certain concepts, as for example that of causality, cannot be deduced from the material of experience by logical methods."[365] Likewise, Einstein realized that he could not deduce the concept of absolute time from the material of experience. Consequently, he replaced it with a new concept of time based on plausible empirical procedures.

Einstein greatly admired Hume and Mach for their critical philosophies. But another important aspect of scientific thought consisted of a speculative

[362] Besso to Einstein, 12 October to 8 December 1947, *Correspondance*, p. 386; translation A. Martínez.

[363] Einstein to Besso, 1952, *Correspondance*, p. 391; translation by A. Martínez. Likewise, he recalled in 1942: "Kant interested me always but I do not believe that he influenced me greatly—if not in the way of opposition. But I was impressed by Mach's works and still more by Hume's philosophical works." Einstein to Arnold S. Nash, 8 August 1942, Einstein Archives, item 17-439.

[364] Einstein to Besso, 6 March 1952; *Correspondance*, p. 464.

[365] Einstein, "Autobiographisches" (1946); trans. Schilpp, p. 13.

and creative element. On this matter, Einstein once commented to Besso, in 1917, that Mach's critical approach, by itself, "cannot create anything living; it can only eliminate that which is rotten."[366] In 1952 he elaborated his views on Mach:

> With conviction he had held onto the standpoint, that these concepts, even the most fundamental, receive authorization only from the empirical, that they are in no way *logically* necessary. He had thereby worked especially to sanitize, as he had made clear, that the most important physical problems are not of a mathematical-deductive kind, but are such that refer to the fundamental concepts. I see his weakness therein, that he more or less believed that science consists of an "ordering" of empirical materials; thereby he had misjudged the freely constructive element in the formation of concepts. He believed, to a certain extent, that theories arise through *discovery* and not through *invention*.[367]

Like other physicists, Einstein advanced by proposing and analyzing physical hypotheses. "His imagination," reportedly, "conjured up the various approaches. His intuition told him when he was on the right track. It was that way with Relativity—a step at a time."[368] For years he struggled to develop concepts, and combinations of concepts, that would appropriately account for the physical relationships between light and moving bodies. This manipulation of hypotheses and concepts was not restricted univocally by experience. Available experiments did not suffice to simply decide which notions were necessarily true. Theoretical speculations were guided and restricted by experimental knowledge. But experiments alone did not compel the choice of particular concepts. As he explained to Solovine, "The concepts can never exactly be derived logically from experience against any objection. . . . Moral: if one does not sin at all against reason, one generally does not arrive at anything."[369]

---

[366] Einstein to Besso, 13 May 1917, *Correspondance*, p. 114; translation by A. Martínez.
[367] Einstein to Besso, 1952, *Correspondance*, p. 391; translation by A. Martínez, emphasis in the original.
[368] Michelmore, *Profile* (1962), p. 44.
[369] Einstein to Solovine, 28 May 1953, Solovine and Einstein, *Lettres* (1956), p. 128.

Throughout his life, Einstein emphasized the essentially artificial character of the fundamental elements of physical theories. He argued that creativity consisted of the "free play" of concepts. This license to playfully manipulate, modify, and rearrange concepts was encouraged by Einstein's childlike approach to physics. He once explained to James Franck: "When I asked myself, how it came to be that I in particular found the theory of relativity it seemed to me to lie in this: The normal adult person does not think about space-time problems, he has already done all that there is to think about that in early childhood. By contrast, I was so slow in my development that I first began to wonder about space and time when I was already grown up and hence naturally I penetrated deeper into the problem than a child."[370] In the discussion meetings with his friends, Einstein had cultivated a childlike enthusiasm for fundamental questions. He had also liked to analyze concepts by how they develop in children, according to Solovine.

A key aspect that distinguished Einstein was his great trust in his physical intuition. By intuition, he meant precisely knowledge derived by contact with experience, the accumulated understanding of phenomena. Thus, in the process of scientific creativity, he argued that "the really valuable factor is *intuition!*"[371] In the end, the irony of his achievement was that even though Einstein began his analyses on the basis of his physical intuitions, he eventually concluded on a solution that was profoundly counterintuitive.

Such factors and others helped Einstein carry out his investigations. And such factors are not unique or original to this case; other scientists throughout history have also employed and benefited from such methods. Yet another aspect that distinguished Einstein from his peers was the consistent obstinacy with which he worked on problems. In an interview, Hans Albert characterized his father, noting that "his major trait definitely is his stubbornness," identifying it as a major reason why Einstein succeeded in the struggle that led to his relativity theory.[372] Einstein's concern over the prob-

---

[370] Quoted in James Franck to Carl Seelig, 16 July 1952, ETH Bibliothek, item Hs 304:637; translation by A. Martínez.
[371] Moszkowski, *Conversations* (1921), p. 96. See also Einstein, "H. A. Lorentz, Creator and Personality," *Mein Weltbild* (1934); reprinted in *Ideas and Opinions* (1954), pp. 73–76.
[372] Kornitzer, "Einstein Is My Father" (1951), p. 136.

lems of light and motion became an obsession. A sister of Michele Besso once asked Einstein why Besso had not made comparable achievements:

> "Herr Professor," she asked . . . "this I really meant to ask you for a long time—why hasn't Michele made some important discovery in mathematics?"
>
> "But, Frau Bice," said Einstein, laughing "this is a very good sign. Michele is a humanist, a universal spirit, too interested in too many things to become a monomaniac. Only a monomaniac gets what we commonly refer to as results."[373]

[373] Bice Rusconi and Albert Einstein, November 1947, quoted in Niccolò Tucci, "The Great Foreigner. A Reporter at Large," *New Yorker* 23 (22 November 1947), 43–57.

CHAPTER SEVEN

Text and Equations
Elements of Einstein's Kinematics

Einstein's paper "On the Electrodynamics of Moving Bodies" ascribed key importance to kinematics and, in particular, to the concept of simultaneity. He divided the paper into two sections: the "Kinematic Part," the subject of the present chapter, and the "Electrodynamical Part."[1]

Already we identified several physicists whose works influenced Einstein during the process of reformulating kinematics. Some of those influences were not explicit in his paper of 1905, an acceptable practice at the time: "It was not unusual for *Annalen* authors to skimp on references," but "Einstein pushed this style to the extreme in the relativity paper."[2] In it, the only names that Einstein explicitly mentioned were Maxwell, Lorentz, Hertz, and Doppler, because of their works in electromagnetism and optics. Also, somewhat indirectly, Einstein referred to Euclid, Newton, and Descartes, by noting methods that they originated. Einstein's incursion into kinematics

---

[1] A. Einstein, "Zur Elektrodynamik bewegter Körper," *Annalen der Physik* (Leipzig) **17** (1905), 891–921; reprinted in *The Collected Papers of Albert Einstein*, ed. John Stachel, vol. 2, (Princeton: Princeton University Press, 1989). Several translations of Einstein's paper are available, including the following: H. A. Lorentz, A. Einstein, H. Minkowski, and H. Weyl, *The Principle of Relativity*, trans. W. Perret and G. B. Jeffery with notes by Arnold Sommerfeld (New York: Dover, 1952); Anna Beck (trans.) and Peter Havas (consultant), *The Collected Papers of Albert Einstein*, vol. 2 (Princeton: Princeton University Press, 1989); Arthur I. Miller, *Albert Einstein's Special Theory of Relativity: Emergence (1905) and Early Interpretation (1905–1911)* (Reading, Mass.: Addison-Wesley, 1981; repr., New York: Springer Verlag, 1998). John Stachel, ed., with assistance from Trevor Lipscombe, Alice Calaprice, Sam Elworthy, *Einstein's Miraculous Year: Five Papers That Changed the Face of Physics* (Princeton: Princeton University Press, 1998). Note that the translations by Perrett and Jeffrey and by Miller have some slight but significant defects.

[2] Miller, *Albert Einstein's Special Theory of Relativity*, p. 83.

stemmed from his extensive studies of electromagnetic theory, owing to his dissatisfaction with asymmetric theoretical descriptions of electromagnetic and optical effects, effects that actually exhibit the relativity of motion. He questioned traditional concepts of kinematics because he had honed the conflict between mechanics and electrodynamics to the traditional addition rule of velocities and the apparent invariance of the speed of light. Lacking faith in the complete validity of Maxwell's theory, Einstein sought general kinematic principles that would facilitate symmetric descriptions of the phenomena of electrodynamics and the optics of moving bodies.

Like Ampère, Pearson, Hertz, and others, Einstein felt that the science of kinematics had been neglected and that analysis of its basic concepts could clarify physics. In contradistinction with Lorentz, Einstein sought to ground electrodynamics on kinematics instead of statics. He analyzed motion on the basis of the concept of inertial systems advanced by Neumann, Lange, and James Thomson, even if perhaps he was not acquainted directly with their writings. By emphasizing the need to establish specific frames of reference for describing motions, Einstein gave prominence to the idea of relativity chiefly as a guiding element of his kinematics. He had studied the works of various writers who emphasized the relativity of motion, including Jules Violle, August Föppl, Karl Pearson, Ernst Mach, and Henri Poincaré. Conversations with his friends Michele Besso, Moritz Solovine, and Conrad Habicht, encouraged Einstein to analyze the conceptual elements of physics.

In the spring of 1905, Einstein suddenly devised a solution to the theoretical problem of formulating a relativistic account of electrodynamics and optics. He realized that the concepts of time and signal velocity are intimately connected if they are defined in terms of measurement procedures. This connection had been recognized previously by other theorists including Poisson, Lange, James Thomson, Calinon, and Henri Poincaré, and there is evidence that Einstein might have been acquainted at least with Poincaré's analysis of the matter. Like Poincaré, Einstein defined time intervals and the simultaneity of distant events in terms of the motion of light signals. But Einstein went beyond his predecessors by realizing that the logical circularity between the concepts of time and of velocity facilitated a reformulation of kinematics in which measurements of time vary with respect to observers in relative motion. Einstein denied significance to the concept of *absolute simultaneity*. He was acquainted with the works of several leading theorists, Kant, Schopenhauer, Mach, Pearson, and Poincaré, who in one way or another had denied the significance of *absolute time.*

Simultaneity in 1905

Einstein simplified electromagnetic theory in terms of new kinematic concepts. Poincaré independently carried out a similar analysis.[3] Other theorists, however, generally discussed the question of the relationship of mechanics to electromagnetism by focusing on concepts of dynamics, particularly mass and force. Presumably, Einstein was encouraged by the writings of Ampère, Pearson, and others to analyze the elementary notions of kinematics. In the introduction to his 1905 paper, Einstein summed up his outlook: "The theory to be developed is based—like all electrodynamics—on the kinematics of the rigid body, since the assertions of any such theory concern the relations among rigid bodies (coordinate systems), clocks, and electromagnetic processes. Insufficient consideration of this circumstance is the root of the difficulties with which the electrodynamics of moving bodies presently has to contend."[4] In short, Einstein believed that the outstanding problems in electromagnetic theory were a consequence of physicists' neglect of kinematics.

Ampère and his followers had construed the science of dynamics as based on the science of kinematics. Accordingly, Einstein realized that the dynamics of electrical particles should be based on kinematics. Like Hertz, he treated motion as more fundamental than the concept of force. With Einstein's work, electrodynamics thus became incorporated into classification schemes that gave a primary role to the science of motion. Previously, leading theorists such as Lorentz had attempted to reduce problems in electrodynamics to simpler problems in electrostatics. This paralleled the way in which earlier physicists had based dynamics, and mechanics in general, on statics, the science they considered the most fundamental. But in Einstein's work, the concepts of neither force nor electrical charge appeared as fundamental. Einstein diverged from Lorentz's approach by basing the dynamics of electrical particles on a new kinematics that described the motion of all material bodies.

[3] For a comparison of the works of Poincaré and Einstein, see Olivier Darrigol, "The Mystery of the Einstein-Poincaré Connection," *Isis* **95** (2004), 614–626.
[4] Einstein, "Zur Elektrodynamik bewegter Körper," p. 892.

In Einstein's paper of 1905, the science of statics was not fundamental. One reason was that the concept of force seemed inessential for the description of motions. Another reason was that bodies that are static or moving inertially seemed equivalent. Experimentally, it was meaningless to talk about motion or rest independent of any reference frame. Thus, in the introduction to his paper, Einstein asserted "the conjecture that to the concept of absolute rest there corresponds no properties of the phenomena, neither in mechanics, nor in electrodynamics," and accordingly, his viewpoint would not involve the concept of an "absolutely resting space."[5]

Einstein divided the Kinematic Part of his paper into five sections. The first two sections consist of a conceptual analysis involving not much mathematics. They introduce the fundamental principles and concepts of his approach. Section 1 is titled "Definition of Simultaneity."

Like earlier kinematics, Einstein's approach was based on the concept of motion relative to inertial frames of reference. The opening sentence of section 1 is: "Let us consider a coordinate system in which the equations of Newtonian mechanics hold."[6] With this brief statement, he touched on the tradition of ideas advanced by Neumann and others, namely, to specify the conditions under which Newton's equations may be observed to be valid. Einstein alluded to the sort of reference frame that Ludwig Lange had named an "inertial system." Einstein would have been acquainted with Lange's conception, because he had read Mach's book, which discussed it. Unlike Lange, though, Einstein did not describe any procedure to identify such a reference frame.

For his purposes, incidentally, the definitional procedure proposed by Lange would have been adequate. As commentators have argued, "Lange's definition of an inertial frame—which, by the way, is very close to Thomson's—is much more appropriate to Einstein's needs than the one that he himself, somewhat carelessly, gives and was therefore appositely prefixed by Max von Laue to his masterly exposition of Special Relativity."[7] Roberto Torretti, a philosopher of science, made this comment on the grounds that the condition that Newton's equations are valid in the coordinate system is

---

[5] Ibid., pp. 891–892.
[6] Ibid., p. 892.
[7] Roberto Torretti, *Creative Understanding: Philosophical Reflections on Physics* (Chicago: University of Chicago Press, 1990), p. 46.

"a condition blatantly at odds with the subsequent development of the [1905] paper."[8] Einstein's special relativity theory *cannot* be derived by assuming the validity of Newton's mechanics. (For example, in Newton's scheme the speed of light is not constant, whereas in Einstein's scheme it is.) For this same reason, presumably, a footnote was added to the Teubner edition (1913) of Einstein's paper, indicating that the condition that "the equations of Newtonian mechanics hold" is valid only "to the first approximation."[9] However, this clause suffers from a serious drawback. As Torretti explained, it only avoids the inconsistency between Einstein's definition of inertial system and his theory "at a rather high price, for the concept of a stationary frame thereby becomes irretrievably ambiguous."[10] That is, if the concept of inertial system were defined in terms that are only *approximately* valid, then any subsequent use of this concept would lack the exactitude necessary to formulate a mathematically precise kinematics.

Hence, one might be inclined to believe, as Torretti suggests, that Einstein was rather "careless" in his specification of an inertial system, and thus one might follow Laue in replacing Einstein's statement with a definition such as Lange's. However, there is an explanation by which Einstein need not be construed as having been careless. We can well surmise that at the outset of his argument Einstein assumed the validity of Newton's mechanics to enable him to show subsequently that it was actually inadequate.

Einstein's discussion of kinematics served two main purposes. It served to introduce a new approach for the analysis of motion, but it did not serve merely as a feasible *alternative* to traditional kinematics. It also demonstrated, by criticizing its notion of time, that the basis of Newton's mechanics lacked some justification. By specifying procedures for the measurement of time intervals, Einstein showed that the traditional conception of time underlying Newton's mechanics lacked empirical justification. Einstein did not prove the impossibility of specifying any procedure by which the traditional concept of time could be substantiated, yet he wrote as though no such procedure could be secured. In discussing the question of simultaneity later in this chapter, we will review evidence to show how Einstein's original arguments were framed in the context of Newton's

[8] Ibid., p. 286.
[9] Miller, *Albert Einstein's Special Theory of Relativity*, p. 371.
[10] Roberto Torretti, *Relativity and Geometry* (Oxford: Pergamon Press, 1983), p. 294.

mechanics. Thus, we may understand why the initial discussion concerned systems in which "the equations of Newtonian mechanics hold."[11]

In Einstein's analysis, nothing distinguishes material bodies as being really stationary instead of in uniform rectilinear motion. There is only the free decision to identify a given reference frame by the *name* "system at rest." Einstein offset the literal meaning of this expression by enclosing it repeatedly in quotation marks.[12]

With the concept of a coordinate system, he proceeded to explain how to describe motion mathematically: "If we wish to describe the motion of a material point, we give the values of its coordinates as functions of time." This way of introducing the basic elements of physics had been employed by Gustav Kirchhoff in his influential *Lectures on Mathematical Physics: Mechanics,* of 1876. After a brief section discussing the aim of mechanics, "the science of motion," Kirchhoff began his first lesson by explaining how to describe the motion of a material point mathematically by reference to a rectangular coordinate system. He noted that the location of the point at any one instant would be specified by the coordinates x, y, z, and that each coordinate had to be determined as a function of t, the duration of the motion.[13] He then proceeded immediately to formulate the equations that transform x, y, z into the coordinates x', y', z', of the same point in a second coordinate system. This specification of how the coordinates of an event in one coordinate system relate to those of the same event in another coordinate system was not a common way to introduce the subject of mechanics in Kirchhoff's day. But it was the approach followed by Einstein, three decades later. He studied Kirchhoff's work well before 1905. Furthermore, it seems that Einstein, like other physicists, regarded Kirchhoff's work as authoritative: he often referred to Kirchhoff's *Lectures* in his scientific papers.

---

[11] Harvey R. Brown argues that inertial frames in special relativity are not identical to inertial frames in Newton's mechanics and, therefore, that "Einstein was wrong in the 1905 paper to identify his inertial frames with Newton's," in Harvey R. Brown, *Physical Relativity: Space-Time Structure from a Dynamical Perspective* (Oxford: Oxford University Press, 2005), p. 88. But this criticism is inappropriate once we acknowledge that Einstein's initial statements were made within the context of Newton's mechanics—that he was specifying coordinate systems within Newton's mechanics.

[12] Einstein, "Zur Elektrodynamik bewegter Körper," pp. 892, 897, 903.

[13] Gustav Kirchhoff, *Vorlesungen über Mathematische Physik: Mechanik* (1876), 3rd ed. (Leipzig: B. G. Teubner, 1883), p. 2.

Yet, Einstein's presentation deviated from that tradition as he elaborated the meaning of the variable t in the equations of motion:

> If we wish to describe the *motion* of a material point, we give the values of its coordinates as functions of time. Now it is important to see that such a mathematical description first gains physical meaning when one is quite clear as to what we understand here by "time." We have to take into consideration that all our judgments in which time plays a role are always judgments about *simultaneous events*. For example, if I say, "That train arrives here at 7 o'clock," this roughly means, "The pointing of the small hand of my clock to 7 and the train's arrival are simultaneous events."[14]

With these words, Einstein initiated the analysis of the concept of simultaneity that entailed the formulation of his new kinematics. He presented a method of connecting the times of physical events occurring at different places. He deviated from the approaches to kinematics of Kirchhoff, Hertz, Ludwig Boltzmann, and others by analyzing how theoretical quantities are defined in terms of physical measurement procedures. This sort of analysis had been advocated and carried out by various theorists, including Mach, Poincaré, and Emil Cohn.

Peter Galison has argued that Einstein's reference to clocks and a train was not merely an imaginative illustration for abstract arguments but referred in fact to contemporary technologies of timekeeping and railways with which Einstein presumably was well acquainted as an expert examiner of patents in Bern, steps away from the train station.[15] Thus, it might seem anomalous that Einstein referred to clocks and a train when discussing kinematics, so it might seem that *he* had done so, by contrast to other physicists, precisely because he was a patent examiner. This impression might even be stronger if one construes kinematics as the purely abstract science of motion. However, kinematics proper was hardly pure; writers on physical kinematics who were not patent examiners regularly referred to clocks, trains, and other technologies. For example, William Kingdon

[14] Einstein, "Zur Elektrodynamik bewegter Körper," p. 892.
[15] Peter L. Galison, *Einstein's Clocks, Poincaré's Maps: Empires of Time* (New York: W. W. Norton, 2003), pp. 29–34.

Clifford, in his lessons on kinematics (published posthumously in 1878) illustrated the relativity of displacements by discussing the motion of a man walking on a moving railway carriage.[16] As another example, Édouard Collignon in 1873 illustrated the concept of average velocity by reference to the departure and arrival times of a train at stations in Paris and Creil.[17] Likewise, while discussing the kinematics of material points in 1902, Paul Appell explained time-space diagrams tracking the relative speeds of trains and their arrival times at multiple stations.[18] Also, in lessons on kinematics from 1902 to 1905, Appell and James Chappuis illustrated the relativity of motion by referring to a train in uniform motion.[19] Such examples show that Einstein's references to clocks and a train, nonspecific commonplace technologies, were not at all unusual in a discussion on kinematics. By that time, imagery of trains had become a common alternative when illustrating uniform rectilinear motion, and the relativity of motion, in addition to the older tradition, since the 1600s, of alluding to a ship cruising on calm waters.[20]

At any rate, Einstein stated that a straightforward determination of "time" at any one place can be made by placing a clock at that place and assigning clock readings to nearby events—thus achieving a specific temporal ordering of these events. But then a difficulty remains concerning the exact dating of events at places distant from that point. Hence, another observer with an identical clock must be stationed at a distant point to as-

[16] William Kingdon Clifford, *Elements of Dynamic: An Introduction to the Study of Motion and Rest in Solid and Fluid Bodies*, part 1: *Kinematic* (London: Macmillan, 1878), pp. 4–5.

[17] Édouard Collignon, *Traité de Mécanique*, Part 1: *Cinématique* (Paris: Libraire Hachette et Cie., 1873), pp. 14–15.

[18] Paul Émile Appell, *Cours de Mécanique à l'usage des Candidats à l'École Centrale des Artes et Manufactures* (Paris: Gauthier-Villars, 1902), pp. 70–73.

[19] Paul Émile Appell and James Chappuis, *Leçons de Mécanique Élémentaire à l'usage des Élèves des Classes de Première C et D, conformément aux Programmes du 31 mai 1902* (Paris: Gauthier-Villars, 1905), pp. 55–56.

[20] Of course, some writers continued to use the image of a ship to illustrate the relativity of motions, such as H. A. Lorentz in his book on visible and invisible motions. Yet in that book Lorentz, too, made brief allusions to a railroad and a locomotive. H. A. Lorentz, *Zichtbare en onzichtbare bewegingen* (1901); German translation: *Sichtbare und unsichtbare Bewegungen; Vorträge auf Einladung des Vorstandes de Departements Leiden der Maatschappij tot nun van't algemeen im Februar und März 1901*, edited from the Dutch original by G. Siebert (Brauschweig: F. Vieweg und sohn, 1902), pp. 18, 25, 71.

sign "times" to the events there.[21] Having done this for a pair of clocks located at distant points A and B, Einstein argued that there remained the task of coordinating a common time for these distant clocks. He claimed that this could not be done without using a "further assumption." To that end, he proposed "establishing *by definition* that the 'time' required by light to travel from A to B equals the 'time' it requires to travel from B to A."[22] It is then plausible to synchronize the two clocks, that is, to establish a "time" common to both, by means of the following operational prescription:

Let a ray of light start at the "*A* time" t_A from A toward B, let it at the "*B* time" t_B be reflected at B in the direction of A, and arrive again at A at the "*A* time" t'_A.

According to this definition the two clocks are synchronous if

$$t_B - t_A = t'_A - t_B.$$

This equation of time intervals, along with the assumption that light takes the same time to travel from A to B as from B to A, constitutes the definition of simultaneity, given the procedure described by Einstein. For him, the concept of simultaneity was justified *essentially* by the requirement that it be formulated in terms of physical operations. "As long as this requirement is not satisfied," he later emphasized, "I allow myself to be deceived as a physicist (and of course the same applies if I am not a physicist), when I imagine that I am able to attach a meaning to the statement of simultaneity."[23]

---

[21] At least since the early 1890s Einstein had read about the synchronization of distant clocks by using electrical signals, because he read the books by Aaron Bernstein, *Naturwissenschaftliche Volksbücher: Wohlfeile Gesammt-Ausgabe* (Berlin: Franz Duncker, 1869); see vol. 4, pp. 88–98.

[22] Einstein, "Zur Elektrodynamik bewegter Körper," p. 894, emphasis in the original. Miller's translation in *Albert Einstein's Special Theory of Relativity* (1981, p. 394) omits the italics.

[23] Albert Einstein, *Über die Spezielle und die Allgemeine Relativitätstheorie* (Brauschweig: Friedrich Vieweg & Son, 1917); *Relativity: The Special and the General Theory*, authorized translation by Robert W. Lawson (New York: P. Smith/H. Holt, 1931; repr., New York: Crown, 1961), p. 22.

By applying this definition of simultaneity, distant clocks can be synchronized by means of the following procedure. At a given instant t_A, marked on a clock at A, a ray of light is sent out from A towards B. This ray is then reflected at B and travels back to A. The designation t'_A then stands for the reading of the same clock at A at the instant the light ray returns. Suppose, for example, that this second reading is at ten seconds ($t'_A = 10$) and that the signal was first emitted when the clock at A marked zero ($t_A = 0$). Then, according to the assumption that light takes the same amount of time to travel from A to B as from B to A, the definition of simultaneity states that the light ray arrived at B in half the duration it took for the round trip. This can be expressed formally by solving for the value t_B in the preceding equation:

$$t_B = 1/2\left(t'_A - t_A\right).$$

Therefore, the clock at B may be adjusted, or "synchronized," with the one at A simply by setting it to mark, say, five seconds ($t_B = 5$), when light gets to B after having been emitted once again from A when the A-clock reads zero seconds.

This was essentially the same synchronization procedure that was very briefly sketched by Poincaré in his paper of 1900, which he applied to systems in relative motion to note that it leads to Lorentz's local time—as Einstein proceeded next to do as well, painstakingly. We do not know whether Einstein read that paper or Poincaré's other relevant papers by May 1905. Yet many writers have discussed this question at length.[24] On this matter, we can consider just two suggestive quotations, which had not been brought to light. First, quoting the following passage from Schopenhauer's writings seems appropriate:

> It may sometimes happen that a truth, an insight, which you have slowly and laboriously puzzled out by thinking for yourself could easily have been found already in a book; but it is a hundred times more valuable if you have arrived at it by thinking for yourself. For only then will it enter your thought-system as an integral part and living member, be perfectly and firmly consistent with it and in accord with

[24] E.g., Darrigol, "The Mystery of the Einstein-Poincaré Connection."

all its other consequences and conclusions, bear the hue, color and stamp of your whole manner of thinking, and have arrived at just the moment it was needed . . . a truth won by thinking for oneself is like a natural limb: it alone really belongs to us.[25]

Indeed, Einstein's analysis of simultaneity came to be seen as distinctive of his style of physics. Thus, one can well come to own certain ideas that one has painstakingly elucidated and secured, even if one has perhaps come across similar ideas along the way. Still, it is also worth noting that at least once, by 1930, Einstein acknowledged: "My peace of mind is often troubled by the depressing sense that I have borrowed too heavily from the work of other men."[26] (He did not specify whose works he was thinking about.)

At any rate, in mid-1905 Einstein explained his definition of simultaneity to Josef Sauter, a colleague at the Swiss patent office. Fifty years later, Sauter recalled the conversation:

Before any other theoretical consideration, Einstein remarked on the necessity of a new definition of "synchrony" of two identical clocks distant from one another; to fix these ideas, he told me, suppose one of the clocks is on a tower at Bern and the other on a tower at Muri (the ancient aristocratic annex of Bern). At the instant when the clock of Bern marks noon exactly, let a luminous signal depart from Bern in the direction of Muri; it will arrive at Muri when the clock at Muri marks a time noon + t; at that moment, reflect the signal in the direction of Bern; if on the moment when it returns to Bern the clock in Bern marks noon + 2t, we will say that the two clocks are synchronized.[27]

---

[25] Arthur Schopenhauer, *Parerga und Paralipomena* (1851); selections reissued in Schopenhauer, *Essays and Aphorisms*, trans. R. J. Hollingdale (London: Penguin Books, 2004), pp. 90–91.

[26] Albert Einstein, untitled reflections, in Albert Einstein, John Dewey, James Jeans, et al., *Living Philosophies* (New York: Forum, 1930; repr., New York: Simon and Schuster, 1931), 3–7, p. 3; no translator noted.

[27] Josef Sauter, "50 Jahre Relativitätstheorie," reprinted in Max Flükiger, *Albert Einstein in Bern: Das Ringen um ein neues Weltbild; Eine dokumentarische Darstellung über den Aufstieg eines Genies* (Berne: Paul Haupt Berne, 1974), p. 156; translation by A. Martínez.

With this simple illustration, Einstein explained the procedure that gave meaning to the notion of the simultaneity of distant events.

Next, Einstein argued that measurements of simultaneity vary among observers in relative motion. Section 2 of his 1905 paper, "On the Relativity of Lengths and Times," began by postulating two principles. As is well known, these were the "principle of relativity" and the "principle of the constancy of the speed of light." Historians and philosophers have analyzed the origin and meaning of these principles extensively, so here we need only summarize their function in the development of Einstein's kinematics. He first formulated the principle of relativity:

> 1. The laws by which the states of physical systems undergo changes are independent of whether these changes of state are referred to one or the other of two coordinate systems moving relatively to each other in uniform translational motion.[28]

In this form, the relativity principle was a declaration that for any equations to be deemed "laws of physics" they must be applicable in the same way in all inertial systems. As discussed earlier, various theorists, for centuries, had entertained notions of relativity, that is, of the physical equivalence among systems in rectilinear uniform motion relative to one another. Nonetheless, the idea of the relativity of motion was questioned in the late nineteenth century, because the ether could be interpreted as a *unique* frame of reference, since Maxwell's "laws" of electromagnetism seemed to be valid exactly only relative to the ether.

Einstein began his 1905 paper with a discussion of the effect experimentally discovered by Faraday in 1831: that the relative motion between a magnet and a conducting wire induces an electric current in the wire. In both cases, the same relative motion induces an identical electrical current in the wire. In support of his principle of relativity, Einstein also alluded to the "unsuccessful attempts" to discover any motion of the Earth relative to the ether. Hence, he affirmed the relativity of all inertial motions on the basis of well-known experimental observations, which he generalized to the broad "conjecture that to the concept of absolute rest there correspond

---

[28] Ibid., p. 895.

no properties of the phenomena."[29] He proceeded to "raise this conjecture (whose content will now be referred to as the 'Principle of Relativity') to a postulate, and moreover introduce another postulate which is only apparently irreconcilable with the former."

Einstein did not allude to any experimental evidence in support of his second postulate, the "constancy of the speed of light." Nonetheless, physicists of the time could justify it on various grounds. Despite the use of accurate measurement devices, no significant variations in the speed of light in free space had been ascertained experimentally. Moreover, in accordance with the widely accepted wave theory of light, light was supposed to travel with the same speed in all directions relative to the ether, and independently of any motion of its source. Likewise, in Maxwell's electromagnetic theory of light, the speed of light in free space appeared as a constant. Hence, regarding the conjecture that light has the same speed in all directions, Poincaré in 1898 claimed, "The postulate, at all events, resembling the principle of sufficient reason, has been widely accepted by everybody."[30] In his 1905 paper, Einstein did not make definite assumptions about the constitution of electromagnetic radiation, whether it consisted of waves or particles, but he declared the idea of a luminiferous ether "superfluous." His new approach could be accepted as consistent with the wave theory at least insofar as each asserted a constancy of the speed of light in free space. In this sense it was a widely held assumption that light traveled with a constant speed. Einstein employed this idea of "constancy" by requiring that the speed of light be constant not explicitly relative to the ether but relative to at least one reference system.

Einstein first described his second postulate simply as the claim that "light is always propagated in empty space with a definite speed c which is independent of the state of motion of the emitting body."[31] This statement was rather vague because, contrary to Einstein's approach, it referred the speed of light to "empty space" rather than to any specific reference body.

[29] Einstein, "Zur Elektrodynamik bewegter Körper," p. 891.
[30] Henri Poincaré, "La mesure du temps," *Revue de Metaphysique et de Morale* 6 (January 1898), 1–13; reissued in Poincaré, *Foundations of Science: Science and Hypothesis, The Value of Science, Science and Method,* trans. George Bruce Halsted (New York: Science Press, 1913), p. 233.
[31] Einstein, "Zur Elektrodynamik bewegter Körper," p. 892.

But this shortcoming was remedied in section 2 of the 1905 paper, where he formulated the principle exactly:

> 2. Any ray of light moves in the "resting" coordinate system with the definite speed c, which is independent of whether the ray was emitted by a resting or by a moving body. Consequently,
>
> $$\text{speed} = \frac{\text{light path}}{\text{time interval}},$$
>
> where time interval is to be understood in the sense of the definition in § 1.[32]

Here Einstein established the constancy of light relative to a specific coordinate system, designated as the "resting" system. As historian Arthur I. Miller pointed out, "The universality of the second principle is perhaps clouded by Einstein's use in 1905 of the term 'resting system.'"[33] At first sight, it might seem that Einstein intended the constancy of the speed of light to be valid only in one frame of reference, a unique system at rest. It is clear, however, that the principle applied to any inertial system. Einstein formulated the postulate in terms of a single system to convey clearly what it means for the speed of light to be "constant" within such a system. Otherwise, as Torretti notes, the principle "remains hopelessly imprecise if the system to which it is referred to is not defined beforehand."[34] The very concept of velocity is more easily understood if explained in terms of a given system than if it related abstractly to all systems at once.

Another reason why Einstein formulated the light principle in relation to a specific coordinate system "at rest" was that he essentially had extracted this principle from the electrodynamics of Hendrik Antoon Lorentz. In his 1905 paper, Einstein did not trace the origin of the light principle to Lorentz's theory explicitly, but he did acknowledge Lorentz clearly in later publications. In one work Einstein remarked, "The fundamental assertion of Lorentz's theory that every light ray in a vacuum always propagates (at least with respect to a certain coordinate system K) with the definite con-

[32] Ibid., p. 895.
[33] Miller, *Albert Einstein's Special Theory of Relativity*, p. 202.
[34] Torretti, *Relativity and Geometry*, pp. 55, 49.

stant speed c, we will call the *principle of the constancy of the speed of light.*"[35] Of course, Einstein did not agree with Lorentz's entire account of electrodynamics; in particular, he dispensed with the conception of the stationary ether. However, Einstein extracted the light principle from Lorentz's theory as a fundamental assertion, valid also in his own theory.

Einstein referred to one arbitrary system as the "resting system" merely "to distinguish this coordinate system verbally from others,"[36] that is, from other equivalent systems also noted in his 1905 paper. He cast his analysis initially in terms of this so-called stationary system but shortly thereafter introduced another system moving relative to the first, which he deemed "the moving system." Thus, "that light (as required by the principle of the constancy of the speed of light, in combination with the principle of relativity) is also propagated with a speed c when measured in the moving system."[37] Therefore the light principle extends to indefinitely many systems by virtue of the relativity principle according to which the same laws apply to all inertial systems.

Because the principle of relativity refers to the laws of physics, and, as it seems feasible to construe the constancy of the speed of light as one such law, it might appear that the invariance of the speed of light was only a special case of the relativity principle. Einstein's decision to present the constancy of the speed of light as an independent principle might seem unnecessary. Thus, the physicist Percy Williams Bridgman wrote: "It would seem to be mostly a historical accident that two postulates were used instead of one. . . . [By] implication . . . the second postulate is superfluous."[38] Indeed, the laws of physics that describe electric and magnetic phenomena involve a term identified as the speed of light, and hence it might appear that its invariance is simply a direct consequence of the relativity principle

---

[35] Albert Einstein, "Die Relativitätstheorie," *Die Kultur der Gegenwart. Ihre Entwickelung und ihre Ziele,* ed. Paul Hinneberg, part 3, sec. 3, vol. 1, *Physik,* ed. Emil Warburg (Leipzig: Teubner, 1915), pp. 703–713, quotation on p. 706; reprinted in *The Collected Papers of Albert Einstein,* vol. 4, *The Swiss Years: Writings, 1912–1914,* ed. Martin J. Klein, A. J. Kox, Jürgen Renn, and Robert Schulmann (Princeton: Princeton University Press, 1995), p. 539. See also, "Zum Relativitäts-Problem," *Scientia* 15 (1914), 337–348; especially pp. 339–340; reprinted in *Collected Papers,* vol. 4, pp. 611–612.

[36] Einstein, "Zur Elektrodynamik bewegter Körper," p. 892.

[37] Ibid., p. 899.

[38] P. W. Bridgman, *A Sophisticate's Primer of Relativity,* Second edition (Middletown, Conn.: Wesleyan University Press, 1962), p. 114.

joined to such laws. However, such conjectures do not represent the logic of the matter properly, nor do they explain Einstein's original development of the subject. The second postulate had to stand on its own because otherwise it would not be altogether necessary that the equations describing electromagnetic phenomena should be regarded as laws and thus apply in the same form in all inertial systems.

Einstein did not base his special theory on the relativity principle together with Maxwell's laws of electromagnetism. Instead, he based it on the relativity principle conjoined with the light principle, and this connection in turn validated Maxwell's equations as "laws." Without independent specification of the constancy of the speed of light, one could instead deduce from the relativity principle alone that Newton's equations of mechanics are to be deemed "laws," whereas the equations that describe electromagnetic phenomena would not be laws, as these would vary among inertial systems.

Einstein's resolve to retain the equations of electromagnetism as "laws" by postulating the light principle (and thus having to revise Newton's equations) was not based on faith in the validity of Maxwell's equations. Actually, "by the time [Einstein] wrote the relativity paper, he no longer regarded Maxwell's theory as universally valid."[39] Einstein's own account of the matter is quite clear. In 1952, for example, he wrote to Max von Laue, commenting on the latter's treatment of relativity and electromagnetic theory in his physics textbook: "When one looks over your collection of proofs of the special theory, one becomes of the opinion that Maxwell's theory is unquestionable. But in 1905 I already knew certainly that Maxwell's theory leads to false fluctuations of radiation pressure and, with it, to an incorrect Brownian motion of a mirror in a Planckian cavity. In my view, one could not get around ascribing to radiation an atomistic structure which, of course, does not fit into the frame of Maxwell's theory."[40] In short, in Einstein's work, the constancy of the speed of light could not be said to be a consequence of the relativity principle. Hence, in 1912 Einstein affirmed that, from the very start, "I knew that the principle of the constancy of the speed of light was something quite independent of the relativity postulate."[41]

---

[39] Stachel, *Collected Papers of Albert Einstein*, vol. 2, p. 265.

[40] Albert Einstein, quoted by Gerald Holton, in *Thematic Origins of Scientific Thought: Kepler to Einstein* (Cambridge, Mass.: Harvard University Press, 1973), pp. 201–202.

[41] Albert Einstein, quoted in Stachel, *Collected Papers of Albert Einstein*, vol. 2, p. 263.

But then the question remains: on what grounds did Einstein come to believe that the value of the speed of light was a fundamental constant? Some writers have supposed that Einstein became certain of the validity of the light principle owing to experiments in optics. Yet closer analysis reveals that Einstein based this postulate essentially on theoretical arguments intimately connected to an analysis of the fundamental concepts of kinematics. As early as 1895, he believed that electromagnetic theory implied variations in the speed of light, and he contemplated carrying out experiments to detect such variations. As Einstein wrote in retrospect, "The difficulty that had to be overcome was in the constancy of the speed of light in vacuum which I first thought I would have to give up. Only after groping for years did I notice that the difficulty rests on the arbitrariness of the kinematical fundamental concepts."[42] We will see that Einstein's principle of the constancy of the speed of light was intimately connected to his analysis of simultaneity.

Once Einstein had synchronized clocks in the "stationary system," the argument that simultaneity is relative followed from applying the same procedure to synchronize clocks in a system moving relative to the first. Before we discuss this, however, it is important to explain why Einstein designated the specified procedure of establishing clock synchrony as a *"definition"* of simultaneity. Why did he stress that the idea that light takes the same amount of time to travel equal distances in opposite directions is an assumption? Why did he establish this as a definition, rather than asserting it as necessarily or experimentally true? Did experiments demonstrate the constancy of the speed of light in any direction?

Consider the question: did the famous experiment of Albert Michelson and Edward Morley prove that light travels with the same speed in opposite directions, say, from point A to point B, and then from B to A? No, their experiment only compared (and indirectly) the times light takes to complete *round trips* between points. It did not involve measurement of times light takes to travel one-way journeys in specific directions. It was not designed to investigate unidirectional propagation of light signals. Hence, the physicist Arthur Stanley Eddington later noted, "Strictly speaking the Michelson-Morley experiment did not prove directly that the speed of light was constant in all directions. The experiment compared the times of a

[42] Miller, *Albert Einstein's Special Theory of Relativity*, p. 137; Einstein Archives, 2-070.

journey 'there-and-back.'"[43] Other physicists also explained that the experiment "merely leaves us at liberty to assume isotropy."[44] Interferometer experiments of the type performed by Michelson and Morley do not provide information on the equality of one-way velocities of light. The equality was not an experimental fact, but an extrapolation from the known equality of round-trip times of light propagation.

In all experiments on the speed of light, what was really measured were round-trip average speeds and not one-way velocities. Whenever the value of a one-way velocity of light was ascertained, it was based on the fundamental assumption that all rays of light, in all directions, have the same speed relative to the observer's system. Thus, already in 1898 Poincaré argued that both terrestrial and astronomical measurements of the speed of light presuppose by convention the equality of the speed of light in different directions.[45]

The underlying difficulty is that, to measure the velocity of a signal moving in a single direction, one needs to know the times at which the signal passes the end points of a given distance, and these times are meaningful only if they are obtained from *synchronized* clocks. In this way, the measurement of any one-way velocity presupposes a knowledge of distant simultaneity. But the synchrony of clocks at distant places can be established or verified only by means of signals of known velocity. Einstein noted, "It would thus appear as though we were moving in a logical circle."[46] As the philosopher of science Hans Reichenbach put it: "To determine the simultaneity of distant events we need to know a velocity, and to measure a velocity we require knowledge of the simultaneity of distant events."[47]

This logical circularity in the definition of the concepts of time and velocity had been considered in the nineteenth century by a few theorists, such as Siméon Denis Poisson, Auguste Calinon, and Ludwig Lange. Following Calinon, Poincaré emphasized the issue in his article of 1898, "The Measure of Time." This same realization was important to Einstein in leading him to

---

[43] Arthur Stanley Eddington, *The Mathematical Theory of Relativity* (1923; repr., Cambridge: Cambridge University Press, 1957), p. 19.

[44] Henri Arzeliès, *Relativistic Kinematics* (Oxford: Pergamon Press, 1966), p. 85.

[45] Poincaré, "La mesure du temps" (1898); reissued in Poincaré, *Foundations of Science*, p. 232.

[46] Einstein, *Relativity* (1917), p. 23.

[47] Hans Reichenbach, *The Philosophy of Space and Time* (Berlin: Gruyter, 1927); trans. M. Reichenbach and J. Freund (New York: Dover, 1958), p. 126.

formulate his new kinematics. Thus, the solution he found to the problems of mechanics and electrodynamics, after his crucial discussion with Besso in May 1905, followed suddenly from the idea that "time is not absolutely defined, and there is an inseparable relationship between time and signal speed. That previous extraordinary difficulty was solved completely by that insight, for the first time."[48]

Therefore, Einstein could not claim that the speed of light is constant in opposite directions on the basis of any experiments. In the words of Eddington, "The measured velocity of light is the average to-and-fro velocity . . . there is a deadlock . . . which can only be removed by an arbitrary assumption or convention. The convention actually adopted is that (relative to the observer) the velocities of light in the two opposite directions are equal."[49] It is thus understandable that in his longest exposition of relativity theory, Einstein emphasized: "That light requires the same time to traverse the path $A \longrightarrow M$ as for the path $B \longrightarrow M$ is in reality neither a *supposition nor a hypothesis* about the physical nature of light, but a *stipulation* which I can make of my own free will in order to arrive at a definition of simultaneity."[50] This is also why in his relativity paper Einstein emphasized that "*by definition* the 'time' required by light to travel from A to B equals the 'time' it requires to travel from B to A." Einstein did not state that the constancy of the one-way speed of light was a matter of empirical fact but only that, given a signal going from A to B and back to A, then

$$\frac{2AB}{t'_A - t_A} = c,$$

where t_A and t'_A are the readings of a *single* clock stationed at A, at the instants the light signal departs from A and returns to A, respectively. This mathematical expression is what is described properly as the constancy of the average round-trip speed, and it is this equation that Einstein deemed to be "in agreement with experience."[51] By contrast, Einstein formulated the

[48] Einstein, "Wie ich die Relativitätstheorie entdeckte," Kyoto lecture (1922), p. 80; translation by F. Kanaya and A. Martínez.
[49] Arthur Stanley Eddington, *The Nature of the Physical World*, Gifford Lectures (1927; New York: Macmillan, 1929), p. 46.
[50] Einstein, *Relativity* (1917), p. 23.
[51] Einstein, "Zur Elektrodynamik bewegter Körper," p. 894.

principle of the constancy of the speed of light mathematically by referring to the times marked by *two* clocks:

$$\text{speed} = \frac{\text{light path}}{\text{time interval}},$$

where "time interval" was to be understood in the sense of his preceding definition, involving clocks at A and B. Accordingly, in a review article of 1907, Einstein stated the postulate of the constancy of the speed of light as

$$\frac{r}{t_B - t_A} = c,$$

where the r stands for the distance AB, and the times are the instants when light reaches each of the clocks, as marked by each clock. This equation represents the one-way speed of light, and *of this* equation Einstein wrote: "It is by no means self-evident that the assumption made here, which we will call 'the principle of the constancy of the speed of light,' is actually realized in nature."[52] For Einstein, the constancy of the speed of light was not an empirical fact but a "concept" or "ordering element" of scientific knowledge.

Furthermore, it is significant that Einstein posited the constancy of the speed of light as a "postulate." Traditionally, a postulate was a fundamental supposition that was not necessarily self-evident. For example, a geometry textbook, studied by Einstein since an early age, characterized a *postulate* as "requiring something to be posited."[53] By contrast, *axioms* (or *common notions*) were names given to statements of obvious and fundamental truths. The geometry textbook characterized axiom as "a self-evident fundamental truth, which is pronounced without proof."[54]

The light principle was not an axiom for several reasons. It was by no means self-evident that the speed of light is independent of the motion of its source at the instant of emission. It was counterintuitive that the speeds

---

[52] Albert Einstein, "Über das Relativitätsprinzip und die aus demselben gezogenen Folgerungen," *Jahrbuch der Radioaktivität und Elektronik* 4 (1907), 411–462; quote on p. 416; reprinted in *Collected Papers*, vol. 2, pp. 433–484.

[53] Theodor Spieker, *Lehrbuch der ebenen Geometrie mit Übungs-Aufgaben für höhere Lehranstalten*, 19th rev. ed. (Potsdam: Aug. Stein, 1890), p. 4.

[54] Ibid., p. 3.

of all light rays, unlike the speeds of other physical signals, would always be constant, in any inertial reference frame. Also, the meaning of the expression "time interval," involved in the light postulate, was not traditional or obvious. Likewise, the assumption of relativity was also a postulate, properly, rather than an axiom. It was a conjecture that not a few theorists and experimentalists doubted.

Years later, however, many writers, and even Einstein himself, did not always distinguish clearly between postulates and axioms. As often happens, one can be careless in the choice of words. Perhaps more important, long familiarity with given propositions tends to make them seem more plausible and admissible, especially in due regard to the growing varieties of experimental evidence.

As someone who actually had experienced firsthand the mental processes involved in conceiving theoretical principles, Einstein argued later in life that "the fundamental principles" of physical science "are free inventions of the human intellect," and he emphasized what he regarded as "the purely fictitious character of the fundamentals of scientific theory."[55] The psychologist Max Wertheimer described how he asked Einstein:

> "How did you come to choose just the velocity of light as a constant? Was this not arbitrary?" . . . Of course it was clear that one important consideration was the empirical experiments which showed no variation in the velocity of light. "But did you choose arbitrarily," I asked, "simply to fit in with these experiments and with the Lorentz transformation?" Einstein's first reply was that we are entirely free in choosing axioms. "There is no such difference as you just implied," he said, "between reasonable and arbitrary axioms. The only virtue of axioms is to furnish fundamental propositions from which one can derive conclusions that fit facts."[56]

In his autobiographical notes, Einstein stated that "one does not easily become aware of the free choice of such concepts, which, through verification and long usage, appear to be immediately connected with the em-

[55] Albert Einstein, *Ideas and Opinions* (New York: Bonanza Books, Crown Publishers, 1954), p. 272.
[56] Max Wertheimer, *Productive Thinking*, enlarged. ed. (New York: Harper & Brothers, 1945), p. 223.

pirical material."[57] Furthermore, as to the possibility of somehow deriving such principles from empirical evidence, he wrote in 1952: "Psychologically the A's [axioms] depend on the E's [experiences]. But there is no logical route leading from the E's to the A's, but only an intuitive connection (psychological)."[58] Accordingly, the foundations of Einstein's kinematics were warranted only indirectly by observation and experiment.

The preceding discussion illuminates the relationship between Einstein's postulate of the constancy of the speed of light and his definition of simultaneity. We will discuss this relationship further, but for the moment we can understand why Einstein called the specified clock synchrony procedure a "*definition*" of simultaneity. For one, Einstein's early geometry textbook explained: "The Clarification or Definition: namely the completely exact specification of a concept, which should be connected with a determined word. This is made through expression of the next higher concept, under which the subject is organized, and the characteristic feature through which it is distinguished from all others."[59] More significantly, the logical relationship between definitions and postulates had been discussed by John Stuart Mill in his *System of Logic*, which was read by Einstein and his friends in Bern.[60] Einstein may have been influenced by Mill's arguments to the effect that scientific definitions do not relate directly to objective matters of fact—although definitions are not fully arbitrary, because they involve postulates that affirm the existence of specific physical entities and relations. Likewise, Calinon argued that the "principal variable t, conveniently chosen," should be adopted "*by definition*" of the equality of simultaneous trajectories.[61] Perhaps more important, Einstein and his friends studied *Science and Hypothesis*, in which Poincaré employed the term definition in a similar way: "The

[57] Albert Einstein, "Autobiographisches" (1946); translated by Paul Arthur Schilpp, in *Albert Einstein: Philosopher-Scientist*, ed. Schilpp, Library of Living Philosophers, vol. 7 (Evanston, Ill.: George Banta, 1949), p. 49.

[58] Albert Einstein, letter to Maurice Solovine, in *Letters to Solovine*, introd. Maurice Solovine (New York: Philosophical Library, 1986), p. 137.

[59] Spieker, *Lehrbuch der ebenen Geometrie*, p. 3.

[60] John Stuart Mill, *A System of Logic, Ratiocinative and Inductive: Being a Connected View of the Principles of Evidence and the Methods of Scientific Investigation* (1846); 8th ed. (New York: Longmans, Green, 1906), e.g., book I, chap. VIII, p. 94.

[61] Auguste Calinon, "Étude critique sur la mécanique," *Bulletin de la Société des Sciences de Nancy*, ser. 2, 7 (1885), 87–180 (Paris and Nancy: Berger-Levrault, 1886), p. 90, emphasis in the original.

principles of dynamics at first appeared to us as experimental truths; but we have been obliged to use them as definitions. It is *by definition* that force is equal to the product of mass by acceleration; here, then, is a principle which is henceforth beyond the reach of any further experiment. It is in the same way by definition that action is equal to reaction."[62] For Poincaré, the simultaneity of distant events was a convention. Feasible conventions were restricted by empirical knowledge, yet seemed to be beyond the reach of experimental confirmation or refutation. Poincaré designated as "conventions" those propositions on which physicists could agree.[63]

Similarly, the "principles" of kinematics, as formulated by Einstein, were rooted in experience. The speed of light seemed to be independent of the speed of its source, and the travel time of light transmitted in one direction seemed no different from the travel time in the opposite direction. By formulating the concept of simultaneity as a definition, one would be admitting that this notion is compatible with experience; that it has a conventional aspect to it; and that, once accepted, it presumably cannot be contradicted by experiments.

Einstein abandoned the idea that some fundamental notions in physics are a kind of transcendental knowledge, as Kant had argued. Over the years, Einstein still continued to appreciate Kant's writings, but he rejected Kant's arguments about *a priori* knowledge. As Einstein commented to the physicist Max Born in 1918: "The 'a priori' I must pare down to 'conventional.'"[64] Accordingly, in a letter of 1924 to the philosopher of science André Metz, Einstein noted that relativity theory involved "conventions" and "physical hypotheses," and he specified one such convention: "simultaneity."[65]

---

[62] Henri Poincaré, "The Classic Mechanics" (1902), originally published in *La Science et L'Hypothèse;* reissued in *Foundations of Science,* Anthology of *Science and Hypothesis, The Value of Science,* and *Science and Method;* authorized translation by George Bruce Halsted (New York: The Science Press, 1913), pp. 92–106, quotation on p. 102, italics in the original.

[63] Scott Walter points out that at least later in life Poincaré clearly did not require conventions to involve consensus among all physicists; see Poincaré, "L'Espace et le Temps," *Scientia* 12 (1912), 159–170; also in *Dernières Pensées* (Paris: Flammarion, 1913).

[64] A. Einstein to Max Born, after 28 June 1918, in Robert Schulmann, A. J. Kox, Michel Janssen, and József Illy, eds., *The Collected Papers of Albert Einstein,* vol. 8, part B, *The Berlin Years: Correspondence 1918* (Princeton: Princeton University Press, 1998), p. 818.

[65] A. Einstein to André Metz, 27 November 1924, Albert Einstein, *Oeuvres choisies 4, Correspondances françaises* (Paris: Éditions du Seuil, 1989), p. 212. Copy of the original

Consider now how Einstein used his definition of simultaneity to derive the relativity of time. For him, the issue of whether clocks in different systems are synchronized was intertwined with the question of comparing the measured lengths of a rigid rod. After all, a method to measure lengths depends on judgments of simultaneity. If these judgments differ, then any measurements based on them will also differ. To measure the length of a moving rod one must ascertain, simultaneously, the locations of its two end points.

Einstein's argument runs as follows. Consider a pair of stationary clocks synchronized in a "stationary system" as described previously. An observer in this system then inquires about the length of a particular rigid rod moving uniformly with a velocity v along a path parallel to the line connecting the stationary clocks. To measure the length of the moving rod, the observer ascertains at what points of the stationary system the two ends of the rod are located *simultaneously,* that is, at a specific instant of time, and then measures directly the distance between these two points in his system. Einstein wrote that the result of this measurement would differ from the result obtained by an observer moving along with the rod and measuring it directly. He anticipated this discrepancy because the application of his definition of simultaneity in reference frames moving relative to one another entailed disagreements between observers concerning the synchrony of clocks.

Einstein's argument crystallized when he turned to consider the synchronization of a second pair of clocks placed on the end points of the moving rod. In this way, Einstein demonstrated the relativity of simultaneity for the first time. He explained that these clocks actually would not be synchronous as judged from the rod if they were synchronized with the stationary system's clocks. Suppose the clocks on the moving rod are synchronized simply by setting each to read the same time marked by a clock in the stationary system at the instants when the moving clocks are positioned exactly adjacent to the ones in the stationary system. Then, according to Einstein, these clocks are "synchronous in the stationary system."[66] One must emphasize that all of the clock readings involved agree with the clocks attached to the "stationary system." Einstein specified this not only by stating that the moving clocks are "synchronous in the stationary system" but also by adding that "'time' here denotes 'time of the stationary system' and

letter in German: Albert Einstein Archives, at the Einstein Papers Project, California Institute of Technology, Pasadena, item 18-255.

[66] Einstein, "Zur Elektrodynamik bewegter Körper," p. 896.

also 'position of hands of the moving clocks situated at the place under discussion,'" and, with regard to the moving clocks, that "their indications correspond at any instant to the 'time in the stationary system' at places where they happen to be."[67] Until this point, Einstein's argument remains within the confines of Newton's mechanics. By contrast, in the kinematics that Einstein was about to formulate, one would not expect the indications of time on "moving" clocks to agree with clocks on the stationary system.

Given the importance of the relativity of simultaneity in the development of modern kinematics, we now review Einstein's original argument. As Arthur I. Miller once noted, "Einstein's demonstration in §2 of the relativity of simultaneity is rarely analyzed."[68] Because Einstein's audience consisted mostly of physicists concerned with electromagnetic theory, it seems that "most readers skipped over Part I with its discussion on the nature of space and time, in order to reach the more familiar material in the electrodynamical §§ 9 and 10."[69] Miller commented, "In the context of the physics of 1905, which strove for a unified field theoretical description of matter in motion, the kinematical part of the relativity paper could only have appeared inelegant, pedantic, and beside the point. And indeed it was for most physicists."[70] This, of course, is not to say that Einstein's paper is rarely read, or that his definition of simultaneity is rarely analyzed. In fact, Einstein's paper is one of the most widely reprinted works in the history of physics and thus has been read many times. Moreover, Einstein's formulation of the definition of simultaneity was a focus of analysis among philosophers of science for decades. Nonetheless, Einstein's 1905 demonstration of the relativity of simultaneity has not been analyzed as much.[71]

---

[67] Ibid.

[68] Miller, *Albert Einstein's Special Theory of Relativity*, p. 204.

[69] Ibid., p. 192.

[70] Arthur I. Miller, *Imagery in Scientific Thought: Creating 20th-Century Physics* (Cambridge, Mass.: Birkhäuser Boston, 1984, and MIT Press, 1986), p. 87.

[71] This lack of analysis of the original argument is an odd gap in the otherwise extensive literature on Einstein's 1905 paper. For example, it is surprising that Torretti's thorough study of 1983 skips exactly over Einstein's demonstration of the relativity of simultaneity. At any rate, a brief analysis of the matter is provided by Miller in his 1981 study, *Albert Einstein's Theory of Relativity*. In addition, the demonstration is summarized sketchily in A. Kopff, *The Mathematical Theory of Relativity* (1921; repr., London: Methuen, 1923); and in Abraham Pais, *"Subtle is the Lord . . .": The Science and the Life of Albert Einstein* (New York: Oxford University Press, 1982). Also, the demonstration is

In Einstein's 1905 paper, the question that facilitated his abandonment of traditional kinematics was whether moving clocks are synchronous as judged by observers traveling along with them. This is determined, according to Einstein's definition of simultaneity, by performing the operational procedure of light-signal synchrony on the clocks on the moving rod. If a light flash departs from one of these clocks at its clock reading of t_A, then reaches the other clock at its particular reading of t_B, then is reflected back in the opposite direction and arrives at the first clock at its time t'_A, these two clocks would be synchronous if

$$t_B - t_A = t'_A - t_B.$$

Einstein next showed that this equation is not valid for observers moving along with the rod. A complication arises because of the rod's state of motion. Its velocity of translation v, in conjunction with the independence of the speed of light from its source, accounts for a difference between the time light takes to travel from one end of the rod to another and the time of the return trip. Letting rAB stand for the length of the moving rod as measured by an observer in the stationary system, a ray of light traveling from one end of the rod to the other would have to move an extra distance $v(t_B - t_A)$ in addition to the rod's length, because the end point B of the rod moves away as the light ray approaches. Then, upon reflection at B the light ray would have to travel a distance *less* than the length of the rod because the end point A approaches the returning light ray. The travel times for the light ray moving in each direction hence are given by

$$t_B - t_A = \frac{rAB + v(t_B - t_A)}{c} \quad \text{and} \quad t'_A - t_B = \frac{-rAB + v(t'_A - t_B)}{-c}.$$

Notice that, as Einstein required, the magnitude of the velocity of light here is the same in opposite directions: c. Thus, Einstein expressed the time intervals that light takes to travel from clock to clock as

$$t_B - t_A = \frac{rAB}{c - v} \quad \text{and} \quad t'_A - t_B = \frac{rAB}{c + v}.$$

illustrated in Banesh Hoffmann, *Relativity and Its Roots* (New York: Scientific American Books, W. H. Freeman, 1983).

These equations are obtained from the preceding ones simply by solving for each time interval. Now, obviously,

$$t_B - t_A \neq t'_A - t_B,$$

that is, as judged by observers moving along with the rod, the clocks on the rod would not be synchronous. Given Einstein's definition of simultaneity, clocks that are synchronous in one frame of reference are not synchronous in another one moving relative to the first. Thus, in ending the second section of his paper, Einstein stated that "we cannot attach any *absolute* signification to the concept of simultaneity, but two events that, viewed from a system of coordinates, are simultaneous can no longer be looked upon as simultaneous events when envisaged from a system which is in motion relatively to that system."[72] This concept became known as "the relativity of simultaneity."

Einstein's analysis of simultaneity was intimately connected to the two principles of his theory. Owing to his principle of relativity, simultaneity could be deemed relative precisely because no reference frame was considered unique. Different determinations of simultaneity in different systems therefore could be deemed equally valid. The connection to the principle of the constancy of the speed of light was no less direct, though perhaps it was not as clear to some readers. For instance, Stanley Goldberg, a historian who studied the early reception of Einstein's theory, wrote that "from the response to the paper, it seems to have been unclear to most readers why the discussion of simultaneity followed the statement of the two postulates."[73] Goldberg identified the second postulate as the root of such ambiguities.

Before dealing with Goldberg's point, his words should be clarified because they do not convey adequately the actual sequence in Einstein's paper. Einstein's first statement of his two postulates appears in his introduction; this is followed by the definition of simultaneity in section 1; the second section then begins with the exact statement of the postulates and ends with the statement of the relativity of simultaneity. Now consider why Goldberg suggested that the sequence in question seemed ambiguous to some readers: "One reason might have been that Einstein did not make a con-

---

[72] Einstein, "Zur Elektrodynamik bewegter Körper," p. 897.

[73] Stanley Goldberg, *Understanding Relativity: Origin and Impact of a Scientific Revolution* (Boston: Birkhäuser, 1984), p. 106.

nection between the second postulate and the definition which followed." This statement is peculiar because this connection is made explicitly in Einstein's paper. Yet, it is presumably representative of similar ambiguities and thus deserves clarification.

The connection between the definition of simultaneity and the second postulate is evident from Einstein's exact formulation of that postulate. Specifically, the speed of light is defined as the ratio of the light path to the time interval (i.e., the distance light travels in a given direction, divided by the time light takes to travel that distance). For this purpose, Einstein immediately added that the "time interval is to be understood in the sense of the definition in §1." Thus, the definition of simultaneity entered directly into the principle of the constancy of the speed of light, for otherwise ambiguity would remain as to how the velocity of light is to be determined.

By Einstein's method, clocks are synchronized using light signals, and hence the speed of light in any direction is fixed by this procedure. The constancy of the speed of light—and, in fact, the determination of any velocity—is contingent on specific synchrony and measurement procedures. In a review of relativity theory of 1907, Einstein explained the constancy of the speed of light by emphasizing: "We now assume that the clocks can be adjusted in such a way that the propagation speed of every light ray in a vacuum—measured by these clocks—becomes everywhere equal to a universal constant c, provided that the coordinate system is not accelerated."[74] Because the definition of simultaneity involves the claim that light signals propagate with equal speeds in opposite directions, this claim enters into the content of the light principle. In this manner, although not explicit in the formulation of the light principle, it is clear that the speed of light is not only independent of its source but also constant irrespective of direction. Thus, the connection between the definition of simultaneity and the light principle is that the former is presupposed by the latter.

The connection between the light principle and the *relativity* of simultaneity remains to be clarified. This connection also may be identified by a straightforward reading of Einstein's paper. Towards the end of section 2, where Einstein investigated the synchrony of moving clocks, he obtained the equations of the relativity of simultaneity by "taking into consideration the principle of the constancy of the speed of light."[75] This operation may be

[74] Einstein, "Über das Relativitätsprinzip" (1907), p. 415; *Collected Papers*, vol. 2, p. 437.
[75] Einstein, "Zur Elektrodynamik bewegter Körper," p. 896.

explained by remarking that the principle simply warrants that the speeds of the light signals sent between moving clocks are independent of the speeds of the sources (in case the sources travel along with the clocks), such that the signals travel at the same speed as those used to synchronize "stationary" clocks. For the stationary system, the "moving" clocks could not be synchronized by the light signals because these clocks were moving during the procedure. Thus, observers on the "stationary" system would claim that the synchronization procedure works on their system but not for the moving clocks. In short, Einstein's conclusion that simultaneity is relative followed from the constancy of the speed of light.

Some writers have claimed that the relativity of simultaneity is a consequence of Einstein's second postulate, while referring not to the second postulate as it appears in section 2 of the 1905 paper but instead to a stronger version of the postulate. This version, commonly used by Einstein and others after 1907, is the statement that the speed of light in a vacuum is the same in all nonaccelerated reference systems. This formulation conveys succinctly the invariance of the speed of light, whereas in Einstein's axiomatic treatment of 1905 this invariance is derived from the light principle in conjunction with the relativity principle. For the present, the importance of the distinction between these two formulations of the light principle (and their relationships to the relativity of simultaneity) is that the stronger version of the principle is *incompatible* with classical physics, and thus it is incoherent to contend that, from that version, the relativity of simultaneity is demonstrated as a problem *within classical physics*. The demonstration that there is something intrinsically wrong with the concepts of classical mechanics was carried out by applying its usual concepts. In this sense, one must understand that the present argument (that the relativity of simultaneity results from "taking into consideration" the light principle) means taking into account the independence of the speed of light from the speed of the source in the "stationary" system. This formulation falls well within the bounds of classical physics, and, from it, the relativity of simultaneity follows. From this relativization of time, Einstein derived his theory. In this way, it properly can be said that relativity theory was reached "through internal criticism of the preceding theory."[76]

Einstein's arguments for the relativity of simultaneity played a crucial role in his formulation of the special theory, as well as in physicists' assess-

---

[76] Torretti, *Creative Understanding: Philosophical Reflections on Physics*, p. x.

ment of the value of his theory over classical mechanics. The importance of the relativity of simultaneity was not only that it constituted a new concept with which to make sense of a problematic situation in physics. It also represented the discovery of a deficiency in the concepts of traditional kinematics. Einstein's accomplishment consisted of a critical demonstration that a fundamental element of the classical scheme, namely, the concept of a common, universal, or "absolute" time, was arbitrary and unwarranted. It remained unsubstantiated even *in principle* in terms of physical processes. This realization, in turn, called for the establishment of a physically meaningful notion of "time" that legitimated the relativistic conception. As Torretti argued, "Einstein's point at the beginning of 'Zur Elektrodynamik bewegter Körper,' § 1, is not that the classical conception of physical time is *wrong* or *inconvenient* and therefore ought to be replaced by his, but rather that classical kinematics does not have a definite notion of time sufficient to determine the relations of simultaneity and succession between distant events. Einstein does not give any reason for *modifying* a given time concept but proceeds to *establish* one where none so definite and far-reaching was yet available."[77] As soon as this deficiency in the classical framework was clearly recognized, Einstein devised a solution to the problems of the current theories of electrodynamics, a solution based on his careful definition of the concept of time.

A Tortuous Path to the Lorentz Transformations

Section 3 of Einstein's 1905 paper, "Theory of Transformation of Coordinates and Times from a Resting System to a System in Uniform Translational Motion Relative to It," consists of a kinematic derivation of the equations that serve to transform values of coordinates from one system to another. Analyses of Einstein's derivation suffer from errors that need correction. The present elucidation clarifies ambiguities that, for decades, deterred even specialists from understanding the derivation. It also exposes difficulties that Einstein's contemporaries may have experienced in attempting to understand his work.

Historians have remarked that the mathematics employed by Einstein was rather simple compared to the analytical methods then employed by leading theorists in electrodynamics. In the words of Russell McCorm-

[77] Ibid., p. 48.

mach, Einstein "was able to carry through a profound critique of the foundations of physics using elementary algebra, differential equations . . . and with that light mathematical equipment he was able to formulate the kinematics of special relativity."[78] But even though Einstein used only algebra and calculus in formulating his theory, his arguments were complex. The mathematical subtleties are best brought to light by reconstructing his derivations in detail. He summarized in only a few equations the results of hundreds of operations. His omission of intermediate steps may have obscured the intelligibility of his kinematics, as has happened with other scientific and mathematical texts throughout history.[79]

Because of its complexity, Einstein's first derivation has not been used in physics textbooks. In the *Annalen der Physik,* it occupies five pages but, once unraveled, occupies approximately thirty pages of text. Einstein presented his derivation in about fifty-five equations, but when worked out explicitly, the derivation involves more than three hundred equations, consisting of roughly five hundred algebraic and differential operations. Einstein's succinct presentation thus allowed his readers to skip many details and to skim the substance of his argument but at the expense of understanding his derivation clearly. Not every step of Einstein's derivation is given here, but for clarity more steps are written and explained than is usually the case in explanations of physicists' works. To distinguish the equations that appear explicitly in section 3 from those that are added here, we identify each of the former by parenthetical numbers preceded by the letter E.

Compared to the conceptual analysis of measurement procedures in the first two sections of Einstein's paper, his formal derivation of the transfor-

---

[78] Russell McCormmach, "Editor's Foreword," *Historical Studies in the Physical Sciences,* ed. McCormmach, vol. 7 (Princeton: Princeton University Press, 1976), xxxvi. McCormmach's expression was later modified to read "ordinary algebra . . . modest mathematical equipment," in Christa Jungnickel and Russell McCormmach, *Intellectual Mastery of Nature,* vol. 2, *The Now Mighty Theoretical Physics, 1870–1925* (Chicago: University of Chicago Press, 1986), p. 337.

[79] For example, even William Rowan Hamilton, with his extraordinary mathematical skills, encountered difficulties as a youth studying Newton's works; he commented that Newton wrote the *Universal Arithmetick* in "the same masterly manner as the *Principia,* and yet in many parts is rendered almost as difficult, by its conciseness and omission of intermediate steps." Letter of 28 September 1823, in W. R. Hamilton, *Life of Sir William Rowan Hamilton,* ed. R. P. Graves (Dublin: Hodges, Figgis & Co., 1882), vol. 1, p. 149.

mation equations was far more abstract. It followed the mathematical tradition of Joseph-Louis Lagrange, Sylvestre François Lacroix, and others who dispensed with geometric diagrams. It followed the descriptive approach espoused by Gustav Kirchhoff rather than explanatory approaches involving models of causes or mechanisms. It involved a profound reliance on formal requirements such as linearity, symmetry, and the theory of functions. It did not involve the methods or concepts of vector theory, although they had been advocated by influential physicists, such as Peter Guthrie Tait, Oliver Heaviside, and August Föppl, to replace the cumbersome methods of "Cartesian" coordinates, especially in electromagnetic theory.

The end result of section 3 of Einstein's paper was a group of four equations. To obtain them, Einstein began by positing two Cartesian coordinate systems, K and k, with rectangular axes X, Y, Z, and Ξ, H, Z, respectively. He identified these systems with rigid bodies, each consisting of three mutually perpendicular rods, as he argued that the meaning of coordinates, lengths, and times should be given by specifications pertaining to rigid bodies and clocks. To distinguish the systems, he identified K as "resting" and k as "moving" (in quotation marks). He then derived the four transformations relating the position and time coordinates, x, y, z, t, of any physical event in K to the coordinates ξ, η, ζ, τ of the same event in k. The transformations expressed the simplest relation between systems in relative motion: the case in which the axes of K are parallel to the axes of k, and the two systems move relative to one another in a straight line with a uniform speed v. By letting K and k be displaced only along the X- and Ξ-axes, the equations he found were

$$\tau = \beta(t - vx/V^2),$$
$$\xi = \beta(x - vt),$$
$$\eta = y,$$
$$\zeta = z,$$

where $\beta = 1/\sqrt{1 - v^2/V^2}$,   and V is the speed of light in empty space, which we henceforth designate by c, as already done in 1905 by some physicists. (For ease of reference to the 1905 paper, all other symbols are in Einstein's original notation.)

These four equations replaced the equations previously used by physicists to transform coordinates between systems in uniform rectilinear relative motion:

$$\tau = t,$$
$$\xi = x - vt,$$
$$\eta = y,$$
$$\zeta = z.$$

These equations were seldom stated explicitly; in particular, there was no need to express the equation relating the time coordinates, because time was assumed to be the same in all reference systems regardless of relative motions. Before the 1880s, physicists used a single variable t for all such systems. The other three equations, by contrast, were stated explicitly at least occasionally, as done by Lorentz, for example, in his 1886 paper "On the influence of the Earth's motion on luminous phenomena."[80] As a result of his research on relative motion in optics and electromagnetics, he advanced a series of modifications to the traditional transformations that eventually led to the equations advocated by Larmor, Poincaré, Einstein, and others.[81] Hence, Poincaré gave the name "Lorentz transformations" to these new equations, although Woldemar Voigt had published equivalent

[80] H. A. Lorentz, "De l'influence du mouvement de la terre sur les phénomènes lumineux," *Verslagen van de gewone vergaderingen der wis-en natuurkundige afdeeling, Koninklijke Akademie van Wetenschappen te Amsterdam* 2 (1886); also in *Archives Néerlander* 21 (1887); reprinted in H. A. Lorentz, *Collected Papers*, 9 vols. (The Hague: Martinus Nijhoff, 1937), vol. 4, pp. 153–214.

[81] H. A. Lorentz, *Versuch einer Theorie der elektrischen und optischen Erscheinungen in bewegten Körpern* (Leiden: E. J. Brill, 1895); reprinted in Lorentz, *Collected Papers*, vol. 5, pp. 1–137. Joseph Larmor, *Aether and Matter: A Development of the Dynamical Relations of the Aether to Material Systems on the Basis of the Atomic Constitution of Matter; including a Discussion of the Influence of the Earth's Motion on Optical Phenomena* (Cambridge: Cambridge University Press, 1900). Hendrik A. Lorentz, "Electromagnetische verschijnselen in een stelsel dat zich met willekeurige snelheid, kleiner dan die van het licht, beweegt," *Koninklijke Akademie van Wetenschappen te Amsterdam. Wis-en Natuurkundige Afdeeling. Verslagen van de Gewone Vergaderingen* 12 (1904), 986–1009; reissued as "Electromagnetic phenomena in a system moving with any velocity smaller than that of light," *Koninklijke Akademie van Wetenschappen te Amsterdam, Section of Sciences, Proceedings* 6 (1904), 809–831. Henri Poincaré, "Sur la dynamique de l'électron," *Comptes rendus de l'Académie des Sciences* [Paris] 40 (1905), 1504–1508; reprinted in H. Poincaré, *Oeuvres*, vol. 9, preface by Louis de Broglie (Paris: Gauthier-Villars, Éditeur-Imprimeur-Libraire, 1954), 489–493. Henri Poincaré, "Sur la dynamique de l'électron," *Rendiconti del Circolo matematico di Palermo* 21 (1906), 129–176; reprinted in *Oeuvres* 9 (1954), pp. 494–550.

equations in 1887.[82] In 1909 the simpler and older transformation equations were named the "Galilean transformations" by Philipp Frank.[83] What distinguished the new transformations in Einstein's work in comparison to the equivalent equations in the earlier work of other physicists was that Einstein introduced such transformations by means of general kinematic arguments, rather than introducing them exclusively for the solution of problems in optics and electrodynamics.

For simplicity, Einstein derived the four transformation equations given only the relative motion of the two systems along the X-axis and $\varXi$-axis. Thus, only the relation between the coordinates x and ξ, and between t and τ, would be expected to vary. To visualize the systems in relative motion, we may suppose that at the initial time t their coordinate axes coincide. Einstein began: "First of all it is clear, that the equations must be *linear* on account of the properties of homogeneity that we attribute to space and time."[84] This requirement can be expressed by the equations:

$$\tau = a_{11}x + a_{12}y + a_{13}z + a_{14}t,$$
$$\xi = a_{21}x + a_{22}y + a_{23}z + a_{24}t,$$
$$\eta = a_{31}x + a_{32}y + a_{33}z + a_{34}t,$$
$$\zeta = a_{41}x + a_{42}y + a_{43}z + a_{44}t.$$

That is, each coordinate of k is a function of the coordinates of K and four undetermined constants. At any one time, any particular value of a coordinate in K corresponds to only one value of a coordinate in k (otherwise, had Einstein allowed the equations to be quadratic, then to each coordinate value of K there could correspond two coordinate values of k, as, for example, $\eta = y^2$ yields two possible values for η).

By the homogeneity of space and time, Einstein presumably meant that no locations or directions in physical space are distinct or privileged and

[82] W. Voigt, "Ueber das Doppler'sche Princip," *Königliche Gesellschaft der Wissenschaften und der Georg-Augusts-Universität zu Göttingen. Nachrichten* (1887), 41–51; reprinted in *Physikalische Zeitschrift* **16** (1915), 381–386.

[83] P. Frank, "Die Stellung des Relativitätsprinzips im System der Mechanik und der Elektrodynamik," *Sitzungsberichte der Kaiserlichen Akademie der Wissenschaften. Mathematisch-Naturwissenschaftliche Klasse* [Vienna], sec. IIa, **118**, no. 4 (1909), 373–446.

[84] Einstein, "Zur Elektrodynamik bewegter Körper," p. 898; emphasis in the original.

that time likewise has a certain uniformity. For instance, if one were to place two identical measuring rods end to end, anywhere in space and at any time, all observers would agree that the result would be twice the length of one such rod. All points fixed in one nonaccelerated system move with the same velocity relative to another nonaccelerated system, and the transformations between systems must be indifferent to the choice of origin for either system.[85] In any case, Einstein did not offer such specific arguments to justify the linearity of the transformation equations.

Einstein proceeded: "We set

$$x' = x - vt, \tag{E1}$$

so it is clear that to a point resting in the system k belongs a definite system of values x', y, z, independent of time."[86] Equation (E1) is identical to the Galilean transformation $\xi = x - vt$.[87] The primed notation, as in x', was used occasionally by some writers to designate coordinates in an additional coordinate system.[88] So it might seem that Einstein had introduced a third coordinate system, because he used x' instead of ξ for the coordinate corresponding to x. Accordingly, in his analysis of Einstein's paper, Miller argued that Einstein had introduced a third set of coordinates, an "intermediate Galilean system," in addition to K (x, y, z, t) and k (ξ, η, ζ, τ). Miller claimed that Einstein related the coordinates in K and k "through an auxiliary set of space and time coordinates in k $(x', y'$ $(= y)$, z' $(= z)$, t' $(= t))$ whose spatial portion x', y, z transforms according to the Galilean transformations; but

---

[85] For a rigorous discussion of the connection between homogeneity and linearity, see Torretti, *Relativity and Geometry* (1983), pp. 71–76. Torretti systematically clarifies the matter because he notes that at first "it is not at all clear that the linearity of the Lorentz transformations can follow from this fact alone" (p. 72), the homogeneity of space and time. Torretti shows that the linearity can be justified on physical grounds by appealing to the principle of inertia. Also, Olivier Darrigol suggests (personal communication) that by homogeneity Einstein presumably meant that a straight line should transform into a straight line, and that a rectilinear uniform motion should transform into another of the same kind.

[86] Einstein, "Zur Elektrodynamik bewegter Körper," p. 898.

[87] Equation (E1) has been identified as a transformation; see, e.g., Torretti, *Relativity and Geometry*, p. 58.

[88] See, for instance, H. A. Lorentz, *Lehrbuch der Differential- und Integralrechnung nebst einer Einführung in andere Teile der Mathematik* (1882), revised by G. C. Schmidt (Leipzig: J. A. Barth, 1900), p. 83.

every time coordinate is relativistic."[89] Likewise, Roberto Torretti claimed that Einstein introduced "an auxiliary coordinate system," although Torretti argued that it would be wrong to describe it "as a Galilei coordinate system."[90] Despite such claims, there is hardly anything in Einstein's paper to indicate that he introduced such an auxiliary system. Einstein did not refer to such a system, nor did he introduce the terms y', z', t'. Moreover, he used the term x' alongside the terms x, y, c, v, and t, which he repeatedly identified as values of system K, as he emphasized, for example, that t "always denotes a time of the resting system." Einstein did introduce, explicitly, a third coordinate system later in the paper. But only (E1) can be interpreted as suggesting that a third system was involved from the outset.

The interpretation of x' as indicative of a third system creates difficulties for the validity of Einstein's arguments. Above all, how could Einstein expect to *derive* new transformation equations on the basis of a traditional transformation? This problematic question is of basic importance for understanding the transition from classical kinematics to Einstein's kinematics. Also, why would Einstein assume the validity of his principle of the constancy of the speed of light only *midway* through his derivation? Moreover, why would he tacitly assume outright that $y' = y$ and $z' = z$, without explanation, whereas he then spent pages demonstrating that $\eta = y$ and $\zeta = z$, rather than likewise assuming these relationships? Finally, toward the end of his derivation, when Einstein explicitly did introduce the terms x', y', z', t' as coordinates of a system, he used them as the coordinates of a system *identical* with K. So why would he use this very notation if he had first used it to designate coordinates of system k?

Such problems disappear once we realize that (E1) need not be interpreted as a transformation equation and that it may have an entirely different meaning. Specifically, it can describe the uniform motion of an object departing from a coordinate x', and moving along the X-axis of K with a velocity v, such that it is located at a coordinate x at time t. This relation is a basic equation in kinematics: $x = x' + vt$, sometimes stated as $x = x_0 + ut$, as for example in Kirchhoff's *Mechanik* (a work studied by Einstein), where it

---

[89] Miller, *Albert Einstein's Special Theory of Relativity* (1981), p. 196; see also p. 213. The claim that x', y', z' are coordinates of the moving system also was made in Stachel, *Einstein's Miraculous Year* (1998), p. 160, n. 2.
[90] Torretti, *Relativity and Geometry*, pp. 57–58.

designated the rectilinear motion of a point just before Kirchhoff discussed the subject of transformation equations.[91] In this context, Einstein's use of (E1) may be reinterpreted. He wrote: "We set $x' = x - vt$, so it is clear that to a point resting in the system k belongs a definite system of values x', y, z, independent of time." This "point resting in the system k" moves at a velocity v in K; because it would depart from the position x', y, z and reach the position x, y, z, after a time t, the value x would vary with time, whereas x' would not. Thus (E1) can be understood as describing the uniform rectilinear motion of a point along the X-axis of K. In this sense, it does not imply a transformation between reference frames. Yet this interpretation, too, is defective, as we will see.

In any case, the point "resting" in k is described in terms of the values x', y, z of the system K. But why did Einstein not describe the position of the moving point with the values x, y, z instead? By employing the value x' instead of x, he could hope to simplify his derivation of the function $\tau(x'$, y, z, $t)$, because then all terms, x', y, z, are constant, that is, "independent of time." Could Einstein have obtained the same transformation equations had he used x instead of x' throughout?

To ascertain the form of the transformation equations, Einstein began by seeking the relation between t and τ: "We first determine τ as a function of x', y, z, and t. To this end, we have to express in equations, that τ is nothing other than the embodiment of the data of clocks resting in system k, which have been synchronized according to the rule given."[92] Thus, he based his derivation of the transformation equations on his procedure for synchronizing clocks: "From the origin of the system k a light ray was sent out at the time τ_0 along the X-axis toward x', and from there at the time τ_1 is reflected back to the coordinates-origin, where it arrives at the time τ_2; thus it must then be: $\frac{1}{2}(\tau_0 + \tau_2) = \tau_1$." Note that because Miller interpreted x' as a coordinate of k, he stated that Einstein's X in the passage just quoted "was an oversight" and that it should have been Ξ instead, "for consistency."[93] But if we interpret x' otherwise, it is not necessary to change Einstein's X; the light ray emitted from the origin of k is described with respect to K.

To relate the values of the function τ to the coordinate values of system

---

[91] Kirchhoff, *Vorlesungen über Mathematische Physik: Mechanik* (1883), p. 4.

[92] Einstein, "Zur Elektrodynamik bewegter Körper," p. 898.

[93] Miller, *Albert Einstein's Special Theory of Relativity,* p. 397.

K, Einstein expressed τ_0, τ_1, and τ_2 in terms of the corresponding values of x, y, z, and t of K:

$$\frac{1}{2}\left[\tau\left(0,0,0,t\right)+\tau\left(0,0,0,t+\frac{x'}{c-v}+\frac{x'}{c+v}\right)\right]=\tau\left(x',0,0,t+\frac{x'}{c-v}\right). \qquad \text{(E3)}$$

To explain (E3) we must clarify an ambiguity: the expressions $c-v$ and $c+v$ seem to indicate variations in the speed of light. Miller, for example, interpreted them in this way.[94] Yet Einstein's analysis led to the conclusion that the speed of light relative to k should be c. Thus, some writers later complained that he could not rightly derive this conclusion by letting light in k propagate at speeds of $c-v$ and $c+v$.[95] Likewise, these expressions are not speeds of light relative to the "resting" system, because Einstein specified from the beginning that the speed of light in K was c. So what do these expressions mean?

The question is answered by clarifying the algebraic analysis Einstein presented in section 2 of his paper where he used the equations

$$t_B-t_A=\frac{rAB}{c-v} \qquad \text{and} \qquad t'_A-t_B=\frac{rAB}{c+v}. \qquad \text{(E0)}$$

Here are the apparent variations in the speed of light. As demonstrated before, these equations are simplifications of the equations

$$t_B-t_A=\frac{rAB+v(t_B-t_A)}{c} \qquad \text{and} \qquad t'_A-t_B=\frac{-rAB+v(t'_A-t_B)}{-c}.$$

Here the speed of light is constant: c. Thus we see that the expressions $c-v$ and $c+v$ did not enter Einstein's argument as speeds of light relative to K or

[94] Ibid., p. 209. Moreover, throughout the decades some writers mistakenly criticized Einstein's demonstration of the relativity of simultaneity and his derivations of the Lorentz transformations, claiming that Einstein had violated his postulate of the constancy of the speed of light by employing speeds other than c in his arguments.

[95] E.g., Hans Strasser, *Einsteins spezielle Relativitätstheorie: eine Komödie der Irrungen* (Bern: Bircher, 1923); G. H. Keswani, "Origin and Concept of Relativity, Part II," *British Journal for the Philosophy of Science* **16** (1965), 19–32; Karl Stiegler, "On Errors and Inconsistencies in Einstein's 1905 Paper 'Zur Elektrodynamik bewegter Körper,'" *Proceedings of the XIIIth International Congress of the History of Science, Moscow, 1971* (Moscow: Naouca, 1974), sec. 6, pp. 53–63.

k, but as *the rates with which light approaches moving points relative to K.* That is, $c - v$ is the rate of light approaching clock *B*, and $c + v$ is its rate of approach toward clock *A*. Such expressions do not imply any transformation but rather the usual vectorial addition of velocities within a single frame, which is valid exactly in both traditional and Einstein's kinematics. This important distinction between the addition of velocities in one system and the transformation of velocities between two systems is often neglected.[96]

Einstein used equations (E0) and (E1) to establish the value of each time coordinate in (E3). As stated, he needed to ascribe definite values to the coordinates in *K* of the three events: the incidence of a light ray on clock *A*, then on clock *B*, and then back to *A*, given that the two clocks are moving along the *X*-axis. So we have

$$\frac{1}{2}\left[\tau(x_0,y_0,z_0,t_0)+\tau(x_2,y_2,z_2,t_2)\right]=\tau(x_1,y_1,z_1,t_1). \tag{1}$$

Because (E1) is valid for any time *t*, Einstein used the same symbol *t* to designate the arbitrary time of departure of the light ray from *A*—that is, he set

$$t_0 = t.$$

By letting $x' = rAB$, equations (E0) serve to establish the values of t_1 and t_2. The time t_0 corresponds to t_A and, likewise, $t_1 = t_B$. Thus, Einstein found that the time of arrival of the light ray at *B* is

$$t_1 = t + \frac{x'}{c-v},$$

and hence the total time taken by the light to return to clock *A* is

$$t_2 = t + \frac{x'}{c-v} + \frac{x'}{c+v}.$$

Miller's procedure[97] to obtain these results for t_0, t_1, and t_2 differs from the one here, because instead of using Einstein's equation (E1), he used

---

[96] Some thorough books do not neglect this important distinction. See, e.g., Henri Arzèlies, *Relativistic Kinematics* (Oxford: Pergamon, 1966), pp. 142–143.
[97] Miller, *Albert Einstein's Special Theory of Relativity*, p. 209.

$x = x' + vt_1$. Anyhow, given these results for t_0, t_1, and t_2, Einstein wrote (E3). If we expect (E3) to agree with (1), then it should give the values of the function τ in terms of the values of the coordinates in K for the three events in question. However, (E3) does not agree with (1).

Consider again the process of synchronization of moving clocks as observed from the "resting" system K. One clock is attached to the origin of system k, and the other clock is a distance away on the Ξ-axis, such that both clocks move uniformly with k, along the X-axis of K. If the coordinate systems coincide at the instant when the light ray is emitted, the ray departs from the origin of both K and k. It then travels to the distant clock and is reflected back to the first. It thus returns to the origin of k, because this process aims to synchronize the moving clocks. Because the origin of k has moved away from the origin of K, then the x-coordinate of the returning light ray cannot be 0 in K. Thus, it is problematic to expect that $x_2 = 0$ is the coordinate value in K for the location of the ray on its return. Likewise, if we interpret Einstein's x' as the location of the ray on its arrival at clock B, we also run into difficulties, because the light ray did not travel only a distance x' to B but an additional distance because the clock moved away.

In short, the problem is that whereas the values given for the time coordinates t_0, t_1, t_2 are values referred to K, the values for the coordinates x_0, x_1, x_2 seem to refer to k. Normally, a statement of coordinates in parenthesis, such as (x_1, y_1, z_1, t_1), was understood to designate values of the coordinates in a *single* system. In contrast, Einstein seems to have mixed values from two different systems. This ambiguity poses interpretive difficulties. We can eliminate them by disregarding x' and determining τ (x, y, z, t) strictly in terms of x. At this point, however, there is still one way out that makes sense of (E1)—namely, to give x' another interpretation, making it stand neither for a coordinate of the moving system nor for the fixed initial coordinate of a moving point.

Suppose that at the initial time t_0, the origins of K and k do not coincide but instead are separated by a distance vt_0. In this case, if we let x_A represent the varying position of clock A, we have

$$x_A - vt = x'_A = 0.$$

That is, for any time t we define a constant, $x'_A - 0$, that describes the separation of clock A from the moving origin of k along the X-axis of K. (Because clock A is attached to the origin of k, that constant is 0.) Likewise, if x_B represents the varying positions of clock B,

$$x_B - vt = x'_B = x'.$$

That is, for any time t we define a constant $x'_B = x'$ that describes the constant separation of clock B from the moving origin of k. Accordingly, for the times of emission of the light ray from A, its arrival at B, and its return back to A, we have:

$$x_A - vt_0 = x'_A = 0,$$
$$x_B - vt_1 = x'_B = x',$$
$$x_A - vt_2 = x'_A = 0.$$

The values 0, x', 0, serve to explain the first terms in each parenthesis in (E3). This explanation implies that we may understand (E1) as meaning that $x'_s = x_{clock} - vt$, where x'_s is any constant expressing the separation (in K) between any fixed clock on k and the origin of k at any given time. Thus, Einstein's definition of x' as describing a point at rest in k with a value independent of time, should refer not just to one point but to any point fixed on k, in particular, to the two points where the clocks are attached. Hence, to be more precise about his notation, Einstein could have stated that he sought a function $\tau(x'_s, y, z, t)$, rather than just $\tau(x', y, z, t)$, because the latter corresponds only to the events at clock B. Likewise, his statement that the light ray travels "along the X-axis toward x', and from there" is reflected back, should not be interpreted literally as meaning that x' is the coordinate of the light ray when it arrives at clock B, because that coordinate is actually $x_B = x' + vt_1$. Rather, to make sense of his expressions, we have to understand him as meaning to say that the light ray is reflected from the constant point of k where clock B is fixed.

Note that to interpret the text consistently, we have gone a slight distance away from giving it a completely literal reading. Only by doing so can we arrive at a consistent explanation of Einstein's analysis. Hence, we may surmise that the subtle ambiguities involved may have caused some difficulties in the understanding of his 1905 argument, especially for readers not sufficiently careful in kinematics, and perhaps for anyone who did not have a good sense beforehand of the end result of the derivation.

Regardless, we set such considerations aside to continue analyzing the derivation. Next, Einstein reformulated (E3) as a partial differential equation. He wrote, "If x' is chosen infinitesimally small,"[98]

$$\frac{1}{2}\left(\frac{1}{c-v}+\frac{1}{c+v}\right)\frac{\partial\tau}{\partial\tau}=\frac{\partial\tau}{\partial x'}+\frac{1}{c-v}\frac{\partial\tau}{\partial t}.$$ (E4)

To understand how to obtain equation (E4) from (E3), we may proceed as follows. Einstein sought differential equations that would express the dependence of the function τ on x', y, z, and t. To do so, he had to find the dependence of each of the values of τ, that is, τ_0, τ_1, τ_2, on each of the corresponding values of x', y, z, and t. He first sought to express their functional dependence on x'. We may begin by making an infinite series expansion of the function τ_1, for example, according to Taylor's theorem,

$$\tau_1 = \tau(x',t+hx') = \tau(x',t) + \frac{\partial\tau}{\partial t}\frac{hx'}{1} + \frac{\partial^2\tau}{\partial t^2}\frac{(hx')^2}{1\cdot2} + \frac{\partial^3\tau}{\partial t^3}\frac{(hx')^2}{1\cdot2\cdot3} + \ldots,$$

where we have let h stand for $1/(c-v)$. We expect, as usual, that with each successive term of this Taylor polynomial its summed total converges toward the value of the function τ_1, as its limit, which is why we assert the equality.[99] Now, by differentiating the function and the series with respect to x',

$$\frac{\partial}{\partial x'}\tau(x',t+hx') = \frac{\partial}{\partial x'}\tau(x',t) + \frac{\partial}{\partial x'}\left(\frac{\partial\tau}{\partial t}\frac{hx'}{1}\right) + \frac{\partial}{\partial x'}\left(\frac{\partial^2\tau}{\partial t^2}\frac{(hx')^2}{2!}\right) + \ldots,$$

we get

$$\frac{\partial\tau(x',t+hx')}{\partial x'} = \frac{\partial\tau}{\partial x'} + \frac{\partial\tau}{\partial t}h + \frac{\partial^2\tau}{\partial t^2}x'h^2 + \ldots.$$

[98] Einstein, "Zur Elektrodynamik bewegter Körper," p. 899.

[99] Taylor's theorem is explained in most introductory calculus books. For a contemporary account see, e.g., Lorentz, *Lehrbuch der Differential- und Integralrechnung* (1900), pp. 346–365. Einstein may have first learned it from the 4th edition (1869) of H. B. Lübsen's textbook (among Einstein's books); e.g., see Lübsen, *Einleitung in die Infinitesimal-Rechnung (Differential- und Integral-Rechnung). Zum Selbstunterricht. Mit Rücksicht auf das Nothwendigste und Wichtigste* (Leipzig: Friedrich Brandstetter, 1862), pp. 43–50, 115–118, 136–138, etc.

Because we assume x' to be infinitesimally small and, likewise, because the large value of c makes the increasing powers of h extremely small, we neglect terms of second order and higher, to get

$$\frac{\partial \tau_1}{\partial x'} = \frac{\partial \tau}{\partial x'} + \frac{\partial \tau}{\partial t}\left(\frac{1}{c-v}\right),$$

giving to h its value. By carrying out the same procedure for the other side of (E3), and by expressing the results jointly, we get

$$\frac{1}{2}\left[(0) + \frac{\partial \tau}{\partial t}\left(\frac{1}{c-v} + \frac{1}{c+v}\right)\right] = \frac{\partial \tau}{\partial x'} + \frac{\partial \tau}{\partial t}\left(\frac{1}{c-v}\right),$$

which Einstein wrote as

$$\frac{1}{2}\left(\frac{1}{c-v} + \frac{1}{c+v}\right)\frac{\partial \tau}{\partial t} = \frac{\partial \tau}{\partial x'} + \frac{1}{c-v}\frac{\partial \tau}{\partial t}, \tag{E4}$$

and which may be simplified as follows:

$$\frac{1}{2}\left(\frac{c+v}{c^2-v^2} + \frac{c-v}{c^2-v^2}\right)\frac{\partial \tau}{\partial t} = \frac{\partial \tau}{\partial x'} + \frac{c+v}{c^2-v^2}\frac{\partial \tau}{\partial t}$$

$$\frac{\partial \tau}{\partial x'} + \frac{v}{c^2-v^2}\frac{\partial \tau}{\partial t} = 0. \tag{E5}$$

Immediately after this result, Einstein stated: "It is to be noted, that instead of the coordinates-origin we could have chosen any other point as starting-point of the ray, and the equation just obtained is therefore valid for all values of x', y, z."[100] This meant, for example, that if the light ray had been emitted from the point $(q + vt, r, s)$ instead of the point $(vt, 0, 0)$, then the expression

$$\frac{1}{2}\left[\tau(0,r,s,t) + \tau\left(0,r,s,t+\frac{x'}{c-v} + \frac{x'}{c+v}\right)\right] = \tau\left(x',r,s,t+\frac{x'}{c-v}\right),$$

if $x' = x_B - vt - q$, would still result in equation (E5).

[100] Einstein, "Zur Elektrodynamik bewegter Körper," p. 899.

Einstein obtained equation (E5) by seeking a partial-differential expression of the dependence of the function τ on the term x'. This equation also conveys the dependence of the function τ on the variable t. Now he had to ascertain the other partial-differential equations expressing the dependence of τ on the variables y and z. Equation (E5) characterizes the propagation of light along the Ξ-axis of k, in terms of values defined relative to the system K (namely, x', y, z, t, v, and c). Einstein proceeded to analyze the propagation of light along the other axes of k, in terms of values also defined relative to K. Hence he argued: "An analogous consideration—applied to the H- and Z-axes—yields, when we note that as observed from the resting system light always propagates along these axes with the speed $\sqrt{c^2 - v^2}$:

$$\frac{\partial \tau}{\partial y} = 0 \tag{E6}$$

$$\frac{\partial \tau}{\partial z} = 0.\text{"[101]} \tag{E7}$$

Again, readers could get the mistaken impression that light propagates in K at a speed other than c, but the ambiguity is explained as before. Anyhow, to understand how Einstein obtained equations (E6) and (E7), we may proceed as follows. Part of this procedure was traced by Miller in his study of Einstein's paper.[102]

Consider the description, in terms of K, of a light signal moving upward along the H-axis of k. Suppose, as before, that k moves uniformly with the velocity v along the X-axis of K. At the time instant t_0, the H-axis coincides with the Y-axis of K. At that very instant, t_0, a light signal is emitted from the origin, the point (x_0, y_0, z_0) of K, upward along H. It arrives at the point (x_1, y_1, z_1) at the time t_1, and then is reflected down to the point (x_2, y_2, z_2) at

[101] Ibid., p. 899. The word *Geschwindigkeit* is often translated as "velocity," though, strictly speaking, in many cases this choice leads to expressions that are literally false— for example, that Einstein claimed that all light rays relative to an inertial system have the same velocity, when really, only the magnitude of their velocities can be considered constant, that is, their "speed."

[102] Miller, *Albert Einstein's Special Theory of Relativity*, p. 198.

the time t_2. Two clocks are attached to the H-axis, at the coordinates y_0 and y_1 of K. At the times t_0, t_1, t_2 of K, the clocks attached to H mark the times τ_0, τ_1, τ_2. As before, these clocks would be synchronous if

$$\tfrac{1}{2}\,(\tau_0 + \tau_2) = \tau_1.$$

This equation again can be restated in terms of the coordinates x, y, z, t of K:

$$\frac{1}{2}\left[\tau(x_0,y_0,z_0,t_0) + \tau(x_2,y_2,z_2,t_2)\right] = \tau(x_1,y_1,z_1,t_1). \tag{1}$$

To find the specific values of the coordinates of K corresponding to τ_0, τ_1, τ_2, we first note that, because the light ray departs from the origin of K at the time t, then

$$
\begin{aligned}
x_0 &= 0, \\
y_0 &= 0, \\
z_0 &= 0, \\
t_0 &= t.
\end{aligned}
$$

Subsequently, the light signal arrives at the point $(x_1,\,y_1,\,z_1)$ at the time t_1. At this time, the H-axis has moved a distance $v(t_1 - t_0)$ along the X-axis. Thus,

$$x_1 = v(t_1 - t_0).$$

To express the value of t_1 we need to realize that, relative to K, the light ray traveled the diagonal length $c(t_1 - t_0)$ up to the coordinate y, thus forming a right triangle with the lengths y and $v(t_1 - t_0)$, such that

$$\left[c(t_1 - t_0)\right]^2 = \left[y\right]^2 + \left[v(t_1 - t_0)\right]^2.$$

If we solve for $(t_1 - t_0)$,

$$(t_1 - t_0) = y\big/\sqrt{c^2 - v^2}.$$

This last expression illustrates Einstein's single comment concerning the analysis of light propagating along the H-axis, that it travels "with the speed $\sqrt{c^2 - v^2}$." Now, given the values of t_1 and t_0, we may write:

$$x_1 = vy/\sqrt{c^2 - v^2},$$
$$y_1 = y,$$
$$z_1 = 0,$$
$$t_1 = t + y/\sqrt{c^2 - v^2}.$$

The specific values of the function τ_2 corresponding to (x_2, y_2, z_2) at the time t_2 are obtained directly from considering that the light ray takes the same amount of time to go back down the H-axis as to go up:

$$\left(t_2 - t_1\right) = \left(t_1 - t_0\right)$$

$$t_2 = 2t_1 - t_0,$$

and, likewise, the light ray travels an extra distance $v(t_2 - t_1)$ along the X-axis, identical to the distance x_1, as it returns down to the coordinate $y = 0$. Therefore:

$$x_2 = 2vy/\sqrt{c^2 - v^2},$$
$$y_2 = 0,$$
$$z_2 = 0,$$
$$t_2 = t + 2y/\sqrt{c^2 - v^2}.$$

Finally, all of the values for the coordinates may be entered into (1):

$$\frac{1}{2}\left[\tau(0,0,0,t) + \tau\left(\frac{2vy}{\sqrt{c^2 - v^2}}, 0, 0, t + \frac{2y}{\sqrt{c^2 - v^2}}\right)\right] = \tau\left(\frac{vy}{\sqrt{c^2 - v^2}}, y, 0, t + \frac{y}{\sqrt{c^2 - v^2}}\right). \quad (2)$$

This expression gives the values of the function τ in terms of the values of coordinates in K for the three different events: the emission of a light signal, its arrival at y, and its return to the point of emission. As Miller noted in his analysis, we can expand both sides of this expression and take y as infinitesimally small, and so forth, to then arrive at Einstein's (E6). But notice that in (2), in contrast to (E3), all the values refer to coordinates in K. We can arrive at (E6) by a simpler procedure if we do not use coordinate expressions for one of the four values in each of τ_0, τ_1, τ_2, that is, by using Einstein's x' instead of the x-coordinate. Because x' is defined as $x - vt$, then for each value of x, we have:

$$x' = x_0 - v\left(t_0\right) = 0,$$
$$x' = x_1 - v\left(t_1 - t_0\right) = 0,$$
$$x' = x_2 - v\left(t_2 - t_0\right) = 0.$$

Thus, for the expression

$$\frac{1}{2}\left[\tau\left(x'_A, y_0, z_0, t_0\right) + \tau\left(x'_A, y_2, z_2, t_2\right)\right] = \tau\left(x'_B, y_1, z_1, t_1\right),$$

instead of (2), we get:

$$\frac{1}{2}\left[\tau\left(0,0,0,t\right) + \tau\left(0,0,0,t + \frac{2y}{\sqrt{c^2 - v^2}}\right)\right] = \tau\left(0,y,0,t + \frac{y}{\sqrt{c^2 - v^2}}\right).$$

In this case, we proceed just as before, by doing a series development of each side of the equation, such that, for example, the function τ_1 is expanded in a convergent series of positive integral powers of ky,

$$\frac{\partial\tau(y, t + ky)}{\partial y} = \frac{\partial\tau}{\partial y} + \frac{\partial\tau}{\partial t}\left(\frac{1}{\sqrt{c^2 - v^2}}\right),$$

where $k = 1/\sqrt{c^2 - v^2}$, and the terms of second and higher orders have been neglected. Hence,

$$\frac{1}{2}\left[(0) + \frac{\partial\tau}{\partial t}\left(\frac{2}{\sqrt{c^2 - v^2}}\right)\right] = \frac{\partial\tau}{\partial y} + \frac{\partial\tau}{\partial t}\left(\frac{1}{\sqrt{c^2 - v^2}}\right),$$

$$\frac{\partial\tau}{\partial y} = 0. \tag{E6}$$

Thus, we get, after much calculation, one of the results Einstein stated in a single line.

Einstein's expression for the independence of the τ-function on the term z is justified in the same way. The round-trip propagation of a light ray along the Z-axis of k as it moves along the X-axis of K is exactly like the propagation of light along the H-axis. Thus, by applying his definition of simultaneity to clocks attached to the Z-axis of k,

$$\frac{1}{2}\left[\tau\left(0,0,0,t\right)+\tau\left(\frac{2vz}{\sqrt{c^2-v^2}},0,0,t+\frac{2z}{\sqrt{c^2-v^2}}\right)\right]=\tau\left(\frac{vz}{\sqrt{c^2-v^2}},0,z,t+\frac{z}{\sqrt{c^2-v^2}}\right),$$

or the simpler expression, if we use x' instead of the x-coordinate, leading to

$$\frac{\partial\tau}{\partial z}=0. \tag{E7}$$

The differential equations (E5), (E6), (E7) express the dependence of τ on x', y, z, and t. Einstein now had to ascertain the exact algebraic form of the function $\tau(x, y, z, t)$. Immediately after stating equations (E5-7), he wrote, "From these equations it follows, since τ is a *linear* function:

$$\tau=a\left(t-\frac{v}{c^2-v^2}x'\right), \tag{E8}$$

where a is a function $\varphi(v)$ presently unknown, and where for brevity it is assumed that at the origin of k, for $\tau = 0$ is $t = 0$."[103] Although his words indicate $a = \varphi(v)$, this is not quite the relation he later used, so he should have stated instead that $a = \varphi(v)b$, as we will see.

Einstein gave no further indication of how he obtained the algebraic expression of τ, equation (E8), from the three preceding differential equations. No explanation is found in the secondary literature either, but one way to deduce (E8) is as follows. Given the requirement of linearity, and seeing that the τ-function is independent of y and z, we may expect the form of the function to be

$$\tau=at+gx',$$

[103] Einstein, "Zur Elektrodynamik bewegter Körper," p. 899.

which in (E5) gives:

$$\frac{\partial(at+gx')}{\partial x'}+\frac{v}{c^2-v^2}\frac{\partial(at+gx')}{\partial t}=0$$

$$g=\frac{-v}{c^2-v^2}a.$$

So that now we have

$$\tau=at+\left(\frac{-v}{c^2-v^2}a\right)x',$$

which Einstein wrote as

$$\tau=a\left(t-\frac{v}{c^2-v^2}x'\right). \tag{E8}$$

At this juncture, you may note that because (E8) is just an algebraic equation stemming from (E3), Einstein did not really need to employ the differential calculus but could have obtained the same result more simply, purely by means of linear algebra. Indeed, many later derivations of the Lorentz transformations, by Einstein and others, did not involve any detour through differential analysis.

Regardless, now having (E8), Einstein argued: "With the help of this result it is easy to determine the quantities ξ, η, ζ, by expressing in equations that light (as required by the principle of the constancy of the speed of light, in combination with the relativity principle) also propagates with a speed c when measured in the moving system." For a ray of light sent out at the time $\tau=0$ in the direction of increasing ξ:

$$\xi=c\tau, \tag{E11}$$

or

$$\xi=ac\left(t-\frac{v}{c^2-v^2}x'\right).^{[104]} \tag{E12}$$

[104] Ibid.

He simplified this equation by expressing t in terms of x': "The light ray moves relative to the origin of k, as measured in the resting system, with the speed $c - v$, so that:

$$\frac{x'}{c-v} = t."$$ (E13)

But why could Einstein assign this value to t when he had given it a different value in (E1)? He could do so because (E1) is valid for *any* value of t. Therefore, he was free to assign a new value to t, by setting $x = ct$ in (E1), from which (E13) follows. He then inserted "this value of t in the equation for ξ":

$$\xi = ac\left(\frac{x'}{c-v} - \frac{v}{c^2-v^2}x'\right)$$

$$\xi = a\frac{c^2}{c^2-v^2}x'.$$ (E14)

To obtain the equations relating the two other coordinates, η and ζ, to the coordinates y and z, Einstein said, "In an analogous manner we find, by consideration of light rays moving along the two other axes, that:

$$\eta = c\tau = ac\left(t - \frac{v}{c^2-v^2}x'\right)."^{[105]}$$ (E15)

As observed from the system K, a ray of light propagating along the H-axis of system k would travel with a speed $\sqrt{c^2-v^2}$ in a diagonal path, as explained previously, such that,

"where $\dfrac{y}{\sqrt{c^2-v^2}} = t; \quad x' = 0."$ (E16), (E17)

If we substitute these values into (E15),

$$\eta = a\frac{c}{\sqrt{c^2-v^2}}y.$$ (E18)

[105] Ibid., p. 900.

Likewise, for a light ray moving along the Z-axis of k,

$$\zeta = a \frac{c}{\sqrt{c^2 - v^2}} z. \tag{E19}$$

Einstein next simplified (E14), (E18), and (E19) by eliminating c from each numerator:

$$\xi = a \frac{x'}{1 - v^2/c^2}, \qquad \eta = a \frac{y}{\sqrt{1 - v^2/c^2}}, \qquad \zeta = a \frac{z}{\sqrt{1 - v^2/c^2}}.$$

He also simplified the equations by finally "substituting for x' its value," namely, $x' = x - vt$, as he had stated at the start of the analysis, so that

$$\xi = a \frac{x - vt}{1 - v^2/c^2};$$

and for the τ-function (E8),

$$\tau = a \left(t - \frac{v}{c^2 - v^2} (x - vt) \right),$$

$$\tau = a \left(\frac{c^2 t - v^2 t}{c^2 - v^2} - \frac{vx - v^2 t}{c^2 - v^2} \right),$$

$$\tau = a \frac{t - vx/c^2}{1 - v^2/c^2}.$$

Einstein provided no explanation or justification for his next step; he did not even explicitly state it. Miller notes that, "without prior warning," Einstein assigned a value for the undetermined term a, namely

$$a = \varphi(v) \sqrt{1 - v^2/c^2},$$

that is, $a = \varphi(v)b$ as stated earlier. Instead of discussing this point, he simply wrote the transformation equations as

$$\tau = \varphi(v)\beta(t - vx/c^2), \tag{E20}$$

$$\xi = \varphi(v)\beta(x - vt), \qquad\qquad\text{(E21)}$$
$$\eta = \varphi(v)y, \qquad\qquad\text{(E22)}$$
$$\zeta = \varphi(v)z, \qquad\qquad\text{(E23)}$$

where

$$\beta = 1 \Big/ \sqrt{1 - v^2/c^2}. \qquad\qquad\text{(E24)}$$

This tacit step eliminated the denominators in the equations for η and ζ and introduced the term $\beta = 1\big/\sqrt{1-v^2/c^2}$ into the equations for ξ and τ. Miller asked: "But why did Einstein make this replacement? It seems as if he knew beforehand the correct form of the set of relativistic transformations."[106] Certainly his first published derivation cannot be construed to be his first investigation of the problem. Miller offered a few historical conjectures to explain how Einstein found the final form of the transformations (in part from Einstein's study of Lorentz's paper of 1895, which, however, lacked the exact transformations), but such arguments need not be reviewed here. Einstein wanted to isolate the term $\varphi(v)$ to then show that it is irrelevant. But the important point is that Einstein did not introduce the value of a in the explicit manner in which he derived the values of the other terms. In view of this unsubstantiated step, readers of Einstein's derivation who did not have a predetermined notion of the final form of the transformation equations might well have been puzzled by his introduction of a and the β-factor.

With the preliminary set of transformation equations, Einstein proceeded to furnish "the proof that the principle of the constancy of the speed of light is compatible with the relativity principle," that is, "to prove that any light ray, measured in the moving system, propagates with the speed c, if, as we have assumed, this is the case in the resting system." He demonstrated this claim by arguing that if a wave of light propagates as a spherical surface in one coordinate system, then it also propagates as a spherical surface

---

[106] Miller, *Albert Einstein's Special Theory of Relativity*, p. 212. Likewise, Fölsing comments: "At an opaque point in his deduction he [Einstein] introduces, without any warning or explanation, a slight mathematical operation whose purpose becomes obvious only if the desired result is already known. This underhand device, by means of which he rather forcibly 'computes his way' to the Lorentz transformations, deprives the deduction of some of its elegance and stringency." Albrecht Fölsing, *Albert Einstein: A Biography*, trans. Ewald Osers (New York: Viking/Penguin, 1997), p. 188.

in the other coordinate system. He wrote: "At the time $t = \tau = 0$, when the coordinates-origins of both systems coincide, a spherical wave is emitted therefrom, which spreads out with the speed c in the system K. If (x, y, z) is a point just reached by this wave, then

$$x^2 + y^2 + z^2 = c^2 t^2 ."$$

(E25)

To transform this equation, Einstein first had to restate the transformation equations by isolating the terms x, y, z, and t. From equations (E20) and (E21), we obtain

$$\frac{\tau}{\varphi(v)\beta} + \frac{vx}{c^2} = t, \qquad \frac{\xi}{\varphi(v)\beta} + vt = x.$$

Substitution of these values of t and x into each other gives

$$\frac{\tau}{\varphi(v)\beta} + \frac{v}{c^2}\left(\frac{\xi}{\varphi(v)\beta} + vt\right) = t, \qquad \frac{\xi}{\varphi(v)\beta} + v\left(\frac{\tau}{\varphi(v)\beta} + \frac{vx}{c^2}\right) = x,$$

$$\frac{\tau + \xi v/c^2}{\varphi(v)\beta} = t - \frac{v^2}{c^2}t, \qquad \frac{\xi + v\tau}{\varphi(v)\beta} = x - \frac{v^2}{c^2}x,$$

$$\frac{\tau + \xi v/c^2}{\varphi(v)\beta(1 - v^2/c^2)} = t, \qquad \frac{\xi + v\tau}{\varphi(v)\beta(1 - v^2/c^2)} = x,$$

and because $\beta = 1 / \sqrt{1 - v^2/c^2}$,

$$\frac{\tau + \xi v/c^2}{\varphi(v)}\beta = t, \qquad \frac{\xi + v\tau}{\varphi(v)}\beta = x,$$

(3), (4)

and,

$$\eta / \varphi(v) = y, \qquad \zeta / \varphi(v) = z.$$

(5), (6)

Substitution of these values into the equation of the sphere in K gives

$$\left(\frac{\beta(\xi + v\tau)}{\varphi(v)}\right)^2 + \left(\frac{\eta}{\varphi(v)}\right)^2 + \left(\frac{\zeta}{\varphi(v)}\right)^2 = c^2\left(\frac{\beta(\tau + \xi v/c^2)}{\varphi(v)}\right)^2,$$

343

and it follows that

$$\xi^2 + \eta^2 + \zeta^2 = c^2\tau^2. \tag{E26}$$

Einstein concluded, "The wave under consideration is therefore also a spherical wave with the expansion-speed c when viewed in the moving system. Herewith is shown that our two fundamental principles are mutually compatible." Note that, algebraically, it would have been simpler to enter equations (E20–24) directly into equation (E26), but Einstein started from (E25) with good reason—because he wanted to *derive* the spherical propagation of light in k, having only assumed it in K.[107]

A logical ambiguity about Einstein's derivation of (E26) was pointed out and clarified by Robert B. Williamson in 1977. Note that to obtain (E26) Einstein relied on equations (E20–24), which stemmed from the equations $\xi = c\tau$, $\eta = c\tau$, $\zeta = c\tau$; but these latter equations already signify the constant and hence spherical propagation of light in system k. Williamson noted that it seems "puzzling" that Einstein seemed to digress from his derivation of the transformation equations to prove the spherical propagation of light when he had already assumed it beforehand.[108] Yet Williamson explained that the logical circularity in question is not vicious because Einstein was not thus attempting to prove "the truth" of the spherical propagation of light but just to show the consistency of the principle of relativity with the principle of the constancy of the speed of light. He had first postulated that light propagates with a constant speed in the "resting" system, and then he required that it also propagates with the same constant speed in the "moving" system. He then demonstrated that the two requirements are mutually compatible, despite what one might expect on the basis of the traditional rule for the transformation of velocities between systems in uniform rectilinear relative motion.

Einstein still had to find the value of the function $\varphi(v)$. Because he was

---

[107] In later derivations, rather than derive the invariance of the spherical propagation of light, Einstein assumed it. For discussions of this important difference, see Robert B. Williamson, "Logical Economy in Einstein's 'On the Electrodynamics of Moving Bodies,'" *Studies in History and Philosophy of Science* 8 (1977), 46–60; or also, Harvey R. Brown and Adolfo Maia Jr., "Light-Speed Constancy versus Light-Speed Invariance in the Derivation of Relativistic Kinematics," *British Journal of the Philosophy of Science* 44 (1993), 381–407.

[108] Williamson, "Logical Economy," p. 50.

satisfied with the form of the transformation equations, he just had to prove that $\varphi(v) = 1$. For this purpose, he introduced "a third system of coordinates K', which relative to the system k is in parallel-translational motion parallel to the Ξ-axis such that its coordinate origin moves with velocity $-v$ on the Ξ-axis." Again in the interest of simplicity, Einstein indicated, "At the time $t = 0$ let all three coordinates-origins coincide, and for $t = x = y = z = 0$ let the time t' of the system K' equal zero." Because he had finished using the term x', he used it anew by designating the coordinates of K' as x', y', z'. He related the coordinates of K' to those of k with the transformation equations

$$t' = \varphi(-v)\beta(-v)(\tau + \xi v/c^2), \tag{E28}$$
$$x' = \varphi(-v)\beta(-v)(\xi + v\tau), \tag{E29}$$
$$y' = \varphi(-v)\,\eta, \tag{E30}$$
$$z' = \varphi(-v)\,\zeta, \tag{E31}$$

as if he had derived them analogously to those obtained before, having changed only the orientation of the relative velocity. From equations (E20–E24), he could also give the transformation equations relating system k to K as:

$$\tau + \xi v/c^2 = \varphi(v)t/\beta,$$
$$\xi + v\tau = \varphi(v)x/\beta,$$
$$\eta = \varphi(v)\,y,$$
$$\zeta = \varphi(v)\,z.$$

By substituting these last four equations into the previous four, and setting $\beta(-v) = \beta$, he obtained the transformation equations relating K' to K:

$$t' = \varphi(v)\varphi(-v)\,t, \tag{E32}$$
$$x' = \varphi(v)\varphi(-v)x, \tag{E33}$$
$$y' = \varphi(v)\varphi(-v)y, \tag{E34}$$
$$z' = \varphi(v)\varphi(-v)z, \tag{E35}$$

He thus concluded that these systems are identical: "Since the relations between x', y', z' and x, y, z, do not contain the time t, so the systems K and K' rest with respect to each other, and it is clear that the transformation from K to K' must be the identity transformation." That is, $t' = t, x' = x, y' = y, z' = z$, such that

$$\varphi(v)\varphi(-v) = 1. \tag{E36}$$

By concluding that systems K and K' are identical, Einstein effectively deduced that the speed of K relative to k is $-v$ (when the speed of k relative to K is v). By contrast, in later derivations, he and other writers allowed themselves to simply assume this symmetry, that is, the reciprocity of the mutual translational velocities.

Einstein continued: "We pursue now the meaning of $\varphi(v)$. We fix our attention on that portion of the H-axis of system k that lies between $\xi = 0$, $\eta = 0$, $\zeta = 0$ and $\xi = 0$, $\eta = l$, $\zeta = 0$. This portion of the H-axis is a rod moving perpendicularly to its axis with velocity v relative to system K, whose ends possess in K the coordinates:

$$x_1 = vt, \quad y_1 = l/\varphi(v), \quad z_1 = 0, \tag{E43--45}$$

and

$$x_2 = vt, \quad y_2 = 0, \quad z_2 = 0." \tag{E46--48}$$

From (E44), we see that $\varphi(v) = l/y_1$. Hence, Einstein wrote: "The length of the rod, measured in K, is therefore $l/\varphi(v)$; hence the meaning of the function φ is given. From reasons of symmetry, it is now evident that in the resting system the measured length of a given rod, moving perpendicular to its axis, can depend only on the speed but not on the direction and the sense of the motion. So the length of the moving rod measured in the resting system does not change if v and $-v$ are interchanged. Consequently:

$$l/\varphi(v) = l/\varphi(-v), \tag{E49}$$

or

$$\varphi(v) = \varphi(-v)." \tag{E50}$$

By substituting this value of $\varphi(-v)$ into (E36), we obtain

$$\varphi(v)\varphi(v) = 1,$$
$$[\varphi(v)]^2 = 1,$$
$$\varphi(v) = \pm 1.$$

Einstein disregarded the negative root, to write

$$\varphi(v) = 1, \tag{E51}$$

and he thus obtained finally the long-sought transformation equations

$$\tau = \beta(t - vx/c^2), \tag{E52}$$
$$\xi = \beta(x - vt), \tag{E53}$$
$$\eta = y, \tag{E54}$$
$$\zeta = z, \tag{E55}$$

where

$$\beta = 1 / \sqrt{1 - v^2/c^2}. \tag{E24}$$

Because we are focusing on the ambiguities in Einstein's expressions, consider again his use of the term x'. We raised the question of whether Einstein used a third coordinate system from the beginning of his analysis, and whether such a system was described by the laws of traditional kinematics. Presumably, Einstein introduced the auxiliary term x' to simplify his derivation, but did it really perform this function?

To clarify this issue, we now reconstruct Einstein's derivation without introducing the x' term. Let us see what we derive instead of equation (E3) if we establish the exact values of the coordinates in K at the times of emission, reflection, and return of the light ray traveling between the clocks in k. Consider again the expression:

$$\frac{1}{2}\left[\tau(x_0, y_0, z_0, t_0) + \tau(x_2, y_2, z_2, t_2)\right] = \tau(x_1, y_1, z_1, t_1). \tag{1}$$

Each of the values of the coordinates in K can be established as follows. At the time t_0, if the origins of K and k coincide, and a light ray is emitted, its coordinates in K are $x_0 = y_0 = z_0 = 0$ and $t_0 = t$. At this time, the point ξ (where a clock is fixed to the Ξ-axis) is located at a distance x from the origin of K. As the light ray travels along the X-axis, ξ moves away with velocity v. Once the ray traverses the distance x, the point ξ is no longer there, and the ray must traverse an extra distance $v(t_1 - t_0)$ to reach ξ. Thus, the coordinates in K of the signal upon its arrival at ξ are

$$x_1 = x + v(t_1 - t_0),$$
$$y_1 = 0,$$
$$z_1 = 0,$$
$$t_1 = t + (x/(c-v)).$$

At this time t_1, the light signal and the point ξ coincide at a distance $x + v(t_1 - t_0)$ from the origin of K. Now the ray travels in the opposite direction, while the origin of k moves toward it with the velocity v. Because at the time t_1 the origin of k was at a distance x from the ray, the ray traverses less than this distance to meet it. Thus, the coordinates in K of the ray when it reaches k's origin are

$$x_2 = v(t_1 - t_0) + v(t_2 - t_1),$$
$$y_2 = 0,$$
$$z_2 = 0,$$
$$t_2 = t + (x/(c-v)) + (x/(c+v)).$$

These values can be restated by expressing $(t_1 - t_0)$ and $(t_2 - t_1)$ in terms of x, v, and c:

$$(t_1 - t_0) = x/(c-v),$$
$$(t_2 - t_1) = x/(c+v).$$

These two equations are equivalent to equations (E0). So, the successive values of the x and t coordinates in K are

$$x_0 = 0,$$
$$t_0 = t,$$
$$x_1 = x + vx/(c-v),$$
$$t_1 = t + (x/(c-v)),$$
$$x_2 = vx/(c-v) + vx/(c+v),$$
$$t_2 = t + (x/(c-v)) + (x/(c+v)).$$

These values can be simplified, but the preceding form conveys their physical significance. These values can now be substituted into equation (1), so that we obtain

$$\frac{1}{2}\left[\tau(0,0,0,t)+\tau\left(\frac{vx}{c-v}+\frac{vx}{c+v},0,0,t+\frac{x}{c-v}+\frac{x}{c+v}\right)\right]=\tau\left(x+\frac{vx}{c-v},0,0,t+\frac{x}{c-v}\right),\ (11)$$

instead of Einstein's expression:

$$\frac{1}{2}\left[\tau(0,0,0,t)+\tau\left(0,0,0,t+\frac{x'}{c-v}+\frac{x'}{c+v}\right)\right]=\tau\left(x',0,0,t+\frac{x'}{c-v}\right).\quad\text{(E3)}$$

Are the two expressions equivalent? Einstein's formulation is algebraically simpler, because of his use of x', but conceptually it is ambiguous for the reasons mentioned. To further distinguish the two approaches, we seek the difference between the next result Einstein obtained and what follows otherwise.

To obtain the differential equation expressing the dependence of τ on x and t, as in Einstein's approach, we may make a series expansion of each term in τ. For example, because $x + (vx/c - v) = x(c/c - v)$, then for τ_1:

$$\tau(xhc,t+hx)=\tau(xhc,t)+\frac{\partial\tau}{\partial t}\frac{hx}{1}+\frac{\partial^2\tau}{\partial t^2}\frac{(hx)^2}{2!}+\ ...,$$

where we have let $h = 1/(c - v)$. Now, if we differentiate the function and the series with respect to x, we obtain

$$\frac{\partial\tau(xhc,t+hx)}{\partial x}=\frac{\partial\tau}{\partial x}hc+\frac{\partial\tau}{\partial t}h+\frac{\partial^2\tau}{\partial t^2}xh^2+\ ....$$

If we neglect terms of second order and higher, we find

$$\frac{\partial\tau_1}{\partial x}=\frac{\partial\tau}{\partial x}\left(\frac{c}{c-v}\right)+\frac{\partial\tau}{\partial t}\left(\frac{1}{c-v}\right).$$

By carrying out the same procedure for the other side of equation (11), we obtain

$$\frac{1}{2}\left[(0)+\frac{\partial\tau}{\partial x}\left(\frac{2vc}{c^2-v^2}\right)+\frac{\partial\tau}{\partial t}\left(\frac{2c}{c^2-v^2}\right)\right]=\frac{\partial\tau}{\partial x}\left(\frac{c}{c-v}\right)+\frac{\partial\tau}{\partial t}\left(\frac{1}{c-v}\right),$$

so that we now obtain

$$\frac{\partial \tau}{\partial x} + \frac{v}{c^2} \frac{\partial \tau}{\partial t} = 0. \tag{12}$$

Compare this result to Einstein's:

$$\frac{\partial \tau}{\partial x'} + \frac{v}{c^2 - v^2} \frac{\partial \tau}{\partial t} = 0. \tag{E5}$$

From equation (E5), he obtained the algebraic expression for τ (E8). But following the same procedure, we derive from equation (12) the different equation,

$$\tau = \beta \left(t - vx/c^2 \right) \tag{13}$$

(where the value of β has not yet been determined), instead of Einstein's equation,

$$\tau = a \left(t - \frac{v}{c^2 - v^2} x' \right), \tag{E8}$$

which may be restated by including the value he gave to x', that is, $x' = x - vt$:

$$\tau = a \left(t - \frac{vx - v^2 t}{c^2 - v^2} \right).$$

Notice how equation (13) resembles the transformation equation for τ that Einstein only subsequently obtained: $\tau = \beta(t - vx/c^2)$.

The only distinction between (13) and the final transformation is that we have yet to establish that β has the same value in both. This great similarity indicates that, from this point onward, the algebraic derivation of the τ-function in terms of x instead of x' leads more directly to the results sought by Einstein. All that remains is to show that this approach agrees with Einstein's by completing the derivation of the transformation equations. Again, we proceed in direct analogy to Einstein's 1905 procedure.

To express the quantities ξ, η, ζ in terms of x, y, z, we need to express in equations that light propagates with a speed c when measured in the moving system. For a light ray emitted at the time $\tau = 0$ in the direction of increasing ξ, $\xi = c\tau$, so that from equation (13) we obtain

$$\xi = \beta(ct - vx/c).$$

Because x corresponds to the distance light travels in K in a time t, $x = ct$, so that

$$\xi = \beta\left(x - vt\right).$$

To obtain the equations relating η and ζ to y and z, we proceed in an analogous manner by considering rays moving along the two other axes. First, for the H-axis,

$$\eta = c\tau = \beta c\left(t - vx/c^2\right). \tag{14}$$

Now, *as observed from the system K*, a ray of light propagating along the H-axis of k travels a diagonal path with a speed $\sqrt{c^2 - v^2}$, such that when $(y/\sqrt{c^2 - v^2}) = t$, $x = vt$, because the H-axis has moved a distance vt as the ray travels to the point η. If we substitute this value of x into the expression for η, equation (14), we find

$$\eta = \beta ct\left(1 - \frac{v^2}{c^2}\right).$$

We substitute the value of t to find

$$\eta = \beta y\sqrt{1 - v^2/c^2}.$$

The same procedure for the light ray transmitted along the Z-axis yields an equivalent result, so that the resulting four equations are

$$\tau = \beta\left(t - vx/c^2\right),$$
$$\xi = \beta\left(x - vt\right),$$
$$\eta = \beta y\sqrt{1 - v^2/c^2},$$
$$\zeta = \beta z\sqrt{1 - v^2/c^2}.$$

These four equations now suggest that $\beta = 1/\sqrt{1-v^2/c^2}$, which may be demonstrated by using the same arguments Einstein used to establish that $\varphi(v)$ = 1, or any simpler arguments showing, say, that $\eta = \gamma$. Hence, we arrive again at the final form of the transformation equations.

We have confirmed that the term x' was not essential for Einstein's analysis. His choice to express the dependence of the function τ on x' simplified the initial steps of his analysis, because it is a bit more complicated to differentiate τ with respect to x because the position of the light ray varies with the changing values of t. Instead of performing every calculation with the coordinates x_0, x_1, x_2, Einstein simply carried the auxiliary term x' throughout the derivation to reintroduce its value at the end. Nonetheless, it also is correct to evaluate the function τ with respect to x on an equal footing with the terms y, z, t, and, after differentiation, the procedure even leads more directly to the desired results, especially because it yields the value of the β-factor directly, rather than requiring the introduction of the function $a = \varphi(v)b$.

The role of the term x' involves sufficient ambiguities to suggest that Einstein's derivation may have confused even careful readers of his paper, because even later physicists misinterpreted the role of x'. Because equation (E1) is formally identical to a transformation equation, readers could easily misunderstand Einstein's analysis. For example, even Torretti misconstrued (E1) as a transformation. Likewise, Miller misconstrued x' as a "Galilean coordinate" of system k and, accordingly, he misinterpreted the expressions $c - v$ and $c + v$ as speeds of light relative to k, as though Einstein had relied on the Galilean transformation of velocities in his analysis. He also construed Einstein as having employed x' to avoid discussing the relativistic effect of length contraction from the outset.[109] It would then seem that Einstein derived new transformation equations by assuming as true the traditional transformations and violating his own notion of the invariance of the speed of light. On such grounds, anyone who approached Einstein's analysis with a critical attitude could easily be confused or reject it as incoherent. But contrary to such interpretations, the term x' does not correspond to a Galilean or quasi-Galilean coordinate of system k. Instead, it may designate the constant separation between the two clocks as judged not from k but from K.

[109] Miller, *Albert Einstein's Special Theory of Relativity,* p. 209.

Yet even in this light, Einstein's analysis involved additional peculiarities that could engender confusion. For example, in equation (E1) the term x' might seem to be the initial position of a material point in its equation of rectilinear motion. Also, equation (E3) could be misinterpreted as consisting of coordinate values of k, whereas the arguments in each parenthesis belong to K, and only three in each, namely the y, z, t terms, are coordinates. Einstein compounded this ambiguity by talking in one place about x' as if it were the position of the light ray when it reaches clock B. Such sloppiness of expression and notation was quite common in papers on theoretical physics at the time, but in the demonstration of radically novel claims, it could hardly help their intelligibility. Yet Einstein gave more than one meaning to some of the symbols he employed, for example, to x, x', and t. By avoiding exact expressions for the positions of a light ray along the X-axis, his analysis scarcely distinguished between the concepts of position and length. For the most part, he distinguished explicitly the concepts of time intervals and instants (single time coordinates), but he hardly drew analogous distinctions for the concepts of space. Also, he did not distinguish explicitly between concepts of distance and displacement, or between speed and velocity—distinctions that stemmed from vector theory and had been introduced precisely to clarify the representation of physical quantities. Thus, Einstein's derivation focused on a clarification of the concept of time, on the basis of measurement procedures and algebraic analysis, but it admitted various ambiguities in the representation of other kinematic concepts.[110]

Moreover, his derivation suffers from some deficiencies. Consider again the factor β in the transformation equations. This factor is crucial, because it is a telltale feature that distinguishes the Lorentz transformation from the Galilean transformation. But Einstein did not derive this factor exactly but introduced it rather freely by setting the value of the variable a to obtain the form of the transformations that he deemed to be correct. In addition to the β-factor, the only other deviation from the Galilean transformations is the term $-xv/c^2$ in the time transformation. It seems awkward that most of the analysis in Einstein's long derivation simply results in the introduction of this small algebraic term (although physically the

---

[110] Alberto A. Martínez, "The Neglected Science of Motion: The Kinematic Origins of Relativity" (Ph.D. dissertation, University of Minnesota, December 2000; UMI, 2001).

term is of crucial importance because it involves first-order effects in velocity v, whereas the β-factor concerns second-order effects). Again, his use of differential analysis was an unnecessary detour. Thus, the amount of analysis does not seem commensurate with the formal simplicity of its results.

Einstein's derivation lacked mathematical elegance; its degree of abstraction along with its ambiguities serves to illustrate why many of his contemporaries had difficulties understanding and accepting his kinematics. Back then, readers well trained in theoretical physics constituted a small minority of physicists. Many steps of his derivation could seem debatable or imprecise to readers who took his every expression literally, or to those who did not already know or accept the final result, the transformation equations and their utility.

The structure and details of the derivation engender the impression of Einstein groping tortuously to construct it, having first discovered that the relativity of simultaneity could provide a kinematic justification for the transformations that he knew to work in optics and electromagnetism. To be sure, the general approach and novel concepts made his derivation well worthwhile. He demonstrated that the Lorentz transformations could be deduced from simple kinematic assumptions, along with the postulate of the constancy of the speed of light, irrespective of the exact validity of the rest of contemporary electromagnetic theory.

Einstein's first derivation included several other virtues. For example, Einstein derived the transformation equations without making any hypotheses about the constitution of matter or of intermolecular forces. The transformations were best suited to the solution of problems in electrodynamics, yet Einstein did not base them on the presumed validity of a contemporary electromagnetic theory or on Maxwell's equations. His derivation also was independent of whether light is presumed to consist of waves or particles, as Einstein used mainly the concept of light ray, appropriate to both conceptions. Moreover, Einstein's derivation involved aspects that show both the logical economy of his thought and its conceptual roots. In particular, rather than postulating the invariance of the speed of light in nonaccelerated frames of reference, Einstein postulated the constancy of the speed of light in a single frame and then derived its invariance. Such expository devices paved the transition from a physics based on the privileged reference frame of the stationary ether to one based on observational and formal symmetries.

Strange Consequences

The Kinematic Part of the 1905 paper has two more sections. In them, Einstein introduced several consequences of his theory. Because our study focuses on the elements of kinematics and not on its theorems, we highlight only some key points of the remaining two sections. These sections, anyhow, are shorter and simpler than the preceding ones. Moreover, the reader may consult Arthur I. Miller's book for further analysis.

Section 4 is titled "Physical Meaning of the Equations Obtained concerning Moving Rigid Bodies and Moving Clocks." Section 2 had provided a somewhat qualitative demonstration of the relativity of lengths and times. Section 4 provides a more exact mathematical account of the metric relations among moving systems.

Einstein began by positing a rigid sphere attached to the "moving" system k. He described this sphere relative to k with the equation

$$\xi^2 + \eta^2 + \zeta^2 = R^2,$$

where R is the radius of the sphere. The same terms that earlier were used for coordinates, ξ, η, ζ, are used here as the components of the sphere. These terms imply that the sphere is at rest at the origin of system k. Using the transformation equations (E53), (E54), (E55), one can convert each of the terms ξ, η, ζ, into terms of the "stationary" system K. We obtain

$$\left(\frac{x-vt}{\sqrt{1-v^2/c^2}} \right)^2 + \left(y \right)^2 + \left(z \right)^2 = R^2.$$

Note that one may well change the symbol R, used already for k, to another proper to K. Einstein wrote this equation in a simpler form by setting $t = 0$ as

$$\frac{x^2}{\left(\sqrt{1-v^2/c^2} \right)^2} + y^2 + z^2 = R^2.$$

Rather than setting $t = 0$, one could obtain the same relationship as follows. Consider instead the equation of the sphere expressed in terms of coordinate intervals:

$$\left(\xi_2-\xi_1\right)^2+\left(\eta_2-\eta_1\right)^2+\left(\zeta_2-\zeta_1\right)^2=R^2.$$

By transforming each coordinate, the terms vt drop out, and the resulting equation would be valid not only for $t = 0$ but for any time t. Also, we may rename each interval by a single letter, such as $y_2 - y_1 = y$, where the single term now clearly denotes a length. In any case, the result is the same as Einstein stated: the figure that is measured to be a sphere in k has instead the form of an ellipsoid relative to K. Bodies that appear as rigid in one system appear as contracted when judged from a system that moves relative to the first. Or, similarly, "moving" bodies contract in the direction of relative motion between reference frames. Thus, for the case illustrated, Einstein explained that, "whereas the Y- and Z-dimensions of the sphere (and therefore of every rigid body of arbitrary form) do not appear modified by the motion, the X-dimension appears shortened in the ratio $1:\sqrt{1-v^2/c^2}$, i.e., the greater the value of v, the greater the shortening." In sum, Einstein thus derived the effect of "length contraction" that had previously been assumed by Lorentz and others as a hypothesis. Einstein noted that, for velocities v greater than that of light, "our deliberations become meaningless." For this reason, but above all for dynamical reasons, Einstein argued that in his theory the speed of light functioned as a limit, a maximum possible velocity for any physical signals. (He later showed that to accelerate a body beyond the speed of light would require an infinite amount of force.)

Next, Einstein demonstrated the relative behaviors of moving clocks. He posited a clock attached to the origin of system k and asked, "What is the rate, of this clock, when viewed from the resting system?" He related the time τ of the clock in k to the time t in system K by the transformation (E52)

$$\tau = \frac{t-vx/c^2}{\sqrt{1-v^2/c^2}}.$$

Because the clock is positioned at the origin of k, its position in time relative to K is given by $x = vt$. Substituting this value of x into equation (E52), Einstein obtained

$$\tau = t\sqrt{1-v^2/c^2} = t-\left(1-\sqrt{1-v^2/c^2}\right)t.$$

As he explained, "The time marked by the clock (viewed in the resting system) is slow by $1-\sqrt{1-v^2/c^2}$ seconds per second." Again, Einstein thus derived this result without using auxiliary hypotheses. And again, the effect would be reciprocal: clocks in system K would seem slow when judged from k. Measures of time, just as those of length, could vary systematically among systems in relative motion.

Einstein proceeded to mention briefly a "peculiar consequence." He argued that two clocks synchronized in K would no longer be synchronized if one were moved relative to the other. In particular, a clock moved toward the other would mark a retarded time—it would lag behind, compared to the other, upon its arrival. Einstein did not then analyze subtleties that stem from the circumstance—that to move a stationary clock it must be subjected to an accelerating force. He did not account for the effect of the acceleration on the rate of the clock. Still, he predicted that because a clock on the equator moves with the rotating Earth more than a clock at the North or South Pole, then the equatorial clock would mark less time than the polar clocks. However, his prediction was mistaken, because it did not account for the different effects of gravity at the equator and the poles, which actually counteract the predicted time dilation.[111]

Finally, the Kinematic Part of Einstein's paper ended with section 5, titled "Theorem of Addition of Velocities." Einstein referred to the traditional rule by one of its older names: "the law of the parallelogram of velocities."[112] Einstein did not use the methods of vector theory in his 1905 paper. He used coordinate algebra systematically and, at most, used the word "vector" merely to describe sets of three coordinates. He did not direct any criticisms against vectorial concepts or methods. Still, he rejected the addition or parallelogram rule.

Briefly put, Einstein derived several algebraic equations to transform velocities in one system into velocities in another system. For example, if a body moves relative to system K with a velocity w, then it appears to move with a velocity V relative to k, such that

$$V = \frac{v+w}{1+vw/c^2}.$$

---

[111] Alex Harvey and Engelbert Schucking, "A Small Puzzle from 1905," *Physics Today* **18**, no. 3 (March 2005), 34–36.

[112] Einstein, "Zur Elektrodynamik bewegter Körper," p. 905.

This formula replaced the usual rule: $V = v + w$. Note, strictly speaking, that here these two expressions are transformation equations. Einstein thus replaced the traditional velocity transformations with new ones. By contrast, the rule for addition of velocities *within a single system* remained the same as in traditional kinematics. For example, two bodies moving in opposite directions, toward one another, one with a velocity v and the other with a velocity w, both relative to K, approach one another, as judged by K, with a total rate of $V = v + w$. In this case, there is no transformation between systems. We already saw this subtlety at play in Einstein's derivation of the coordinate transformations in section 3. Likewise, the vectorial addition concept remained valid within any individual reference frame.

The new transformation equations replaced the addition rule for velocities along a common line and also the "parallelogram" rule for the composition of velocities in arbitrary directions. Thus, Einstein, and independently Poincaré, replaced the well-known equations with new equations. To a limited extent, this replacement of the vectorial concept with algebraic rules constituted a formal rejection of one of the fundamental principles of vector theory. By 1900 the rule for adding velocities vectorially had been regarded as a fundamental principle in the study of motion. The high degree of trust placed by physicists on this concept was voiced by August Föppl in the foreword to his *Introduction to Mechanics* of 1898: "Above all, mechanics cannot dispense with the concept of the geometrical sum of two directed magnitudes, without considerable loss of clarity and orderliness. This is generally acknowledged nowadays."[113] Indeed, the acceptance and development of vector theory had been greatly facilitated by the convenient simplicity of the rule for vector addition.

Broadly speaking, Einstein's rejection of the parallelogram law was a denial of the exact applicability of the traditional geometric concept. It constituted a rejection of the exact validity of the basic equations of traditional kinematics. In a manuscript of 1912, Einstein placed strong emphasis on the intimate connection between the mathematical theory of classical kinematics and the parallelogram rule. He virtually equated the two, writing about the traditional "kinematic equations, *i.e.*, the law of the parallelogram

[113] August Föppl, *Vorlesungen ueber Technische Mechanik*, vol. 1, *Einfuehrung in die Mechanik* (1898; 2nd ed., Leipzig: B. G. Teubner, 1900), p. vi.

of velocities," and again, "the [traditional] transformation equations, or the law of the parallelogram of velocities."[114] By rejecting the traditional rule of velocities, Einstein created a new kinematics.

Einstein's first work on the relativity principle was difficult to understand for many readers. Consider the following examples.[115] In the spring of 1905, Einstein discussed his ideas with his good friend and colleague Michele Besso, who defended Newton's concepts against Einstein's radical conception of time. Afterward, the physicist Josef Sauter was one of the first persons to hear about Einstein's work in depth. Sauter was one of Einstein's co-workers in the Swiss patent office, and because he had studied and published on Maxwell's theory of electromagnetism, Einstein "gave him his notes, which Sauter criticized severely: 'I pestered him for a whole month with every possible objection.'" Despite his criticisms, Sauter facilitated a meeting between Einstein and Paul Gruner, professor of theoretical physics at the University of Bern. Consequently, in 1907 Einstein gave his paper "On the Electrodynamics of Moving Bodies" to Gruner, in support of his candidacy for a faculty appointment at Bern. In the words of Gruner, "I received his essay although the whole theory at the time seemed to me to be highly problematical." The reaction of the faculty was even less favorable, as they "declared the work as inadequate: it was more or less clearly rejected by most of the contemporary physicists." The professor of experimental physics, Aimé Forster, returned Einstein's paper with the remark, "I cannot understand at all what you have written here." To be sure, a few physicists seem to have understood the gist of Einstein's arguments promptly, including Max Planck. But even Planck first wrote to Einstein asking that he clarify certain points in his paper.[116]

After his paper on the electrodynamics of moving bodies was published in the *Annalen der Physik* in 1905, Einstein discarded the original manu-

---

[114] Albert Einstein, manuscript on the special theory of relativity (1912–1914, in *The Collected Papers* vol.4, p. 25.
[115] The quotes that follow are from Carl Seelig, *Albert Einstein: A Documentary Biography,* trans. Mervyn Savill (London: Staples Press, 1956), pp. 73, 88. See also Flückiger, *Albert Einstein in Bern* (1974), pp. 103, 209.
[116] Maja Winteler-Einstein, "Albert Einstein—Beitrag für sein Lebensbild" (1924), quoted in Pais, *"Subtle is the Lord . . . ,"* p. 150.

script.[117] Late in life, Einstein expressed surprise at the complexity of the paper. In 1943 Einstein was drafting a copy of the paper to donate for a fundraising auction. His secretary, Helen Dukas, recounted that "she would sit next to Einstein and dictate the text to him. At one point, Einstein lay down his pen, turned to Helen and asked her whether he had really said what she had just dictated to him. When assured that he had, Einstein said, 'Das hätte sich einfacher sagen können.'"[118]

By comparison to other parts of Einstein's 1905 paper, his derivation of the transformation equations stands out as the most subtle mathematical argument. Later derivations of the same equations, by Einstein and others, were immensely simpler.[119] Hence, we may understand Einstein's remark, "That could have been said more simply," as referring to his derivation of the transformation equations.

Einstein distinguished verbally neither between distances and displacements nor between speeds and velocities. Nor did he distinguish systematically between the concepts of spatial and temporal points as opposed to intervals. In hindsight, his style of expression falls short of the high standards of clarity that had been advocated by Friedrich Haller at the patent office: "to select every word in its most exact sense."[120] This circumstance apparently reflects a nonverbal aspect of the creative process in Einstein's work. Max Wertheimer reported a discussion in which Einstein explained the thoughts that led him originally to his theory: "'These thoughts did not come in any verbal formulation. I very rarely think in words at all. A thought comes, and I may try to express it in words afterward.' When I

---

[117] As Einstein informed Julian Boyd, librarian of the Princeton University Library; see Pais, *"Subtle is the Lord . . . ,"* p. 147. Einstein's written statement of having discarded the original manuscript appears in Flückiger, *Albert Einstein in Bern* (1974), p. 103.

[118] Pais, *"Subtle is the Lord . . . ,"*, p. 147. John Stachel, having also heard the story directly from Helen Dukas, received the impression that it was at two distinct places during the reading that Einstein made this remark to her. We do not know whether he necessarily referred to something in his derivation of the transformations. Incidentally, the *Collected Papers of Albert Einstein*, vol. 2, p. 309, includes notes on three slight corrections that Einstein made to a reprint copy of his 1905 paper, but such corrections are far too minor for us to assume that his later comment to Dukas referred to them.

[119] E.g., Einstein, "Über das Relativitätsprinzip" (1907), pp. 411–462. See also A. Einstein, *Relativity* (1917), appendix I.

[120] Anton Reiser, *Albert Einstein: Biographical Portrait*, with a preface by Albert Einstein (New York: A. & C. Boni, 1930), p. 65.

[Wertheimer] remarked that many report that their thinking is always in words, he only laughed."[121] Similarly, Einstein provided the following reply to Jacques Hadamard for his *Essay on the Psychology of Invention in the Mathematical Field:*

> The words or the language, as they are written or spoken, do not seem to play any role in my mechanism of thought. The physical entities which seem to serve as elements in thought are certain signs and more or less clear images which can be "voluntarily" reproduced and combined.
>
> There is, of course, a certain connection between those elements and the relevant logical concepts. It is also clear that the desire to arrive finally at logically connected concepts is the emotional basis of this rather vague play with the above mentioned elements. But taken from a psychological viewpoint, this combinatory play seems to be the essential feature in productive thought—before there is any connection with logical construction in words or other kinds of signs which can be communicated to others.[122]

Neglecting the verbal focus that might have served to distinguish among various concepts, Einstein relied on a considerably abstract algebraic analysis to obtain the results of his new kinematics. Consequently, Einstein's writings presented interpretive challenges to those who wished to understand his kinematics in detail. Challenging reactions came even from his good friend, the young mathematician Conrad Habicht, who cultivated a "great aversion against the 'unnecessary propagation of printed matter.' A hundred hours transpired the *pro* and *contra*. Later Einstein reckoned 'that with him, in many conversation-walks, the discussions were more animated than all later critics put together.'"[123]

---

[121] Wertheimer, *Productive Thinking*, chap. 7: "Einstein: The Thinking That Led to the Theory of Relativity," p. 184.

[122] Albert Einstein to Jacques Hadamard, (no date noted, 1944), translated in Jacques Hadamard, *An Essay on the Psychology of Invention in the Mathematical Field* (Princeton: Princeton University Press, 1945; repr., New York: Dover, 1954), p. 142; a shorter draft of this letter (dated 17 June 1944) is in the Einstein Archives, item 12-57.10; see also item 1-146.00.

[123] Paul Habicht to Melania Serbu, "Brief eines Jugendfreundes Einstein," 26 October 1943, Einstein Archives, item 39-275.

CHAPTER EIGHT

Critical History
The Algebra of Motion

It would be predictably a pleasant story. And we like success stories. It would be the story of a science that was not a science, repudiated and neglected, and how it grew to high esteem, foremost among the physical sciences. A child of abstraction and experience, that served to solve vexing and far-reaching problems. It would be a story that began with old philosophers in Greek antiquity and ended with a twenty-six-year-old clerk at the Swiss patent office. Problems solved, and we might frame the solutions on the wall and turn our attention to other things.

But let us leave happy endings aside. Instead, we may focus on the negatives. Otherwise, it is easy to foster the illusion that kinematics was completed in the early 1900s. Indeed, nearly all of the elementary kinematics taught in college nowadays had already been formulated a century or more ago. Yet that is not the complete story. Because the science of kinematics continues to have plenty of shortcomings, in this chapter we discuss a selection of concepts and problems that still require critical attention.

In 1905 the latest word in kinematics was Einstein's work. Thereafter, many physicists came to admire it as a relatively permanent contribution to science. But that work was not intended as such. Early on, Einstein argued that his work was not a "closed system"; instead he compared it to a "heuristic principle," concerning only expressions about rigid bodies, clocks, and light signals.[1] From the beginning, his contribution was intended to be a temporary solution. The reasons were several.

[1] A. Einstein, "Bemerkungen zu der Notiz von Hrn. Paul Ehrenfest: 'Die Translation deformierbarer Elektronen und der Flächensatz,'" *Annalen der Physik* **23** (1907), 206–208; quotations on p. 206; also in *The Collected Papers of Albert Einstein*, vol. 2, *The Swiss*

First, it applied only to inertial reference systems. Because it had a limited range of applicability, it became known as the "special" or "restricted" theory. On the one hand, it stemmed from the critical outlook that judged as artificial the idea that the laws of physics should have their simplest form relative to a single undetectable reference system. But, on the other hand, it replaced that privileged system with an infinity of special systems, the inertial systems. Why should we restrict the form of physical theory around one special kind of reference frame? Einstein clearly sensed this problem and struggled to solve it. In hindsight, he commented: "I see the most essential thing in the overcoming of the inertial system, a thing that acts upon all processes, but undergoes no reaction. The concept is in principle no better than that of the center of the universe in Aristotelian physics."[2]

The artificiality of the special theory is evident when we realize the extent to which an inertial system is a concept, a construct. In ordinary practice, the Earth is routinely treated as an inertial system. Strictly speaking, however, it is not. It does not move rectilinearly (it rotates), does not move at a constant speed, and is not free of outside influences and forces. Worse, there is no free body in space that has been identified as a true inertial system. All bodies seem subject to interactions with other bodies, at least by gravity. Moreover, the ideal construction of an inertial system as composed of three orthogonal rigid rods, like the axes of a coordinate system, disregards the notion that one would expect gravitational forces to act among the parts of this construction and, in effect, deform its shape. If the shape of such a construction is deformed, then free particles measured relative to such axes would not appear to move according to the first law of motion, that is, they would not traverse equal spaces in equal times.

A further difficulty concerns the notion of empty space. An inertial system, for example, was said to move rectilinearly in empty space. But experimentally we know of no such thing as truly empty space. Even when engineers manufacture a vacuum, there is a wealth of subatomic particles

---

Years: Writings, 1900–1909, ed. John Stachel (Princeton: Princeton University Press, 1989), pp. 410–412.

[2] A. Einstein to Georg Jaffe 19 January 1954 (Albert Einstein Archives, at the Einstein Papers Project, California Institute of Technology, Pasadena, item 13-405), quoted in John Stachel, "Einstein and the Quantum: Fifty Years of Struggle" (1986), Einstein from 'B' to 'Z' (Boston: Birkhäuser, 2002), pp. 367–402, quotation on p. 393.

that can readily be detected in the apparent "void." The walls of any material enclosure seem to be quite permeable to certain kinds of particles. And, given any such enclosure, even its walls emit radiation. Einstein, for one, gradually came to abandon the concept of empty space; he eventually denied its existence. Some theorists believed that space is densely packed with extraordinarily small particles of various kinds. Others came to believe that space is permeated by a continuous homogeneous medium. Although Einstein had dismissed the concept of the ether in 1905, in favor of the concept of empty space, he later changed his mind. He did not pursue a corpuscular theory of radiation and matter in space, but he espoused instead the field concept as that which is everywhere, a continuous medium. He even came to admit the option of designating that medium by the name "ether," at least so long as one does not ascribe to it a state of motion or rest. In any case, the notion of empty space, so convenient in kinematics and particle dynamics, has been judged to be a concept, to some extent unwarranted, changeable, and perhaps even dispensable.

Next, the special theory of relativity assumes the existence of clocks and measuring rods. It has been a subject of some discussion concerning whether a fundamental physical theory should posit such entities at the very outset or not. Einstein himself believed that it was a provisional stopgap. He believed that, preferably, clocks and measuring rods should emerge from the physical theory as by-products rather than as presuppositions. The original form of the special theory, as including phenomenological elements such as observers and pointer readings on clocks, was rooted partly in the limitation that no theory of matter available yielded definitive results. On the one hand, the special theory was appealing because it harmonized with the idea of studying motion as it appears to observation. On the other hand, the behavior of matter is affected by properties that are not "purely" kinematic, so that physicists do not aspire to understand only that which is observable. Nature itself is scarcely divided according to the divisions of knowledge. The kinematic approach was conceived as a convenient starting point for physical analysis, but in that sense it was always preliminary. Physicists sought to further understand phenomena by analyzing material interactions and invisible motions, invisible processes. In this context, the special theory began by positing clocks and rulers as a starting phase for the analysis of physical processes.

Furthermore, the behaviors of bodies such as clocks and rods depend on their rigidity. Yet the notion of length contraction, whether dynamical or kinematic, cast doubts on the property of rigidity. If moving bodies contract, and clocks are composed of moving parts, how would the rate of any such clock be affected by relative distortions among its parts? Parts of a rotating body, for example, would contract as they moved along certain segments of its cycle but not among others. In the end, a new concept of rigidity had to be carefully formulated, in contradistinction to the traditional notion.

Length contraction was itself a subject of interpretive disagreements. Originally, it was supposed to be a dynamical effect, a rearrangement of the molecules of a body. (Even today, some theorists advocate this account.)[3] But following Einstein's work, increasingly many physicists came to construe the effect as merely an appearance, an artifact of certain measurement conventions. There seemed to be no actual contractions of bodies in themselves, but only varying perspectival relationships. Accordingly, many physicists inferred that length contraction was partly an optical effect. They expected that the effect should be observable in principle. However, notwithstanding the role of observers and light in the kinematics of the special theory, the theory of relativistic optics had to be developed considerably to derive what optical effects would be observed. Consequently, it was shown, for example, that according to relativistic optics, three-dimensional bodies in motion would exhibit not so much contraction as rotation relative to the observer.[4] Meanwhile, physicists and philosophers also analyzed the connection between length contraction and synchronization procedures. They showed that different synchrony conventions would generate different apparent contractions or expansions.[5] Such results seemed to support the conjecture that relativistic effects such as contraction were essentially appearances, by-products of measurement conventions. Yet there were various arguments to the contrary. In 1911, Einstein formulated a thought experiment to show the effect of length contraction irrespective of synchroni-

---

[3] In particular, a well-articulated defense of the dynamical explanation is now advocated by Harvey R. Brown, in *Physical Relativity* (Oxford: Oxford University Press, 2005). For arguments to the contrary, see A. Martínez, "There's No Pain in the FitzGerald Contraction, Is There?" *Studies in History and Philosophy of Modern Physics* **38** (2007), 209–215.

[4] James Terrel, "Invisibility of the Lorentz Contraction," *Physical Review* **116**, no. 4 (1959), 1041–1045.

[5] E.g., see John A. Winnie, "Special Relativity without One-Way Velocity Assumptions: Part I," *Philosophy of Science* **37**, no. 1 (1970), 81–99; and "Special Relativity without One-Way Velocity Assumptions: Part II," *Philosophy of Science* **37**, no. 2 (1970), 223–238.

zation procedures.[6] Two identical rods move toward one another relative to the observer's reference frame. The rods overlap as they pass one another, and when each of their respective end points coincide, marks are made on the observer's reference frame. The resulting length should be shorter, according to the special theory, than the length of each rod at rest. By contrast, according to Newton's theory the lengths should be the same. Such predictions offer the possibility of testing experimentally the kinematics of the special theory. As usual, the technical problem is that the velocities involved would need to be very great, comparable to the speed of light, for any such effect to be observable.

Another aspect of the special theory that drew critical attention concerned its postulates. The theory attained empirically verifiable results by positing principles that Einstein himself characterized as freely chosen constructs of the imagination. Having failed to formulate a theory based on hypotheses about the constitution of matter and radiation, Einstein formulated his theory in terms of overarching principles. Yet he saw this structure as a provisional stage. He hoped that theories of principle would eventually be clarified by way of so-called constructive theories. The latter kind of theories explain phenomena in terms of underlying physical structures and processes. Complicated phenomena, such as clocks and electromagnetic radiation, would be understood in terms of the workings of known or hypothetical simpler constituents. For decades after 1905, numerous physicists labored to analyze or circumvent Einstein's elegant but elected postulates. Some physicists sought to justify the two principles on the grounds of a growing wealth of experimental knowledge. Other physicists worked to devise alternative formulations of relativistic physics based on different principles and assumptions. For example, several attempts were made to derive general forms of the Lorentz transformations without using the principle of the constancy of the speed of light.

The light principle, in particular, received a great deal of critical attention. It was occasionally construed as the root of many counterintuitive consequences of the special theory. In classical kinematics, time and space were constant, whereas the velocities of light, like all velocities, were relative. In the special theory, time and space became relative once one adopted the in-

[6] Albert Einstein, "Zum Ehrenfestschen Paradoxon," *Physikalische Zeitschrift* **12** (1911), 509–510; John A. Winnie, "The Twin-Rod Thought Experiment," *American Journal of Physics* **40** (August 1972), 1091–1094.

variance of the speed of light. Physicists and philosophers analyzed various aspects of the light principle. Some questioned whether the speed of light is really the fastest possible speed of a physical transmission. They speculated about superluminal signals and explored the bearing that such would have on the special theory. A few other writers pondered whether there might be variations in the speed of light over the centuries or even decades.[7]

When Einstein made his first derivation of the Lorentz transformations, it seemed counterintuitive that the speed of light could be independent of the inertial motion of reference frames. It might have then seemed conceivable, perhaps, that the speed of light somehow could also be independent of accelerated motions of frames. But soon Einstein admitted that in accelerated frames light rays have many and variable speeds. In order to formulate a theory that encompassed accelerated reference frames, he found it necessary that the velocity of any light ray is observed to vary if the observer accelerates. Still, he believed, as did most physicists, that astronomical and experimental evidence demonstrated, at least, that the speed of light is independent of the speed of its source.[8] By 1965, however, virtually all the early evidence that had been interpreted as proving decisively the independence of the speed of light had been shown to prove nothing of the sort.[9] Again, however, new evidence was attained to substantiate the light postulate.

Furthermore, another controversy over the light postulate concerns the equality of the speed of light in opposite directions. Poincaré and Einstein had asserted that it was a convention, a stipulated definition. Some physicists agreed. But many others believed that the speed of all light is truly equal in opposite directions. It seemed natural. Why would light propagate in space in one direction differently from another? Actually, there were plausible reasons why it might. In a theory of relative motion, the motion of any body or signal is determined relative to reference frames, not space

[7] E.g., see J. H. Rush, "The Speed of Light," *Scientific American* **193**, no. 2 (August 1955), 66–67.

[8] E.g., see Willem de Sitter, "Ein astronomischer Beweis für die Konstanz der Lichtgeschwindigkeit," *Physikalische Zeitschrift* **14** (1913), 429. Wolfgang Pauli, "Relativitatstheorie," *Encyklopadie der mathemathischen Wissenschaften*, vol. 19 (Leipzig: B.G. Teubner, 1921); *Theory of Relativity*, trans. G. Field (New York: Pergamon, 1958). Robert S. Shankland "Conversations with Albert Einstein," *American Journal of Physics* **31** (1963), 49.

[9] J. G. Fox, "Evidence Against Emission Theories," *American Journal of Physics* **33** (1965), 1–17; A. Martínez, "Ritz, Einstein, and the Emission Hypothesis," *Physics in Perspective* **6**, no. 1 (April 2004), 4–28.

itself. Bullets shot in opposite directions are expected to travel at the same speed relative to their source, but not relative to an observer that moves in the same direction as one of the bullets. Likewise, if light signals travel at the same speed in opposite directions relative to one reference frame, the same two signals might travel at different directions relative to another frame. From early on, it seemed plausible that any such difference in the travel times might be undetectable. For example, let two distant clocks, at rest in a system K, be activated when light rays emitted from a middle point reach them. Suppose that one light ray takes three seconds to reach clock A. Suppose that another light ray, traveling in the opposite direction, takes longer, five seconds, to reach clock B. We might say that clock A was activated two seconds before clock B. But the difference would not be observable in that reference frame K. To see the time indicated by each clock, an observer between them has to wait for light to travel back from each clock face. But the light from clock A would take five seconds, and the light from clock B would take three seconds. Both light rays would return to the middle point at the same time. So both clocks would seem to be activated simultaneously, and the observer would not perceive differences in the travel times in opposite directions. Thus, some theorists have sought to somehow design an experimental method that would test or verify the equality of the speeds of light rays sent in opposite directions. They have not found one.

All the critical questions reviewed thus far were acknowledged and examined by physicists and philosophers to various extents. Rather than continue to review relatively well-known points, we now proceed to others. Of course, from a distance it might seem that there are scarcely problems in the foundations of modern kinematics. Just like in the 1800s, physicists nowadays do not turn to kinematics because of any immediate emergency. Owing to the progress of theoretical physics, we find time to look at the elements of physics with a certain leisure. Again, the aim here is to identify conceptual ambiguities and to try to elucidate them.

Seemingly Impossible Measurements

One place where we might expect to find no problems is in Einstein's analysis of the simultaneity. His analysis was carefully scrutinized by many physicists and philosophers, and most of them, by far, agreed with his conclusions. Still, numerous aspects continue to be debated, such as whether the modern definition of simultaneity is conventional.[10] But we do not analyze such questions here, because the goal is to understand the past in terms of its earlier history. Accordingly, we consider first the transition from classical to modern kinematics.

Einstein's analysis of simultaneity showed not only how to establish a relative kind of simultaneity. It critically suggested that the old notion of invariant simultaneity could not be substantiated within the framework of classical kinematics. Traditional concepts that seemed to entail ways to define simultaneity abstractly, did not serve to substantiate it in practice.

For example, consider one pair of clocks fixed at some distance from one another. And consider another identical pair of clocks, but jointly moving relative to the first, on the same plane. If a perfectly rigid and flat body, parallel to the plane of the clocks, were to be pushed downward such that when it touches the top parts of both pairs of clocks it activates them, one would expect that all four clocks could be synchronized simultaneously. In practice, however, bodies of any measurable length undergo deformations when pushed or pulled, such that one would not be justified in assuming that, as the "flat" body moves, its parts make simultaneous contact with the clocks, synchronizing them all at once. Ideal procedures such as this one could not be instantiated in practice because they depend on the imagined concept of a perfectly rigid body.

Likewise, the concept of an infinite speed could be invoked to define invariant distant simultaneity. If the pairs of clocks are all activated by a single signal traveling infinitely fast, it would reach all of them simultaneously, and they would thus be absolutely synchronized. But in practice no physical transmissions of infinite speed are known to exist, so this procedure is impossible.

Furthermore, suppose that one of the pairs of clocks is truly at rest in space. Then light would propagate at the same speeds in opposite directions

---

[10] For a review of the literature, see Ronald Anderson, I. Vetharaniam and G. E. Stedman, "Conventionality of Synchronisation, Gauge Dependence, and Test Theories of Relativity, *Physics Reports* **295** (1998), 93–180. For a subsequent attempt to deny conventionality, see Hans C. Ohanian, "The Role of Dynamics in the Synchronization Problem," *American Journal of Physics* **72** (2004), 141–148; for criticisms, see A. Martínez, "Conventions and Inertial Reference Frames," *American Journal of Physics* **73** (2005), 452–454, and Alan Macdonald, "Comment on 'the Role of Dynamics in the Synchronization Problem' by Hans C. Ohanian," *American Journal of Physics* **73** (2005), 454–455.

relative to those clocks. Such clocks could be synchronized by light signals sent from the midpoint between them. All other clocks in other reference frames could be adjusted to mark the special time of the unique stationary system. Again, however, the concept of an absolutely stationary reference frame was not substantiated in practice either, so such a procedure was impracticable.

Owing to the apparent nonexistence of any unique reference frame, or of instantaneous signals, many physicists inferred that no experimental method could possibly substantiate the old simultaneity concept. For example, Max Born argued in hindsight that "the result of the experimental researches was that it is impossible to detect motion with respect to the ether by physical means. From this it follows that absolute simultaneity can likewise be ascertained in no way whatsoever."[11] Likewise, Louis De Broglie claimed that "the absence of signals which travel at infinite speed results in the impossibility of verifying the simultaneity of two events occurring at points distant from each other. . . . Central to this reasoning is the fact that no signal can travel with a speed greater than that of light in a void."[12]

Strictly speaking, however, others realized that just because no plausible method is known for substantiating a concept within classical physics, that does not necessarily mean that no such method can be devised. Arthur Stanley Eddington formulated a plausible conversation between a classical physicist and an advocate of the special theory:

> *Physicist.* What you have shown is that we have not sufficient knowledge to determine in practice which are simultaneous events . . . . It does not follow that there is no definite simultaneity.
>
> *Relativist.* That is true, but it is at least possible that the reason why we are unable to determine simultaneity in practice (or, what comes to pretty much the same thing, our motion through the aether) in spite of many brilliant attempts, is that there is no such thing as

---

[11] Max Born, *Die Relativitätstheorie Einsteins und ihre physikalischen Grundlagen gemeinverständlich dargestellt* (Berlin: J. Springer, 1920); reissued as *Einstein's Theory of Relativity*, trans. Henry L. Bose (1924); rev. ed., prepared with the collaboration of Günther Leibfried and Walter Biem (New York: Dover, 1962), p. 228.

[12] Louis De Broglie, "A General Survey of the Scientific Work of Albert Einstein," trans. Forrest W. Williams, in Paul Arthur Schilpp, ed., *Albert Einstein, Philosopher-Scientist* (Evanston, Ill.: Library of Living Philosophers, 1949), pp. 109–127, quotation on p. 113.

absolute simultaneity of distant events. It is better therefore not to base our physics on this notion of absolute simultaneity, which may turn out not to exist, and is in any case out of reach at present.[13]

More generally, consider the problem of proving that a given puzzle has no solution. One philosopher argued: "Can you stop a man from looking for a way . . . if he says just that he has not yet found a way? If you say, 'You see, this doesn't work,' . . . he says, 'I know; I haven't found it.'—We simply *decide* that there isn't a way."[14] For the present discussion, the question is then, Within the context of classical kinematics, is there any plausible way to ascertain the invariant simultaneity of distant events? And, if so, what consequences would that have for the modern kinematics? (By invariant simultaneity, we do not mean an absolute simultaneity but one that depends precisely on the relations among material bodies and signals, but which yet yields common results in frames in relative motion.)

First, though, consider a similar question. Is there any way, within the context of classical mechanics that observers in relative motion would agree that their clocks are synchronized? In the simplest case, let two observers move at a constant uniform relative speed toward one another such that their paths are parallel. At the moment when they are adjacent to one another, let their identical clocks be activated to start marking time. According to classical physics, the two clocks would be synchronized. The same procedure could be used to synchronize two or more clocks on each system. Let there be two identical clocks A and B stationed on a rod at rest in one system, and two other identical clocks C and D on a rod at rest in the other. As the systems approach one another, at a given moment clocks B and C coincide, that is, are immediately adjacent to one another and are activated. Then, as these two clocks recede from one another, the classical physicist would expect them to continue to mark a common time. Thus, clock B would serve to synchronize clock D when they coincide, while C would synchronize A. According to classical mechanics, all four clocks would hence be synchronized. The procedure is completely symmetrical. It is also inde-

---

[13] A. S. Eddington, *Space, Time and Gravitation: An Outline of the General Theory of Relativity* (1920; repr., Cambridge: Cambridge University Press, 1923), p. 12.

[14] Ludwig Wittgenstein, in Cora Diamond, ed., *Ludwig Wittgenstein's Lectures on the Foundations of Mathematics, Cambridge 1939* (Ithaca: Cornell University Press, 1976), p. 226.

pendent of the relative speed of the systems, and it is independent of the separations between the clocks—that is, it would yield the same results irrespective of whether the rods involved have equal lengths.

However, the procedure depends on certain physical hypotheses. It presupposes that identical clock mechanisms always run at the same rate. To test this hypothesis we would need to already possess a means of measuring and comparing times. It also presupposes that the rate of clocks is not affected by their relative motions. This physical hypothesis was rejected by Einstein, and eventually by most physicists. The key problem then is that the classical physicist and the modern physicist would disagree on what procedure serves to synchronize clocks. The former would argue that light signals cannot be employed because the speed of light is independent of the speed of its source and has equal speeds in opposite directions relative only to one system. The modern physicist would claim that the procedure with moving clocks depends on the untested hypothesis that clocks keep a constant rate regardless of relative motions.

Moreover, the procedure in question is supposed to produce invariant clock synchrony, but it does not specify any procedure by which a physicist or observer would be able to *verify* such synchrony. How could one, having supposedly synchronized distant clocks in even one reference frame, confirm whether they are indeed synchronized? The question is, How can one determine whether two distant events, such as the pointing of the clock hands to a certain number on a clockface, happen simultaneously?

Any positive answer to this question would have to abide by certain requirements in order to meet the critical standards that Einstein espoused in 1905. A definition of classical simultaneity would need to be formulated as a plausible empirical procedure that could be carried out in precisely the same way in each reference frame to determine exactly whether two events are simultaneous. For example, Einstein argued that two lightning bolts striking a railroad track would be defined as simultaneous if an observer positioned midway between them sees them happen at the same time. The same definition applies to an observer on the ground, as for an observer on a train. The procedure entails a relative simultaneity, even in the context of classical mechanics. Can a different procedure lead to invariant determinations of simultaneity in the context of classical mechanics?

Two lightning bolts strike a railroad track. If there are identical guns stationed on the tracks at the places where lightning strikes, which are thus immediately triggered, classical mechanics allows that the bullets shot in opposite directions can meet exactly at the midpoint. If they do, the classical physicist would judge that the two lightning bolts were simultaneous. Then, if a train carries identical guns that were fixed at the points where the lightning struck, the corresponding bullets will meet midway between the two locations on the train. Observers on each reference frame would thus employ identical procedures, and they would yet obtain the same result. If the events appear simultaneous for one of them, they also appear simultaneous for the other.

The key difference to Einstein's procedure is simply that light signals are not used and that each system employs its own pair of signals, emitted from sources at rest in that system. Of course, the speed of each bullet is assumed to depend on the speeds of its source, according to the classical law of the addition of velocities. The procedure does not escape the logical circularity that in order to measure a velocity one needs to have knowledge of the simultaneity of distant events and that in order to determine distant simultaneity one would need to know a velocity. Neither classical nor modern kinematics escaped that circularity. (As with Einstein's procedure, the velocities involved are not measured.) Regardless, classical kinematics predicts that the experiment would yield invariant distant simultaneity.

We can apply this procedure in the context of Einstein's arguments of 1905. He assumed that light would move with a constant velocity in the "stationary system," such that a light ray used to synchronize clocks on the rod would travel at the same speed relative to the "stationary system." Instead, other signals could be used, such as mechanical projectiles, that would travel at equal though opposite velocities relative to the system from which they are projected. In the "stationary system," let one bullet be shot from one clock to another clock, and once it reaches the latter clock let another bullet be immediately shot back. One could claim the two clocks are synchronized if

$$\overrightarrow{\Delta t} = \frac{rAB}{w} = \overleftarrow{\Delta t} = \frac{-rAB}{-w}.$$

The same procedure can be carried out on a system moving relative to the first. If the speed of the signals depends on the speed v of their source, as required by classical mechanics, then, relative to the "stationary system," the equations that represent their travel times in opposite directions are

$$t_B - t_A = \frac{rAB + v(t_B - t_A)}{w + v} \quad \text{and} \quad t'_A - t_B = \frac{-rAB + v(t'_A - t_B)}{-w + v}.$$

Solving these equations for each time interval gives

$$t_B - t_A = \frac{rAB}{w} \quad \text{and} \quad t'_A - t_B = \frac{rAB}{w},$$

such that now

$$t_B - t_A = t'_A - t_B$$

instead of Einstein's $t_B - t_A \neq t'_A - t_B$. Hence, even in direct analogy to the procedure described by Einstein in 1905, it is possible to synchronize clocks within classical kinematics.[15] The relationship defined is completely symmetrical. No system is selected as unique, no hypothetical medium is required or suggested, and no signals of either infinitely fast or infinitely slow speeds are involved.

Still, irrespective of any alternate definitions of simultaneity or synchronization procedures, Einstein showed that it was possible to establish a procedure to measure time *without* assuming that relative to all reference frames events occur in the same temporal sequence. He demonstrated that determinations of time and simultaneity could vary depending on the reference frame from which such determinations are made. The value of this argument is that Einstein proceeded from this relativized notion of "time" to devise a new kinematics that systematically accounted for the known facts of mechanics, optics, electromagnetism, and many other branches of science.

Still, that classical physics admits a plausible definition of invariant simultaneity is significant. The procedure for verifying the simultaneity of distant events can be inverted in order to define a procedure for synchronizing clocks. That is, let identical bullets be emitted from the midpoint of

[15] Alberto Martínez, "Understanding Simultaneity" (M. A. Thesis, New York University, Gallatin School, 1995).

each system, in opposite directions toward the clocks. Classical mechanics requires that if such bullets are shot simultaneously from identical mechanisms, they will reach both clocks at the same time. This procedure is especially significant because it *also* serves to synchronize clocks within the context of the special theory. Even in the modern context, identical mechanical processes emitted in opposite directions are supposed to transpire in equal times relative to the closed system on which their sources are stationed. Therefore, if this particular procedure was actually carried out, then both the classical physicist and the modern physicist *would agree* that it would serve to synchronize clocks. Once the clocks were "synchronized" thusly, it would be experimentally testable whether the kinematic predictions of one theory or the other are fulfilled. In particular, two pairs of clocks in two separate systems could serve to test the one-way velocity of light. By measuring the travel times of a single ray of light passing by both pairs of clocks, the special theory predicts that the travel time AB will be equal to the travel time CD. Classical kinematics predicts that they will not be equal.

Of course, most contemporary physicists would disregard such a test as utterly unnecessary because they have no shortage of evidence in support of the special theory. But nearly all such evidence is electromagnetic and dynamical rather than kinematic. By having an experimental setup that is compatible with both theories, it becomes possible to test propositions that are otherwise beyond the reach of experimental testing because they are presupposed by the fundamental measuring conventions of each theory. If the one-way speed of a single ray of light is measured to be constant relative to two reference systems (that do not employ clocks synchronized by light), then that result would constitute experimental evidence that the constancy of the speed of light in one direction is not merely a convention. (Physicists might still then say that the equality of speeds of mechanical transmissions in opposite directions remains an untested convention.) Otherwise, if the same ray were measured to have different speeds, it would show either that the second postulate of the special theory is false or at least that, oddly, mechanical synchronization does not agree with electromagnetic synchronization.

An Old-Fashioned Preference for Coordinates

How did modern kinematics incorporate or neglect basic concepts employed traditionally for the description of motion? To construct a radically new theory of motion, Einstein began by considering "a coordinate system in which the equations of Newton hold."[16] Yet one of the main consequences of Einstein's theory was that it showed that Newton's equations were, strictly speaking, incorrect. For more than two hundred years, physicists had relied on Newton's theory of motion, and many had regarded its equations as examples of the highest degree of certainty to which human knowledge could aspire. Relativity theory changed that. Einstein argued that fundamental concepts of Newton's physics were unjustified. In presenting his arguments, Einstein began by supposing that Newton's physics was correct to then show that it was actually deficient.

Einstein based his original argument on an inspection of the procedures by which physicists could measure motion and time. He argued that position and motion could be determined by using "the methods of Euclidean geometry, and can be expressed in Cartesian coordinates." Again, this statement also seems peculiar because a further major consequence of Einstein's theory was that it convinced physicists eventually that the geometry of Euclid was *not* exactly valid in physics. For more than two thousand years, people had regarded Euclid's geometry as a systematic statement of commonsense properties of all things. For example, it seemed natural to conceive of space as three-dimensional. Yet following the invention of non-Euclidean geometries, physicists occasionally wondered whether the structure of nature really matched the old geometry. In 1906 Poincaré pointed out that physical quantities that are invariant under Lorentz transformations could well be described by a four-dimensional space. Then, in 1908, Hermann Minkowski claimed that this four-dimensional geometry was physically necessary. One could then still consider the geometry of Euclid valid but only within individual partitions of space-time. Consequently, physicists adopted the idea that the traditional geometry was not the best mathematics with which to represent and analyze physical phenomena exactly. Nonetheless, Einstein's original work began by assuming the validity of Euclidean geometry.

Einstein's first relativity paper lies at the juncture between changing conceptions of mathematical physics. The transition from Newton's physics to Einstein's physics has received much attention from scientists, historians, and philosophers. Likewise, the transition in physics from the use of the geometry of Euclid to the use of non-Euclidean geometries has been studied

---

[16] Einstein, "Zur Elektrodynamik bewegter Körper," *Annalen der Physik* **17** (1905), 892.

extensively. With these changes, a new understanding of the relationship between physics and mathematics developed. However, one area of this changing relationship remained unanalyzed. From the initial statements in Einstein's 1905 paper to the many later contributions and commentaries by other writers, it seemed clear that the advance of physics was tied to critical analyses of the concepts of Newton and Euclid. Yet, in the preceding two centuries, the physics of Newton and the geometry of Euclid were commonly related by an auxiliary mathematical language, namely, the analytic geometry of "Cartesian coordinates." Thus, what escaped critical inspection was the role played by the mathematics originated by Descartes.

Einstein's theory led to the critical rejection of basic concepts of Euclid's geometry and Newton's mechanics. Traditionally, these concepts had been held in the highest esteem by philosophers, mathematicians, and scientists who regarded them as examples of logical clarity and perfection. By contrast, for centuries the foundations of algebra had been criticized and viewed with suspicion, as lacking logical foundations. Ironically, relativistic kinematics emerged from traditional algebraic methods.

The study of practically any kind of phenomena—heat, light, electricity—each depended on the analysis of motion. This analysis was commonly carried out in the mathematics of analytic geometry, also known as Cartesian algebra. Yet by the late nineteenth century, several other new mathematical approaches had become available for physicists. In particular, vector algebra had emerged as a popular alternative to coordinate methods. Since the 1880s vector mathematics had been especially useful in the fields of electricity, dynamics, and kinematics. Physicists, led by Heaviside and Gibbs, developed vector methods tailored to the needs of physics. Hence, traditional methods were criticized as being poorly suited for physics, by comparison to vector methods. Thus, some hoped or believed that the use of Cartesian algebra in physics would be replaced entirely by the use of vector algebra. But this did not happen. Any inspection of later textbooks readily shows that although vector methods are employed in physics, their use is commonly restricted to certain areas of physics, while other areas are typically presented in traditional coordinate algebra. In particular, college textbooks introduce vector methods in the opening chapters as fundamental for analysis in the areas of mechanics and electromagnetism, but then in later chapters, without explanation, proceed to employ coordinate algebra for the treatment of relativity theory.

Einstein based his work on a revision of the concepts of kinematics and conceived it with the aim of constructing a theory of the motion of electric particles. Vector methods had been shown to be effective precisely in analyses of motion and of electricity. But Einstein did not use vector methods to formulate relativity theory. He used coordinate algebra. Why?

Perhaps algebra seemed sufficiently easy and trustworthy for research in theoretical physics. In those days, Einstein "still believed that an elementary knowledge of mathematics was a sufficient prerequisite for further work in physics."[17] In his childhood, he had first become acquainted with algebra from his uncle, Jakob Einstein, who had explained its nature in the following words: "Algebra? It's a lazy kind of arithmetic. What you don't know you simply call x and then look for it."[18] In his "Autobiographical Notes," Einstein recalled that before the age of sixteen his studies of "the elements of mathematics" were "truly fascinating," and he characterized his acquaintance with "the basic idea of analytical geometry" and the calculus as "climaxes . . . whose impression could easily compete with that of elementary geometry."[19] He employed the methods of algebra skillfully and confidently. In 1896 he finished course work at the cantonal school in Aarau as best in his class in algebra. Later at the Zurich Polytechnic, he registered for three algebraic courses, a decision that was "unusual for a physical scientist of the period," as noted by historian Lewis Pyenson, "even for one expressing strong interest in mathematics."[20] At the time, by contrast, "Traditional mathematical preparation for physicists consisted of analysis, geometry, and mechanics, each defining a large domain in applied mathematics. Together with arithmetic, algebra constituted one of the most important fields of pure mathematics at the end of the nineteenth century."[21]

Given that algebra was taught mainly as a field of pure mathematics rather than for its applications in physics, why did Einstein pursue the subject? As Pyenson suggests, "To a student who sought powerful mathematical syntheses that also held the promise of being solutions to practical

[17] Anton Reiser, *Albert Einstein: A Biographical Portrait* (New York: Albert & Charles Boni, 1930), p. 52.

[18] Ibid., p. 35.

[19] Einstein, "Autobiographisches" (1946), trans. Paul Arthur Schilpp, in Schilpp, *Albert Einstein: Philosopher-Scientist*, pp. 14, 15.

[20] Lewis Pyenson, *The Young Einstein: The Advent of Relativity* (Boston: Adam Hilger, 1985), p. 21.

[21] Ibid.

problems, algebra might have seemed attractive."[22] Indeed, one branch of modern algebra in particular, group theory, served Einstein well in developing his theory of relativity. Already in 1886, for example, Josiah Willard Gibbs had commented, "Our Modern Higher Algebra is especially occupied with the theory of linear transformations."[23] Accordingly, Minkowski occasionally employed coordinate transformations to derive results in his lectures at the Zurich Polytechnic.[24] Einstein may have been attracted to certain general methods of algebraic analysis. Although he registered for a total of nine courses with Minkowski, his attendance was rare. Einstein recollected, "For me, Minkowski was somewhat difficult to follow, because I often did not see just what was essential."[25] In the words of Einstein's son-in-law, "No one could stir him to visit the mathematical seminars."[26] Minkowski reportedly commented that "in his student days Einstein had been a lazy dog. He never bothered about mathematics at all."[27] In 1912 Einstein himself wrote that until then, in his "simple-mindedness," he had considered the "subtler parts" of mathematics merely as "pure luxury."[28]

After his early childhood trust in geometry dissipated, basic algebraic analysis remained as a reliable standard mathematics for the investigation of fundamental questions in physics. Einstein's distrust for geometric constructions may be illustrated by noting that one of his co-workers at the Swiss patent office in the early 1900s, the physicist Josef Sauter, recalled that in those days Einstein "reproached" Sauter's "predilection for

---

[22] Ibid., p. 22.

[23] Josiah Willard Gibbs, "On Multiple Algebra," *Proceedings of the American Association for the Advancement of Science* 35 (1886), 37–66; reprinted in *The Scientific Papers of J. Willard Gibbs*, vol. 2 (New York: Dover, 1961), p. 96.

[24] Pyenson, *The Young Einstein*, p. 23.

[25] Albert Einstein to Walter Leich, 24 April 1950, Einstein Archives, item 60-252 .

[26] Reiser, *Albert Einstein: A Biographical Portrait*, p. 51. "It should be noted," comments Russell McCormmach, "that the Zurich Polytechnic generally and Minkowski in particular did not attract many students in its advanced mathematics courses." In McCormmach and Christa Jungnickel, *Intellectual Mastery of Nature*, vol. 2, *The Now Mighty Theoretical Physics, 1870–1925* (Chicago: University of Chicago Press, 1986), p. 335.

[27] According to Max Born, in Carl Seelig, *Albert Einstein: A Documentary Biography*, trans. Mervyn Savill (London: Staples Press Limited, 1956), p. 28.

[28] Einstein quoted by Russell McCormmach, "Editor's Foreword," in *Historical Studies in the Physical Sciences*, ed. McCormmach, Seventh Annual Volume (Princeton: Princeton University Press, 1976), p. xxviii.

graphic solutions."[29] In contradistinction, the works of many writers might have fostered Einstein's trust in the validity of algebraic reasoning.[30] In particular, he may have shared Helmholtz's opinion that the difficulties involved in the traditional formulation of the principles of geometry could be avoided by employing "the analytic method developed in modern calculative geometry."[31] Despite Einstein's reservations toward mathematics in general, algebra stands out as a field in which he apparently was quite comfortable.

By the early 1900s, many physicists were aware of the uncertainty elicited in traditional geometry by the non-Euclidean geometries, yet most remained quite unaware that similar developments also had cast doubt on the foundations of algebra. Einstein, in particular, apparently was hardly acquainted with Hamilton's work on quaternions, and only years later did he begin to read Grassmann's work.[32] At least he was acquainted with vector theory, owing to his studies of the works of Föppl, Hertz, and others. By then, vector theory was presented basically as a convenient shorthand for traditional methods of analysis. Most physicists did not construe vector theory as a critical demonstration that the principles of algebra could

[29] Josef Sauter, statement delivered at the conference "50 Jahre Relativitätstheorie" in Bern (1955); reprinted in Max Flückiger, *Albert Einstein in Bern: Das Ringen um ein neues Weltbild; Eine dokumentarische Darstellung über den Aufstieg eines Genies* (Berne: Paul Haupt Berne, 1974), p. 158.

[30] See, e.g., Arthur Schopenhauer, *Die Welt als Wille und Vorstellung*, vol. 2 (1844); *The World as Will and Representation*, trans. E. F. J. Payne, (New York: Dover, 1966), pp. 73, 179. That Einstein read works by Schopenhauer was noted by Rudolph Kayser, Einstein's son-in-law, in Reiser, *Albert Einstein: A Biographical Portrait*, p. 55.

[31] Hermann von Helmholtz, "Über den Ursprung und die Bedeutung der geometrischen Axiome" (1868); *Vorträge und Reden*, 5th ed., vol. 2 (Braunschweig: F. Vieweg, 1884), pp. 1–31; reissued in Helmholtz, *Epistemological Writings*, trans. Malcolm F. Lowe and ed. Robert S; Cohen and Yehuda Elkana, Boston Studies in the Philosophy of Science, vol. 37 (Boston: D. Reidel, 1977), p. 5, or see the longer quotation in chapter 4 of the present work.

[32] See Einstein's letter to Michele Besso, of May 1911, published in Einstein and Besso, *Correspondance, 1903–1955* (Paris: Hermann, 1972), p. 21; also in *Collected Papers of Albert Einstein*, vol. 5, *The Swiss Years: Correspondence, 1902–1914*, ed. Martin J. Klein, A. J. Kox, and Robert Schulmann (Princeton: Princeton University Press, 1993), p. 25. Later in life Einstein expressed admiration for Grassmann's work; see Einstein's letter to Besso, of 30 June–2 July 1941, published in Einstein and Besso, *Correspondance*, p. 361.

be altered. They did not interpret vector algebra as a possible challenge to traditional notions of mathematical truth. It seemed reasonable to pursue or disregard vector methods because they were allegedly equivalent to the methods of analytic geometry. Accordingly, Einstein developed his kinematics without explicitly employing any methods or even notations of vector theory.

The 1905 paper exhibits minimal contact with vector concepts. The "Kinematic Part" did not make a single direct reference to vector concepts, notation, or terminology. The "Electrodynamical Part" at least included the term *vector*, although merely as a word, for it operated entirely with coordinate methods and notation. Decades earlier, Maxwell had employed prominently the notations and concepts, if not the methods, of vectors and quaternions. Likewise, other leading theorists such as Heaviside, Hertz, Föppl, Abraham, and Lorentz had employed vectors further in electromagnetic theory. By contrast, Einstein based his electromagnetic theory on a system of kinematics that did not refer even minimally to vector notions.

It seems plausible to imagine that Einstein formulated relativity theory in coordinate algebra because vector methods were still relatively new, and he simply used the mathematics that was most familiar to him and to the community at large. It seems that relativity theory owed more to conceptual rather than mathematical advances in kinematics. But regardless of the extent to which such factors were involved, there lay an issue of greater significance. Physicists did not merely neglect the use of vector algebra to formulate relativity theory; ordinary vector algebra could not be used to derive relativity theory. This is remarkable because physicists believed that vector algebra and coordinate algebra were equivalent, in that either could be used to obtain the same results. Were they mistaken?

Scarcely anything has been written about the divergence of relativistic kinematics from the vector concepts of Gibbs and Heaviside. Historians have briefly noted that Einstein relied fully on coordinate notations, and continued to do so, at a time when vectors had become commonplace.[33] Some writers have briefly noted the divergence of relativistic kinematics

---

[33] Arthur I. Miller, *Albert Einstein's Special Theory of Relativity: Emergence (1905) and Early Interpretation (1905–1911)* (Reading, Mass.: Addison-Wesley, 1981; repr., New York: Springer, 1998), p. 271; Françoise Balibar, Olivier Darrigol, Bruno Jech, and John Stachel, eds., *Œuvres Choisies d'Albert Einstein*, vol. 2, *Relativités I* (Seuil: Éditions du Seuil, Éditions du CNRS (Mame Imprimeurs), 1993), p. 45.

from vector concepts by observing that in the theory of relativity the traditional "law of addition of velocities" does not apply.[34] By far, most books present the Lorentz transformations in terms of coordinate geometry. Comparatively few also include the transformations formulated in actual vector notation. Writers do not derive the vector formulation of the Lorentz transformations solely from vector concepts; instead, they first derive the coordinate expressions and then modify them into vector form.[35]

Several issues emerge in connection with Einstein's use of the ordinary methods of analysis instead of vectors. First, Einstein's new kinematics entailed a focus on physical quantities that are invariant in different reference frames, a subject for which vectors seemed ideally suited. For example, Edwin B. Wilson observed that "vectors are to mathematical physics what invariants are to geometry."[36] Second, it is noteworthy that Einstein chose not to employ vector methods, although he aimed to solve fundamental problems in electromagnetism, precisely the area where vectors had been proved to be most useful. Furthermore, because Hamilton had construed algebra as the "Science of Pure Time," it seems noteworthy that Einstein developed his new conception of time based essentially on an algebraic analysis. Although he was not familiar with Hamilton's writings, he could have been aware of the association of the concepts of arithmetic or algebra with the notion of time from his studies of the works of Kant, Schopenhauer, Richard Dedekind, and Helmholtz.[37] Moreover, the works of Poin-

---

[34] E.g., see Albert Shadowitz, *Special Relativity* (New York: Dover, 1968), p. 78. See also Banesh Hoffmann, *About Vectors* (New York: Dover, 1966), p. 25.

[35] See, e.g., Yuan Zhong Zhang, *Special Relativity and Its Experimental Foundations,* Advanced Series on Theoretical Physical Science, vol. 4 (Singapore: World Scientific, 1997), p. 25; and C. Møller, *The Theory of Relativity* (Oxford: Clarendon Press, 1952), p. 41.

[36] Edwin B. Wilson, *Vector Analysis* (New Haven, Conn.: Yale University Press, 1901), p. xii.

[37] As mentioned in chapter 5, at an early age Einstein's favorite philosopher was Kant. Therefore, see Kant, *Prolegomena,* §10, quoted by Norman Kemp Smith, in *A Commentary to Kant's Critique of Pure Reason,* 2nd ed. (London: Macmillan, 1923), p. 129; and *Kritik der reinen Vernuft* (1781), *Critique of Pure Reason,* trans. and ed. Paul Guyer and Allen W. Wood, Cambridge Edition of the Works of Immanuel Kant (Cambridge: Cambridge University Press, 1998), pp. 144, 633–636. See also Arthur Schopenhauer, *Vierfache Wurzel des Satzes von zureicheneden Grunde,* translated by Karl Hillebrand as "On the Fourfold Root of the Principle of Sufficient Reason," in *Two Essays by Arthur Schopenhauer,* rev. ed. (London: George Bell and Sons, 1891), chap. VI, §38, p. 156. Schopenhauer, *Die Welt als Wille und Vorstellung,* vol. 2 (1844); chap. 4: "Von der Erkenntniβ

caré, Einstein, and Minkowski engendered the notion of a four-dimensional physical manifold, where one dimension corresponds to time (as suggested earlier by Lagrange), in a way similar to Hamilton's interpretation of the fourth term of a quaternion as representing a scale of values of time.[38] Hamilton believed that the "sciences of Space and Time" were "intimately intertwined," but that ordinary analytic methods were inadequate for the integration of the two, especially in physics.[39] It thus seems ironic that, although vector theory originated partly from Hamilton's critique of algebra and his metaphysical ideas about time, Einstein's critical revision of the physical concept of time emerged from traditional algebraic analysis.

A related issue concerns the sense in which early relativity theory received a vectorial interpretation. After the works of Poincaré and Einstein, theorists reformulated relativistic kinematics in terms of "four-dimensional vectors."[40] This movement was impelled by Minkowski, who pursued a symmetric interpretation of the parameter t analogous to the coordinates x, y, z.[41] In Minkowski's interpretation, the concept of vector summarized

a priori"; translated by Payne in The World as Will and Representation, pp. 34–35, 45, 51, 179, 379. Schopenhauer explicitly includes algebra with arithmetic as constituting the "doctrine of the ground of being in time" only in chapter 7, p. 127. Hermann von Helmholtz, "Zählen und Messen," Philosophische Aufsätze Eduard Zeller zu seinem fünfzigjährigen Doktorjubiläum gewidmet (Leipzig: Fues' Verlag, 1887); translated in Helmholtz, Epistemological Writings, especially pp. 72–77.

[38] Thomas L. Hankins, Sir William Rowan Hamilton (Baltimore: Johns Hopkins University Press, 1980), p. 309.

[39] William Rowan Hamilton, "Account of a Theory of Systems of Rays," presented to the Royal Irish Academy (23 April 1827); reprinted in R. P. Graves, Life of William Rowan Hamilton, vol. 1 (Dublin: Hodges, Figgis, & Co., 1882), p. 229.

[40] Hermann Minkowski, "Die Grundgleichungen für die elektromagnetischen Vorgänge in bewegten Körpern," Nachrichten der Gesellschaft der Wissenschaften zu Göttingen, Mathematisch-physikalische Klasse (1908); pp. 53–111; reprinted in Gesammelte Abhandlungen von Minkowski, vol. 2 (New York: Chelsea, 1967), 352–404. Arnold Sommerfeld, "Zur Relativitätstheorie, I. Vierdimensionale Vektoralgebra," Annalen der Physik, ser. 4, 32 (1910), 749–776. Arnold Sommerfeld, "II. Vierdimensionale Vektoranalysis," Annalen der Physik, ser. 4, 33 (1910), 649–689.

[41] Ibid., and Hermann Minkowski, "Raum und Zeit," lecture delivered at the 80th Naturforscher-Versammlung at Cologne on 21 September 1908; published in Phyikalische Zeitschrift 10 (1909), 104–111; reprinted in Gesammelte Abhandlungen von Minkowski, vol. 2, pp. 432–444; translated by W. Perrett and G. B. Jeffery as "Space and Time," in Einstein, Lorentz, et al., The Principle of Relativity, with notes by Arnold Sommerfeld (New York: Dover, 1952), see particularly pp. 84–90.

coordinate-analytic notions. Previously, vector theorists had advocated the priority of vectors by conceiving them as consisting *fundamentally* of direction and magnitude and only incidentally as expressible in Cartesian coordinates.[42] Therefore, Tait, Heaviside, Föppl, and others sought to replace the so-called Cartesian methods. In Minkowski's work, however, coordinates appeared as fundamental. Moreover, a central idea for Minkowski was the four-part algebraic expression consisting of three spatial scalars and one imaginary temporal parameter,

$$x^2 + y^2 + z^2 + (it)^2,$$

where $it = itc = it(1)$, whereas Hamilton had based his work on a four-part algebraic expression consisting of three imaginary spatial magnitudes and one scalar temporal parameter,

$$ix + jy + kz + t.$$

The similarity of these formalisms led a number of theorists after 1911 to formulate relativity theory in terms of quaternions.[43]

Rather than pursue such developments here, we consider more elementary concepts. Vector theory enabled physicists to systematically distinguish directed and nondirected magnitudes. Such distinctions were lacking in

---

[42] See, e.g., Peter Guthrie Tait, "On the Importance of Quaternions in Physics," *Philosophical Magazine* (January 1890), and "On the Intrinsic Nature of the Quaternion Method," *Proceedings of the Royal Society of Edinburgh* (2 July 1894); both reprinted in Tait, *Scientific Papers* (Cambridge: Cambridge University Press, 1900), vol. 2, documents XCVII and CXVI. Heaviside, *Electromagnetic Theory*, vol. 1 (London: Benn Brothers Ltd., The Electrician Printing and Publishing Company, 1894), chap. III: "The Elements of Vectorial Algebra and Analysis."

[43] E.g., A. W. Conway, "On the Application of Quaternions to Some Recent Developments of Electrical Theory," *Proceedings of the Royal Irish Academy*, A 29 (1911), 1–9; L. Silberstein, "Quaternionic Form of Relativity," *Philosophical Magazine* 23 (1912), 790–809; A. W. Conway, "The Quaternionic Form of Relativity," *Philosophical Magazine*, ser. 6, 24 (1912), 790–809; P. A. M. Dirac, "Application of Quaternions to Lorentz Transformations," *Proceedings of the Royal Irish Academy*, A 50 (1945), 261–270. For a review of the subject and additional references, see J. L. Synge, *Quaternions, Lorentz Transformations, and the Conway Dirac-Eddington Matrices, Communications of the Dublin Institute for Advanced Studies*, ser. A, no. 21 (Dublin: Dublin Institute for Advanced Studies, 1972).

traditional methods. Was this deficiency inherited in the new kinematics? Hermann Grassmann, in support of his novel geometric analysis of lines, argued that in physics "this new analysis is indispensable if one is not to annihilate the intuition by the introduction of coordinates and other apparatus disturbing the treatment, and entangling the method in useless details."[44] Heaviside, too, claimed that "for general purposes of reasoning the manipulation of the scalar components instead of the vector itself is entirely wrong."[45] Moreover, Peter Guthrie Tait argued at length, "The intensely artificial system of Cartesian coordinates, splendidly useful as it was *in its day,* is one of the wholly avoidable encumbrances which now retard the progress of mathematical physics."[46] And likewise, in advocating vectors in kinematics, Paul Appell and James Chappuis complained in 1905 that "the abuse of the methods of analytic Geometry destroys the intuition and the spirit of invention."[47]

In view of such claims, it is remarkable that Einstein's reasoning leading to his new kinematics was based precisely on the analytic manipulation of coordinates. By analyzing basic equations of Einstein's kinematics, we may begin to gauge whether it incorporated ambiguities and "encumbrances" rooted in traditional algebra.

Consider again the mathematical core of modern kinematics, the equations

$$t' = \beta(t - vx/c^2), \quad x' = \beta(x - vt),$$
$$y' = y, \quad z' = z,$$

where

$$\beta = 1 / \sqrt{1 - v^2/c^2}.$$

[44] Hermann Grassmann, *Geometrische Analyse genüpft an die von Leibniz erfundene geometrische Charakteristik* (Leipzig: Weidmann'sche Buchhandlung, 1847); *Geometric Analysis* in *A New Branch of Mathematics, The Ausdehnungslehre of 1844 and Other Works*, trans. Lloyd C. Kannenberg (Chicago: Open Court, 1995), p. 384.

[45] Heaviside, *Electromagnetic Theory,* vol. 1, p. 298.

[46] Tait, "On the Importance of Quaternions in Physics," *Philosophical Magazine* (1890); reprinted in Tait, *Scientific Papers,* vol. 2, emphasis in the original.

[47] Paul Émile Appell and James *Chappuis, Leçons de Mécanique Élémentaire a l'usage des Élèves des Classes de Première C et D, conformément aux Programmes du 31 mai 1902* (Paris: Gauthier-Villars, 1905), p. vi.

Because these equations underlie many others in physics, the factor β (now commonly known as γ) is widespread. It seems to be a distinctive mathematical signature of relativistic kinematics. As indicated earlier, Einstein introduced it rather abruptly in his first derivation of the transformation equations, and he did not indicate its exact physical significance. This deficiency may be remedied by explaining its algebraic origin. Einstein's use of coordinates incorporated one of the main features of Cartesian algebra that was criticized insistently by vector theorists. Specifically, the telltale β factor was rooted in the traditional way of expressing diagonal distances in analytic geometry. This can be shown as follows.

By the Pythagorean theorem, the sides a and b of a right triangle are related to its hypotenuse by $a^2 + b^2 = c^2$. In the study of motion, according to the parallelogram rule, two motions in perpendicular directions would be composed into one by writing

$$(v\Delta t)^2 + (w\Delta t)^2 = (c\Delta t)^2,$$

where $v\Delta t$, $w\Delta t$, and $c\Delta t$ are the distances, or products of speed and time, in each direction (here Δt represents a time interval, a duration). Now, when a motion is analyzed with reference to two inertial frames, we must distinguish how it is measured relative to each frame. If a light ray is projected in a direction perpendicular to a platform, the speed of this ray relative to the platform may be described as $c'\Delta t'$. Meanwhile, from the vantage point of another platform, if the first platform moves at a speed v and covers a distance $v\Delta t$, the same ray of light will seem to traverse a diagonal path $c\Delta t$. Without making key assumptions about the relationship or equivalence of the measurements of time and speed made from either platform, we may distinguish between the values measured in one and the other by writing

$$(v\Delta t)^2 + (c'\Delta t')^2 = (c\Delta t)^2. \tag{1}$$

The diagonal path $c\Delta t$ may be expressed by

$$\sqrt{(v\Delta t)^2 + (c'\Delta t')^2} = c\Delta t.$$

Vector theorists would agree that it was reasonable to designate a path simply as $c\Delta t$, whereas to identify the same path in terms of *a square root of a sum of squares* seemed unnecessarily artificial and inconvenient. Hence, Peter Guthrie Tait observed that vector concepts "enable us by a mere mark to separate the ideas of length and direction without introducing the cumbrous square roots of sums of squares which are otherwise necessary."[48] Yet Einstein's algebraic kinematics essentially involved just this "cumbrous" expression. Notice how readily the β factor follows from expression (1):

$$c'^2(\Delta t')^2 = (\Delta t)^2(c^2 - v^2)$$

$$\frac{c\Delta t}{c'\Delta t'} = \frac{1}{\sqrt{1 - v^2/c^2}}. \tag{2}$$

Hence, the β factor arises entirely out of the algebraic formulation of the Pythagorean expression relating the trajectory of a single ray of light described in two reference frames in uniform relative motion.

Equation (2) also clarifies an algebraic connection between modern and classical kinematics. The special theory requires that the speed of light is invariant. Thus, we must substitute $c = c'$ in equation (2),

$$\frac{c\Delta t}{c\Delta t'} = \frac{1}{\sqrt{1 - v^2/c^2}},$$

such that we get

$$\Delta t = \frac{\Delta t'}{\sqrt{1 - v^2/c^2}},$$

which is just Einstein's well-known "time-dilation" equation. It expresses the relation between intervals of time in different reference frames. Thus, in the algebraic formulation of the ratio of light distances, the mere adoption of the assumption of the invariance of the speed of light yields essential mathematical content of the modern theory of motion. Alternately, the same equation (2) can be used to derive a similarly essential equation of

[48] Peter Guthrie Tait, "Address to Section A of the British Association," *British Association Report*, Edinburgh (3 August 1871); reprinted in Tait, *Scientific Papers*, vol. 1 (Cambridge: Cambridge University Press, 1898), p. 164.

classical kinematics. The old assumption that time is the same in all reference frames translates simply into the requirement that $\Delta t = \Delta t'$. Substituting into (2) gives

$$\frac{c\Delta t}{c'\Delta t} = \frac{1}{\sqrt{1-v^2/c^2}},$$

so that the times cancel out and the expression can be reformulated as

$$c = \sqrt{c'^2 + v^2},$$

or, because the relative speed between the frames is the same, $v = v'$, we may also write

$$c = \sqrt{c'^2 + v'^2}.$$

These two preceding equations are just algebraic expressions of the parallelogram law of the transformation of speeds. Hence, with respect to equation (2), the only algebraic difference between the basic content of classical and that of modern kinematics is whether one assumes the invariance of time or the invariance of the speed of light. So we see that the β factor by itself is not a purely special-relativistic term. It is just the ratio of the distances traversed by a ray of light relative to two reference frames:

$$\frac{c\Delta t}{c'\Delta t'} = \beta.$$

This equation, simply put, significantly illustrates an old tradition in the analysis of motion. It is reminiscent of the argument voiced by d'Alembert and others, that motion should be understood quantitatively not really by "dividing a space by a time" but by comparing motions. Hence relativistic kinematics systematically incorporated ratios of the determination of one motion in one system to its determination in another.

Asymmetries in the Basic Algebra

To further elucidate ways in which Einstein's work inherited aspects of traditional algebraic kinematics, consider now the concepts of position and motion. Writers have extensively analyzed Einstein's definition of time, but scarce attention has been paid to his definitions of position and motion.

To formulate the mathematical concepts of motion, Einstein began with three basic definitions. First, he defined the concept of a coordinate system, as the reference frame relative to which any motion may be ascertained. Second, he defined the concept of the position of a material point at rest. Third, he defined the idea of motion of a material point. Einstein's first specification, the coordinate system, was discussed in the previous chapter. As for position, he defined it as follows: "If a material point is at rest relatively to this coordinate system, its position can be defined relative to it by rigid measuring rods employing the methods of Euclidean geometry, and can be expressed in Cartesian coordinates."[49] Thus he indicated the physical procedure that could be employed to establish the position of a body relative to a reference system. He imagined physical objects to be located at "points" of space corresponding to a coordinate system. For example, he referred to "points at which clocks are located."[50] The location of objects, such as material point particles, would be described by three numbers or variables corresponding to coordinates along three axes X, Y, Z. He stated, for example, that "x, y, z denote the electron's coordinates."[51] As usual, he expressed position in terms of coordinates as three terms inside parentheses, stating for instance that "(x, y, z) is a point."[52] He also included the parameter of time as a fourth term, such that "any system of values x, y, z, t, completely determines the place and time of an event" relative to a specific system.[53]

Einstein next defined a mathematical concept of motion: "To describe the *motion* of a material point, we give the values of its coordinates as functions of time."[54] Later, in his book of 1917, he emphasized the importance of this idea by presenting it as the main aim of the entire science of mechanics: "The purpose of mechanics is to describe how bodies change their position in space with 'time.'" He immediately added: "I should load my conscience with grave sins against the sacred spirit of lucidity were I to formulate the aims of mechanics in this way, without serious reflection

[49] Einstein, "Zur Elektrodynamik bewegter Körper," p. 892.
[50] Ibid., p. 898.
[51] Ibid., p. 917.
[52] Ibid., p. 901.
[53] Ibid., p. 898.
[54] Ibid., p. 892.

and detailed explanations."[55] So he proceeded to analyze the concepts of position, coordinate system, and time, emphasizing that "in order to have a *complete* description of the motion, we must specify how the body alters its position *with time; i.e.* for every point on the trajectory it must be stated at what time the body is situated there."[56]

Despite the apparent simplicity of such statements, several basic mathematical concepts remained ambiguous. Einstein's verbal expressions suggest that, in his conceptualization of motion, the idea of position played a prime role alongside the concept of time. Expressions such as

$$x = ct \quad \text{and} \quad x = vt$$

were understood by Einstein as expressing the functional relationship of a coordinate of position to any instant of time. This interpretation was quite distinct from another traditional definition of the concepts of motion as the ratio of spatial *intervals* over *intervals* of time, for example,

$$c = \frac{(x_2 - x_0)}{(t_2 - t_1)} \quad \text{and} \quad v = \frac{(x_1 - x_0)}{(t_2 - t_1)}.$$

Here, the x_n are coordinates that designate positions along one axis. Because physicists usually preferred succinct notations, the subtle difference between spatial and temporal intervals, as opposed to spatial positions and temporal instants, was often disregarded. The distinction was obscured further owing to the devices of the calculus. Because velocity was conceived as a ratio of infinitesimally small distance and time intervals, these intervals could be confused easily with points, such as single coordinates, instead of being identified as line segments between pairs of coordinates.

Basic notions of the differential calculus facilitated confusion between kinematic concepts. Like vector theory, the differential calculus hardly dealt explicitly with the concept of position. For example, the equation,

$$v = \frac{dx}{dt},$$

[55] Albert Einstein, *Über die Spezielle und die Allgemeine Relativitätstheorie* (Brauschweig: Friedrich Vieweg & Sohn, 1917); *Relativity: The Special and the General Theory*, trans. Robert W. Lawson (New York: Crown, 1961) p. 9, emphasis in the original.
[56] Ibid., p. 10.

merely expresses the ratio of an infinitesimally small line element to an infinitesimally small time interval. It conveys no indication of the positions that bound the infinitesimal interval dx, or of the time instants that bound the infinitesimal duration dt. The concept of position emerges clearly only when the expression in question is defined in terms of coordinates:

$$v = \frac{dx}{dt} = \lim_{\Delta t \to 0} \frac{\Delta x}{\Delta t} = \lim_{t_2 \to t_1} \frac{(x_2 - x_1)}{(t_2 - t_1)}.$$

Physicists and mathematicians could say, of course, that because t_2 approaches t_1 by an infinitesimally small amount, in the limit the two are virtually identical, such that t by itself suffices to designate a single instant of time, and x designates a single point. But that was precisely the root of an ambiguity. By abstractly conceiving spatial lengths or temporal durations as being infinitely small, symbols that designated intervals could be construed as points. Instead of always conceiving of velocity as the ratio of intervals of displacements and times, the notions of calculus allowed physicists to construe "instantaneous velocity" also as a velocity at a single point and instant of time.

The distinction between the concepts of position and of infinitesimal interval may be illustrated as follows. Consider a moving body with a changing velocity, and represent its motion along the X-axis by a curve, such as that depicted in figure 22.

The instantaneous velocity at time t is represented by the tangent to the curve at x, the hypotenuse of the characteristic rectangular triangle. The coordinates x and t give the position of the tangent point. Suppose that dx and dt are infinitesimal intervals bound by pairs of coordinates along the X- and T-axes. Here, $dx/dt = v$, and $(x_2 - x_1)/(t_2 - t_1) = v$, but clearly

$$\frac{x}{t} \neq v.$$

The ratio of an x-coordinate and a t-coordinate is not equal to dx/dt, *except* in the special case where the extension of the hypotenuse coincides with the origin. Thus, coordinates of position are distinct from infinitesimal intervals. The distinction applies equally to accelerated as well as to inertial motions. Therefore, Einstein's use of expressions such as $x = vt$, where x is a

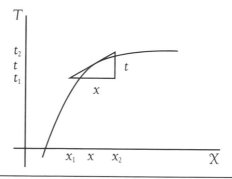

Figure 22. The average speed of an accelerating body

single coordinate, did not constitute generally valid descriptions of motions but only descriptions valid in special cases.

Moreover, Einstein did not adopt explicitly significant distinctions that had been elucidated in the late 1800s. In contrast to vector theorists, he did not operate with the distinction between scalars and vectors. In particular, while he focused his analysis on the exact significance of the concept of time, he did not distinguish between the concepts of speed and velocity and between the concepts of position, distance, and displacement.

To understand the relationship of Einstein's kinematics to the concepts of vector theory, we must answer the following questions. To what extent were the concepts of position, distance, displacement, speed, and velocity present in Einstein's work? What functions did they play in relation to one another? These questions may be approached by analyzing Einstein's expressions in light of the definitions of these concepts as formulated by vector theorists. To aid in understanding mathematical physics, Heaviside argued that one should "practice to take a symmetrically written-out Cartesian investigation and go through it, boiling it down to a vector investigation."[57] To this end, we can review first the basic kinematic distinctions between scalars and vectors.

The concept of speed was defined as a scalar quantity, a magnitude devoid of direction, consisting of the ratio of the two scalars, distance and time:

$$\text{speed} = \frac{\text{distance}}{\text{time interval}}.$$

[57] Heaviside, *Electromagnetic Theory*, vol. 1, p. 163.

It was formulated symbolically, in various ways, all of which meant the same thing:

$$v = \frac{d}{t}, \qquad v = \frac{|\Delta x|}{\Delta t}, \qquad v = \frac{|(x_2 - x_1)|}{(t_2 - t_1)}.$$

Here, the absolute value of a difference of two position-coordinates represents a distance. The concept of velocity, by contrast, was defined as a directed magnitude, a vector, consisting of the ratio of the displacement vector (movement in a specific direction) and the scalar magnitude of time:

$$\text{velocity} = \frac{\text{displacement}}{\text{time interval}},$$

or,

$$\mathbf{v} = \frac{\mathbf{d}}{t}, \qquad \mathbf{v} = \frac{\Delta \mathbf{x}}{\Delta t}, \qquad \mathbf{v} = \frac{(x_2 - x_1)}{(t_2 - t_1)},$$

all of which mean the same thing, and none of which are equal to the expressions for v. Here a displacement may be construed as a change in a position vector, or as the difference between coordinates indicating a positive or negative direction along the X-axis.

Einstein's use of kinematic concepts may now be elucidated in light of these definitions. First, the concepts of distance and displacement are both *intervals*, differences between pairs of quantities. Einstein did not express this mathematically in his derivations of the Lorentz transformations. Distances or displacements were expressed in the same way as positions, namely, by a single coordinate or by three. Einstein could justify this notation partly owing to his reliance on distances or displacements with their starting point at the origin of a coordinate system. For example, he wrote ξ = $c\tau$, instead of $(\xi_2 - \xi_1) = c\tau$, because the light ray under consideration was supposed to be emitted from the origin of the coordinate axis ($\xi_1 = 0$), so that $(\xi_2 - 0) = c\tau$, which naturally seemed equivalent to the simpler expression. Einstein relied on this simplification throughout. Hence, he did not distinguish mathematically between the concepts of position and spatial interval.

Also, he did not designate explicitly duration, in equations such as $\xi = c\tau$, as an interval. To be precise, he could have written $(\xi_2 - \xi_1) = c(\tau_2 - \tau_1)$. Indeed, at the beginning of his 1905 paper, Einstein employed the concept of duration explicitly, by writing expressions where the time interval is bounded by two instants: "$t_B - t_A$." Afterward, he also made this concept of duration explicit by denoting with t, instead of zero, the arbitrary instant of time when a ray of light is emitted from the origins of the coordinate systems. Nonetheless, Einstein proceeded to simplify his analysis by assuming "for brevity" that "at the origin of k, $\tau = 0$ when $t = 0$," and thereafter by stipulating to discuss the propagation of light "emitted at the time $\tau = 0$," as well as $t = 0$.[58] He thus expressed time intervals explicitly, in the initial equations of his 1905 paper, whereas he did not express distances or displacements as intervals.

As for the concepts of speed and velocity, nowhere did Einstein distinguish explicitly between them. At the time, most physicists writing in German did not use different words for these two concepts. Unlike physicists such as Föppl and Hertz,[59] Einstein did not even indicate the distinction by using an expression such as "the magnitude of the velocity" to refer to speed. He only wrote the word *Geschwindigkeit* to mean either velocity or speed. The closest he came to making a distinction between velocity and speed in his 1905 paper was in a footnote where he mentioned a case "when v changes sign without changing its numerical value."[60] Here his expression "numerical value" may be understood to mean speed, and the change in sign signifies a change in orientation.

The ambiguity between the concepts of speed and velocity was incorporated in formulations of the second principle of the special theory. Einstein described the propagation of light by writing:

$$\text{Geschwindigkeit} = \frac{\text{Lichtweg}}{\text{Zeitdauer}},$$

[58] Einstein, "Zur Elektrodynamik bewegter Körper," p. 899.
[59] August Föppl, *Einführung in die Maxwell'sche Theorie der Elektricität* (1894); 2nd ed.: Max Abraham and August Föppl (Leipzig: B. G. Teubner, 1904), p. 9. Heinrich Hertz, *Die Prinzipien der Mechanik* (1894); *The Principles of Mechanics, Presented in a New Form*, authorized translation by D. E. Jones and J. T. Walley (New York: Dover, 1956), p. 54.
[60] Einstein, "Zur Elektrodynamik bewegter Körper," p. 909.

which often has been translated as[61]

$$\text{velocity} = \frac{\text{light path}}{\text{time interval}},$$

The concept of "time interval" poses no ambiguity because Einstein defined it clearly as the scalar expression $(t_2 - t_1)$. But the other two concepts require attention. Since the early 1900s, common practice has been to translate Einstein's statement "die Konstanz der Lichtgeschwindigkeit" as "the constancy of the velocity of light." However, this expression is equivocal, strictly speaking, because light rays in free space do not all travel with the same velocity but only with a constant speed. Thus, Einstein's expression should be translated better as

$$\text{speed} = \frac{\text{light path}}{\text{time interval}},$$

It suggests that "light path" must designate a *distance*, not a displacement, because the equation is true, by definition, only if the scalar concept of speed is equal to the ratio of a directionless line segment and a time interval.

One might say that Einstein used scalar algebra rather than vector algebra—and that Einstein's algebraic theory of motion was based on the treatment of quantities irrespective of direction. But such an interpretation is problematic, because traditional algebra involved several conventions for representing and analyzing directions. Did the concepts of coordinate algebra capture the physical distinctions that had been elucidated with vector algebra?

Consider again some of the algebraic expressions Einstein used in his derivation of the Lorentz transformations. To describe the constancy of the speed of light, he used the following equations:

$$\xi = c\tau, \quad \eta = c\tau, \quad \zeta = c\tau,$$

[61] E.g., in Einstein et al., *The Principle of Relativity*, p. 41; Miller, *Albert Einstein's Special Theory of Relativity*, p. 373.

where τ has the same value in each equation. Each of these equations seems to say

$$\text{coordinate} = \text{speed} \times \text{time},$$

because in the 1905 paper the variables ξ, η, ζ are coordinates. In view of Einstein's definition of motion in terms of *position* and *time*, it seems reasonable to understand the coordinates ξ, η, ζ as specifying a position. Evidently, ξ, η, ζ later played this function so that they could appear in the final form of the transformation equations. In the three preceding equations, however, ξ, η, ζ *cannot* properly be interpreted as positions, because the statement,

$$\text{position} = \text{speed} \times \text{time},$$

is incorrect according to the definition of the concept of speed. A possible explanation: because ξ, η, ζ represent positive or negative values along each axis, they may be understood to designate displacements. But note that the variables ξ, η, ζ should represent *different* displacements. Yet Einstein used the same expression, $c\tau$, in the three equations, without indicating any direction, suggesting that c, again, should be construed as a speed, and $c\tau$ as a distance. But, then, each of the three equations would read

$$\textbf{displacement} = \text{distance},$$

another physically incorrect statement.

How can this ambiguity be resolved? One either could make the equations true by refining them to express the equality of two distances:

$$\text{distance} = \text{speed} \times \text{time};$$

or one could refine them to express the equality of two displacements:

$$\textbf{displacement} = \textbf{velocity} \times \text{time}.$$

For these purposes, Einstein could have employed a more definite notation. For example, the notation introduced by Karl Weierstrass indicates absolute values.[62] Thus, one could relate coordinate-displacements to the distance $c\tau$ by writing:

$$|\xi|=c\tau, \quad |\eta|=c\tau, \quad |\zeta|=c\tau.$$

The absolute value of the coordinates is equal to the distance traveled by light (the product of the speed of light and time). This notation was employed by August Föppl in his *Introduction to the Maxwellian Theory of Electricity*.[63] It would clarify Einstein's equations by restating them in terms of nonnegative quantities. Alternatively, the equations could be restated in terms that specify the directionality of the light rays. In other words, we can distinguish velocities along the three coordinate axes by writing:

$$\xi=c_{\xi}\tau, \quad \eta=c_{\eta}\tau, \quad \zeta=c_{\zeta}\tau.$$

Indeed, Einstein used such subscripts to distinguish velocities in his 1905 paper but only after having derived the transformation equations.[64] Specifically, he wrote:

$$\xi=w_{\xi}\tau, \quad \eta=w_{\eta}\tau, \quad \zeta=0.$$

Likewise, in the electrodynamical part of the paper, he noted that "the vector $(u_{\xi}, u_{\eta}, u_{\zeta})$ is nothing else than the velocity of . . ."[65] So of course he knew how to distinguish velocities symbolically. Considering the variety of vectorial notations in existence by 1905, physicists had a range of choices by which to attempt clear distinctions between the concepts of speed and velocity, and distance or displacement.

But Einstein did not distinguish between the concepts of light speed and light velocity. He treated the *directional* quantities ξ, η, ζ, as equal to the

[62] Karl Weierstrass, *Mathematische Werke*, vol. 1 (Berlin: 1894), pp. 67, 252, quoted in Florian Cajori, *A History of Mathematical Notations*, vol. 2 (Chicago: Open Court, 1929), pp. 123–124.
[63] Abraham and Föppl, *Einführung in die Maxwell'sche Theorie*, 2nd ed., e.g., pp. 6, 11, 48.
[64] Einstein, "Zur Elektrodynamik bewegter Körper," p. 905.
[65] Ibid., p. 917.

directionless quantity $c\tau$. His lack of distinction among the different directions of the propagation of light, along with his use of the same value τ, seem to allow the equivalence:

$$\xi = \eta = \zeta = c\tau.$$

But this seemingly simple equivalence is problematic: the light rays do not move in the same direction, and hence each coordinate cannot be equated to the one value $c\tau$. In short, the algebra underlying Einstein's theory of motion involved the nondirectional concept of *Lichtgeschwindigkeit* alongside the directional concept of coordinate. Was this peculiarity merely an incidental and superficial aspect of his notation, or did it have a substantial bearing on the theory of motion?

This odd aspect of physical mathematics can be exhibited clearly by analyzing the propagation of light along a *single* coordinate axis in accord with the historical development of analytic geometry. Einstein's identical way of representing the motion of light signals along the different directions of three coordinate axes may be simplified to a form that illustrates the ambiguity by means of positive and negative signs. By disregarding two of the three coordinate axes, Einstein, to describe light rays moving in opposite directions along a single axis, could have written the equations

$$\xi = c\tau \quad \text{and} \quad -\xi = c\tau.$$

These equations express directly the peculiar algebraic "constancy": light seems to move with the same velocity in opposite directions. Even though Einstein did not write these equations in his 1905 paper, in 1917 he published a popular book on relativity in which he derived the Lorentz transformations precisely on the basis of such equations representing the motion of two light signals in opposite directions. That derivation is immensely simpler than the first. By then, Einstein knew well that he could derive the transformations purely by using only elementary algebra. He titled the section: "Simple Derivation of the Lorentz Transformations." He explained in the preface: "The author has spared himself no pains in his endeavor to present the main ideas in the simplest and most intelligible form."[66] This derivation was widely reprinted and translated. It has also been employed

---

[66] Einstein, *Relativity*, p. v.

as a pedagogical tool.[67] In the words of one writer, "All the mathematics we need for this [derivation] is a little simple algebra, such as any school boy knows."[68] The derivation is also significant because it yields the transformation equations for x and t without appealing to the propagation of light along axes *transverse* to the direction of the relative motion of the coordinate systems.

The 1917 derivation, describing two rays of light moving away from each other along the X-axis of system K, was based directly on the equations[69]

$$-x = ct, \quad \text{and} \quad x = ct.$$

These equations display a peculiar asymmetry. While the opposite of the displacement x is expressed by $-x$, the opposite of the displacement ct is not expressed by $-ct$. These equations literally contradict one another if both are affirmed simultaneously. Instead, we could expect to describe opposite motions by the equations

$$-x = -ct, \quad \text{and} \quad +x = +ct.$$

If x represents a displacement in the positive direction along the X-axis, then $-x$ represents a displacement of the same magnitude in the negative direction. Likewise, if ct represents the same displacement x, in the positive direction along the X-axis, then $-ct$ represents the displacement of the same magnitude but in the opposite sense. Proof is straightforward. Describe the

[67] Gerald Holton, *Introduction to Concepts and Theories in Physical Science*, 2nd ed., rev. Stephen G. Brush (Princeton: Princeton University Press, 1985), pp. 514–516; Stanley Goldberg, *Understanding Relativity: Origin and Impact of a Scientific Revolution* (Boston: Birkhäuser, 1984), pp. 458–463; Lillian Lieber and Hugh Gray Lieber, *The Einstein Theory of Relativity, Part I: The Special Theory* (Lancaster, Pa.: Science Press Printing Company, 1936), pp. 36–42.
[68] Lieber and Lieber, *The Einstein Theory of Relativity, Part I: The Special Theory*, p. 38.
[69] These key equations appear explicitly in the works of later writers, such as Holton, *Introduction to Concepts and Theories in Physical Science*, p. 514; Goldberg, *Understanding Relativity: Origin and Impact of a Scientific Revolution*, pp. 166, 459; Cornelius Lanczos, *The Einstein Decade (1905–1915)* (New York: Academic Press, Harcourt Brace Jovanovich, 1974; London: Elek, 1974), p. 79; Rolf Hagedorn, *Relativistic Kinematics: A Guide to the Kinematic Problems of High-Energy Physics* (New York: W. A. Benjamin, 1963), p. 5; Shadowitz, *Special Relativity*, p. 77; V. A. Ugarov, *Special Theory of Relativity*, translated from the Russian by Yuri Atanov (Moscow: MIR Publishers, 1979).

motion of light along the positive direction of the X-axis by the equation $x = ct$. Hence, the opposite of the displacement x may be designated by $-x$. Now, designate the velocity of a ray of light moving along that trajectory $-x$ in the same time t as $-x/t = e$. Into this equation, substitute $c = x/t$ to find the relationship of one velocity to the other: $-(c) = e$. Therefore, $-x/t = -c$. In short, *if $x = ct$, then $-x = -ct$*. The opposite of the displacement x is $-x$, and the opposite of the velocity c is $-c$. In the words of Euler, "If z is defined as a function of y, it follows that when y is replaced by $-y$, z must necessarily become $-z$."[70] Accordingly, in his *Introduction to Mechanics*, Föppl noted that "in an equation of mechanics or theoretical physics both sides must have the same dimension and also the same plus or minus signs. An equation in which this is not applied must necessarily be false."[71]

These arguments seem evident, yet they clash with the mathematical analyses that underlay the modern kinematics. For example, *Einstein's 1917 derivation of the Lorentz transformations would not work with the symmetric equations $x = ct$ and $-x = -ct$*. Moreover, the problem cannot be dismissed by simply rejecting any particular derivation as defective, because the problematic equations are intrinsic to what became the algebraic geometry of relativistic kinematics. When Minkowski reformulated Einstein's theory algebraically in terms of a four-dimensional geometry, these equations were embodied as a basic aspect of the structure of space-time diagrams,[72] as conveyed in figure 23.

Here the equations play a fundamental role: they describe the boundaries of a cone representing the propagation of light, which in turn determines other geometric structures.[73] In the diagram, the cross section of the cone is perfectly symmetric, but the algebraic expressions that describe the diagonals are not. Therefore, even though the formalisms of Lorentz, Einstein, Minkowski, and others led to symmetric solutions in countless problems in electrodynamics and in physics in general, they involved certain formal asymmetries in kinematics.

As seen previously, Einstein's analysis focused mainly on the concept of

---

[70] Leonhard Euler, *Introductio in analysin infinitorum* (1748); *Introduction to Analysis of the Infinite*, trans. John D. Blanton, book I (New York: Springer-Verlag, 1988), p. 15.

[71] Föppl, *Einfuehrung in die Mechanik* (1900), p. vi.

[72] Minkowski, "Raum und Zeit."

[73] See, e.g., Hagedorn, *Relativistic Kinematics* (1963), p. 5; Lanczos, *The Einstein Decade*, p. 79; Shadowitz, *Special Relativity*, p. 77.

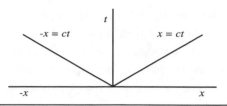

Figure 23. Rays of light emitted in opposite directions, in a space-time diagram

time. Because t was conceived as a scalar, it would seem to be exempt from the directional distinctions inherent in the symbols for speed or velocity, and position, distance, or displacement. For example, in Heaviside's vector notation one could write:

$$t=\frac{x}{c} \quad \text{and} \quad t=\frac{\mathbf{x}}{\mathbf{c}}.$$

The ratio, t, of two scalars was the same as the ratio of the two corresponding vectors because the directional components of the vectors canceled one another.[74] However, vector theorists did not allow an expression such as: $t = (\mathbf{x}/c)$. But because Einstein did not employ vector notation, his algebraic analysis admitted such relations. Numerically, Einstein could equate the concepts of position, distance, and displacement because all three could be represented by positive numbers. Moreover, the concepts of displacement and position could be represented equally by negative numbers. *Arithmetic itself did not capture the distinctions between the physical relations.* And, to this day, it still does not.

The distinctions between the three meanings of the term x did not arise verbally either, because instead of using expressions such as the distance x, the position x, or the displacement x, Einstein used only the words *coordinate, value, quantity,* and *position* in 1905.[75] His lack of verbal clarity persisted

[74] There exist ambiguities regarding the division of vectors. Early writers such as Hamilton and Tait defined the quotient of vectors to be a quaternion, but some writers in the 1900s avoided the subject of the division of vectors by leaving the operation undefined; some even claimed that vectors cannot be divided. In any case, the expression of time as the ratio of two vectors is justified because both vectors have the same direction, and thus their ratio is a scalar; see, e.g., Hoffmann, *About Vectors*, pp. 106–107.
[75] Einstein, "Zur Elektrodynamik bewegter Körper," pp. 898–917.

in his later derivations or the transformations, where he again employed these terms and also the expression the abscissa x.[76]

Against such verbal ambiguities, we should clarify the meaning of Einstein's algebraic expressions. In principle, we would expect that once an algebraic quantity is defined physically, its meaning should remain throughout our manipulations of it. This principle was asserted, for one, by the logician George Boole.[77] Likewise, for example, the engineer Walter R. Browne explained:

[Suppose] we begin by saying, "Let x = the number of acres in the field." We have then defined the term x for the purpose of that particular problem; from henceforward it stands merely as a convenient symbol for the words by which it is defined, and if we take care to preserve its meaning unaltered, we shall solve the problem much more easily and clearly by its aid. . . . It should, however, be noted that in giving several successive definitions to the same word, care should in general be taken to preserve some connection between them. Thus the connection between the innumerable definitions given to x in algebraical problems, is that in every case it expresses the unknown quantity which is the subject of inquiry. Similarly in Co-ordinate Geometry, x expresses the co-ordinate of a point as measured not up but across the paper. If in any particular case we were to reverse this, and call x the co-ordinate measured up the paper, we should be extremely likely to get our work into confusion.[78]

A single algebraic term may be used to designate various concepts, such as position, displacement, distance, and more; but then care must be exercised to ensure that one term is not used simultaneously in mutually incompatible ways. Moreover, the meaning of each term should be explicit. Appropriately, Arthur Stanley Eddington, the one scientist who did more than any

[76] Einstein, *Relativity*, p. 115.

[77] George Boole, *An Investigation of the Laws of Thought, on which are founded the Mathematical Theories of Logic and Probabilities* (London: Walton and Maberley/Cambridge: Macmillan, 1854), pp. 6, 68.

[78] Walter R. Browne, *The Foundations of Mechanics*, reprinted from "The Engineer" (London: Charles Griffin and Co., 1882), pp. 2–4.

other in England to promote Einstein's theory, admitted: "But the mind is not content to leave scientific Truth in a dry husk of mathematical symbols, and demands that it shall be alloyed with familiar images. The mathematician, who handles x so lightly, may fairly be asked to state, not indeed the inscrutable meaning of x in nature, but the meaning which x conveys to him."[79] Accordingly, consider the roots of Einstein's algebraic expressions.

The origins of the asymmetric pair of equations $-x = ct$ and $x = ct$ lie in the algebraic formulation of particular physical theories. At the root of the matter lies a subtle distinction between the mathematical approaches that were used in two branches of physics: mechanics and optics. In mechanics, physicists typically described motions of bodies such as projectiles moving in definite trajectories. In optics, physicists often described the motion of light in indefinitely many directions simultaneously. Einstein's asymmetric equations for the motion of light make no sense from the perspective of mechanics, but they can be derived readily from traditional algebraic optics.

Consider the algebraic equation describing the propagation of an expanding sphere of light, as given by Einstein[80] and many others:

$$x^2 + y^2 + z^2 = c^2 t^2.$$

The sphere of light may be conceived as constituted of indefinitely many rays of light, all emitted from one origin simultaneously and traveling outward at the same speed. The coordinates y and z may be disregarded by considering the propagation rays moving only along the X-axis:

$$x^2 + (0)^2 + (0)^2 = c^2 t^2$$
$$x^2 = c^2 t^2.$$

---

[79] A. S. Eddington, *Space, Time and Gravitation* (Cambridge: Cambridge University Press, 1929), p. vi. See also, A. S. Eddington, *The Mathematical Theory of Relativity* (Cambridge: Cambridge University Press, 1924), pp. 1–7, 240.

[80] See, e.g., Einstein, "Zur Elektrodynamik bewegter Körper," p. 901. Einstein, "Über das Relativitätsprinzip und die aus demselben gezogenen Folgerungen," *Jahrbuch der Radioaktivität und Elektronik* 4 (1907), p. 419. Einstein, "Manuscript on the Special Theory of Relativity" (1912–1914), *The Collected Papers of Albert Einstein*, vol. 4, *The Swiss Years: Writings, 1912–1914*, ed. Martin J. Klein, A. J. Kox, Jürgen Renn, and Robert Schulmann (Princeton: Princeton University Press, 1995), p. 40.

To simplify this expression further, ordinary algebra requires that the square roots of both sides of the equation be extracted as follows:

$$\sqrt{x^2} = \sqrt{c^2 t^2}$$
$$\pm x = ct.$$

This last equation expresses the rule that any quantity has two possible square roots, one positive and one negative. Hence one could write: $-x = ct$, and $x = ct$; the "distance" ct seems to be equivalent to two "displacements" in opposite directions. Thus, algebraic optics already involved the equivalence between nondirectional and directional quantities underlying Einstein's kinematics. Of course, optical theory was subsumed under electromagnetic theory and the analytical devices of calculus, but the essential principle is the same.

Unlike other algebraic operations, oddly, the extraction of square roots is *not* carried out in the same way on both sides of an equation. For centuries mathematicians believed that quantitative statements, such as algebraic equations, remain true so long as any alteration to one side of an equation is also carried out on its other side. However, the equation $\pm x = ct$, for example, does not appear to follow symmetrically from the equation $\sqrt{x^2} = \sqrt{c^2 t^2}$. In particular, why not write instead

$$\pm x = \pm ct$$

as the solution for the second-degree equation? The explanation is not found in the history of physics but in the history of algebra.

The double sign $\pm$ was used since the early days of analytic geometry. It was employed by Descartes in his *Géométrie* of 1637 to designate places in equations where either the plus or the minus sign could be used alternately.[81] The $\pm$ sign came to be used in the solution of second-degree equations such that any positive quantity then has two square roots, one positive and one negative (owing to the rule that minus times minus is a positive). But the double solution of second-degree equations had been explored by algebraists long before the days of analytic geometry. For example, in the

[81] René Descartes, *Oeuvres de Descartes,* ed. Charles Adam and Paul Tannery; also in *Discourse on Method, Optics, Geometry, and Meteorology,* trans. Paul J. Olscamp (Indianapolis: Bobbs-Merrill Company/Howard W. Sams & Co., 1965), e.g., see p. 239.

first chapter of his *Great Art* of 1545, titled "On Double Solutions in Certain Types of Cases," Girolamo Cardano noted that in equations such as "$x^2 = 9$, x is 3 or -3," and he proceeded to investigate the double solutions of more complicated equations.[82] The double solution of second-degree equations led to interpretive problems that played a significant role in the prolonged resistance of mathematicians to completely accepting negative numbers.

By the 1890s, controversy about the significance of negative quantities had faded from the complexion of algebra. In pure analytic geometry, students readily solved equations of the second degree by the straightforward application of basic rules and notations. Thus, for example, as a seventeen-year-old student, Albert Einstein solved the following equation of an ellipse,

$$\tfrac{1}{2}\, x^2 + y^2 = r^2,$$

by writing

$$\text{For } x = 0 \quad y = \mp r$$
$$\text{For } y = 0 \quad x = \mp r\sqrt{2}$$

in his Matura examination at the cantonal school at Aarau in 1896.[83] Such solutions seemed to present no problems in the field of pure analytic geometry, and they were applied directly in analytical physics. Decades earlier, however, mathematicians had pondered the ambiguities in the double solution of second-degree equations.

In the early 1800s, the mathematician Sylvestre François Lacroix played an important role in giving a standard form to algebra. Lacroix was the successor of Lagrange, as professor of analysis at the École Polytechnique, and he authored definitive textbooks on mathematics. At the time, the curriculum of the Polytechnique was so highly mathematical that students often used the catchword "X" as a name for the school itself. And, above all,

---

[82] Girolamo Cardano, *Artis Magnae Sive de Regulis Algebraicis* (1545); *The Great Art, or the Rules of Algebra*, trans. and ed. T. Richard Witmer (Cambridge, Mass.: MIT Press, 1968), p. 10.

[83] Albert Einstein, doc. 23: "*Matura* Examination (C) in Geometry," in *The Collected Papers of Albert Einstein*, vol. 1, *The Early Years: 1879–1902*, ed. John Stachel (Princeton: Princeton University Press, 1987), p. 31.

Lacroix was the ruling authority on the proper manipulation of algebraic signs and symbols such as x.

Lacroix discussed the "ambiguity according to which every equation of the second degree is susceptible to two solutions" in his textbook *Elements of Algebra*. He noted that "it is a general rule that it is necessary to give to the square root of any quantity the double sign $\pm$," and he proceeded to indicate that an equation such as $x^2 = b$ is solved by writing $x = \pm\sqrt{b}$. Lacroix commented on this "general rule" for the solution of square roots:

> Given this rule, one could ask why, x being the square root of x^2, one does not affect x also with the double sign $\pm$? One responds first, with M. Develey (*Algèbre d'Emile*, T. II.), that the letter x having been posed simply without sign (that is, with the sign $+$), as the symbol for the unknown, it is in this state that it is necessary to determine the value, and that inasmuch as one seeks a number x of which the square is b, for example, there are only these two possible solutions: $x = +\sqrt{b}$, $x = -\sqrt{b}$. Therefore, when equally, in solving the equation $x^2 = b$, one writes $\pm x = \pm\sqrt{b}$, and one arranges those signs in all possible ways, as:
>
> $$+x = +\sqrt{b}, \quad -x = -\sqrt{b},$$
> $$+x = -\sqrt{b}, \quad -x = +\sqrt{b},$$
>
> one will not obtain anything more, because by changing the sign of the members of the second equation of each line, one returns to the first.[84]

Lacroix gave two arguments here. First, if we seek the value of the "unknown" quantity x, then it should be expressed simply as "x," not with the $\pm$ sign or with the minus sign. Second, of four possible solutions to the equation $x^2 = b$, the expressions $-x = -\sqrt{b}$ and $-x = +\sqrt{b}$ are equivalent to the equations $x = \sqrt{b}$ and $x = -\sqrt{b}$, respectively. Mathematicians thus were

[84] S. F. Lacroix, *Élémens d'Algèbre, à l'Usage de l'École Centrale des Quatre-Nations;* 11th rev. ed. (Paris: Mme. Ve. Courcier, Imprimeur-Libraire pour les Mathématiques, 1815), p 156; translation by A. Martínez. Lacroix referred to the work of the mathematician Isaac-Emmanuel-Louis Develey: Em. Develey, *Algèbre d'Emile*, 2 vols. (Lausanne: Imprimerie de Hignou, 1805).

aware of the apparent asymmetry in treating only the square root of one side of an equation with the ± sign. It was done only as a convenient matter of notation, because the resulting four equations could be expressed simply as two, or as one: $x = \pm \sqrt{b}$.

Does this algebraic reasoning explain the asymmetric equations used decades later by Einstein? Once $x^2 = c^2t^2$ is solved as four possible equations, further arguments are needed to select two of these four equations as descriptive of motions in opposite directions. To select a pair of equations from among the four,

$$+x = +ct, \quad -x = -ct,$$
$$+x = -ct, \quad -x = +ct,$$

to describe the motion of light in opposite directions, we need to know what the symbols x and ct represent. These letters represent numbers, so the physical meaning of each equation depends on the numerical value of each letter. Whereas in Descartes' work letters always represented only positive numbers, following the work of Johann Hudde in 1659, letters came to stand for negative as well as positive numbers in analytic geometry.[85] Hence, to describe opposite motions, the following pairs of equations are possible, according to the numerical values of x and ct:

1. If x is positive and ct is positive, then $x = ct$ describes motion in the positive direction, and $-x = -ct$ describes motion in the negative direction.
2. If x is positive and ct is negative, then $x = -ct$ describes motion in the positive direction, and $-x = ct$ describes motion in the negative direction.
3. If x is negative and ct is positive, then $-x = ct$ describes motion in the positive direction, and $x = -ct$ describes motion in the negative direction.
4. If x is negative and ct is negative, then $x = ct$ describes motion in the negative direction, and $-x = -ct$ describes motion in the positive direction.

[85] Florian Cajori, A History of Mathematics, 2nd ed. (New York: Macmillan, 1926), p. 178. Also see Cajori, History of Mathematical Notations, vol. 2: Notations Mainly in Higher Mathematics (Chicago: Open Court, 1929), articles 392, 708: pp. 5, 323.

Yet Einstein's kinematics did not involve *any* of these possible pairs of equations to describe the motion of light. Because two of them, namely 1 and 4, can be summarized by the equation $x = ct$, and the other two, namely 2 and 3, can be summarized by the equation $-x = ct$, then one could say that motion in opposite directions may be represented *either* by means of the equation $x = ct$ *or* by means of the equation $-x = ct$. Hence, one may interpret the expression $\pm x = ct$ as signifying algebraically that

$$\text{either } -x = ct \quad \text{ or } \quad x = ct.$$

But, instead, Einstein relied on the validity of both equations simultaneously, that is:

$$-x = ct \quad \text{ and } \quad x = ct.$$

This interpretation of the double $\pm$ sign entered into the formal content of relativistic kinematics. Accordingly, it might seem that one can write:

$$\frac{-x}{c} = \frac{x}{c}.$$

This impossible equation, in a slightly generalized form, was also discussed by Lacroix.

Lacroix considered algebraically two couriers moving along the same line in opposite directions, each traveling a different distance at a different speed.[86] In the *same* time interval, these two couriers would traverse opposite paths x and $-y$ at the speeds b and c, so that

$$\frac{x}{b} = \frac{-y}{c}.$$

Lacroix obtained this equation by modifying the equation that he used to describe the equal duration of two motions in the same direction:

$$\frac{x}{b} = \frac{y}{c}.$$

[86] Lacroix, *Élémens d'Algèbre*, pp. 105–106.

Thus, he first considered only the change in motion by altering only the sign of the path y. Yet he added that "one must observe that the path y is composed by multiples of the space c that is traversed in one hour by a courier," that is, $c = yt$, where $t = 1$, such that c,

being directed in the same sense as the space y, should be equally supposed to be of the same sign – that one has given y: thus one will have

$$\frac{x}{b} = \frac{-y}{-c} \quad \text{or} \quad \frac{x}{b} = \frac{y}{c}.$$

It suffices therefore from a simple change of sign to comprehend the second case of the question within the first; and it is in this way that Algebra yields at the same time the solution of many analogous questions.[87]

Thus, he expressed explicitly the symmetry of signs corresponding to opposite paths and motions. Lacroix's example may be compared to Einstein's formulation simply by considering the special case in which couriers traverse paths of equal lengths at equal speeds, such that $x = y$ and $b = c$, that is,

$$\frac{x}{c} = \frac{-x}{-c}.$$

Again, this relationship was missing in Einstein's use of coordinate algebra.

Lacroix's analysis was not unique. For example, an earlier version was published by Newton in his *Arithmetica Universalis*,[88] and another was discussed later by Augustus De Morgan in his 1831 book, *On the Study and*

---

[87] Ibid., p. 106.

[88] Isaac Newton, *Arithmetica Universalis* (1707); *Universal Arithmetick: or, a Treatise of Arithmetical Composition and Resolution*, trans. Mr. Ralphson and rev. Mr. Cunn (London, 1728), reprinted in *The Mathematical Works of Isaac Newton*, vol. 2, assembled by Derek T. Whiteside, Sources of Science (New York: Johnson Reprint Corporation, 1967), see p. 72.

Difficulties of Mathematics.[89] By the way, this analysis of the motion of two messengers traveling in opposite directions is similar to the famous analysis of simultaneity published by Einstein in his book of 1917. There he considered the motion of light signals emitted from opposite directions and meeting at a midway point as a means to determine the simultaneity of distant events.

Not only did Einstein's expressions clash with the some of the kinematic concepts of vector theory but, surprisingly, they also diverged from some of the kinematic concepts of traditional algebra. Einstein used negative and positive algebraic quantities in a way that had been criticized as mistaken by mathematicians who had given basic analytic geometry its definitive form. The ambiguity originated from the duplicity in the algebraic solution of quadratic equations, which physicists imported into the kinematics of single motions from the analysis of spherical waves in optics.

The inherent equivalence of nondirected and directed magnitudes in modern kinematics requires further explanation. Its historical origin is found in the application of traditional algebraic rules that were not designed for the analysis of motion. The identification of nondirected and directed magnitudes rests on subtle differences between algebra, analytic geometry, and vector theory. A comparative analysis of the same simple expression $x = ct$ suffices to disclose the source of the ambiguity. Thus far, we have considered only the meaning of the algebraic letters, but the question can be refocused by analyzing the meaning of the remaining sign.

Consider first the equation in terms of vector theory:

$$\mathbf{x} = \mathbf{c}t.$$

Here the equality sign indicates that the terms $\mathbf{x}$ and $\mathbf{c}t$ are equivalent in length *and* direction. Vector theorists defined this concept thus: "Two vectors are said to be *equal* when they have the same magnitude and the same direction."[90] Two distinct vectors, say, vectors at different places, would be

---

[89] Augustus De Morgan, *On the Study and Difficulties of Mathematics,* Society for the Diffusion of Useful Knowledge (1831; repr., Chicago: Open Court, 1902), see pp. 112–119.
[90] Wilson, *Vector Analysis,* p. 4. Josiah Willard Gibbs, *Elements of Vector Analysis* (New Haven, Conn.: privately printed, 1881), and "On the Determination of Elliptic Orbits From Three Complete Observations," *Memoirs of the National Academy of Sciences* 4, part II (1889); both reprinted in *The Scientific Papers of J. Willard Gibbs,* vol. 2 (New York: Dover, 1961), pp. 18, 119.

deemed equal only if they had the same length and direction.[91] This defini-tion *also* admitted the possibility that the terms on both sides of the equa-tion were merely different names for the same vector. This latter meaning applies to the equation $\mathbf{x} = \mathbf{c}t$, because physically $\mathbf{x}$ and $\mathbf{c}t$ represent the same thing: a single displacement.

Consider now the case in analytic geometry. There existed a subtle but significant difference in the use of the equality sign. Consider the simplest possible expression relating two coordinates in analytic geometry:

$$x = y.$$

It specifies a diagonal line, and here the equality sign had only one mean-ing: that x and y have the same numerical value. However, although x and y have the same numerical value, they are not identical because x and y repre-sent coordinates *on different axes*. Their equality thus was never complete in analytic geometry. In this context, the equality sign served to compare *only numerical quantities* of x, y, and z.

This was a key difference between vector algebra and analytic geometry. Whereas an *equation* in vector algebra could signify *complete identity*, in ana-lytic geometry *it could not*. In analytic geometry, an equation signified only the numerical relation between distinct coordinates.

Now consider the equation in ordinary algebra:

$$x = ct.$$

Like vector theory, algebra was sufficiently flexible to admit two meanings of the equality sign. One meaning was common to both: equality could mean *complete identity*: that x and ct were just different names for the same thing. The other meaning was foreign to vector theory but common to analytic geometry and algebra: equality could mean that x and ct were identical in quantity only, irrespective of direction.

In Einstein's work, the equality sign may be understood to play two dis-tinct roles. In some places, it could mean that two physical quantities were

[91] In modern vector theory, another definition of equality involves the additional re-quirement that to be considered equal, vectors must have the same starting point, i.e., the same position. See Banesh Hoffmann, *About Vectors* (1966; repr., New York: Dover, 1975), pp. 13–16.

equal merely in absolute magnitude. In other places it would indicate that two expressions constituted different names for the same thing. Owing to this ambiguity, he equated directed magnitudes (e.g., positive and negative coordinates of position, such as ξ, η, ζ) to scalars (e.g., distances, such as $c\tau$). He used the equality sign in an algebraic way, to indicate either identity or mere numerical equivalence. Because of the double meaning of the equality sign, and because Einstein did not present his arguments by graphic methods, it may seem that the modern theory of motion was essentially based on algebra rather than analytic geometry. Algebra provided the flexibility involved in Einstein's derivations. Of course, analytic geometry was a subfield of algebra, and it provided conventional grounds for equating quantities that were not of the same kind.

This issue may be further illustrated by Minkowski's interpretation of the new kinematics. Minkowski reformulated Einstein's algebraic kinematics in graphic analytic geometry. He used the vertical axis of a rectangular coordinate system to represent the parameter t.[92] He then could represent the equation $x = ct$ by a diagonal line, as depicted in figure 24.

This graphical representation of t was quite analogous to the ordinary use of the Y-axis in analytic geometry. Even if x and t had the same numerical value, they did not represent the same thing. The constant c served to equate values of x with values of t by introducing speed as conversion units. For example,

$$x \text{ meters} = c\frac{\text{meters}}{\text{seconds}} \times t \text{ seconds}.$$

Still, the term c did not involve direction. The expression ct could designate only a distance, whereas x could be interpreted as a distance, a position, or a displacement. Therefore, there existed a tacit distinction between x and ct.

Minkowski's analytic-geometric interpretation facilitated the simultaneous graphical representation of distinct concepts. On the one hand, ct could be equated to x, as illustrated, while, on the other, ct also could be equated to $-x$. Minkowski and his followers also equated the term x to the expression vt, simply on the grounds of the traditional definition of $v = x/t$.[93] This use

[92] Minkowski, "Raum und Zeit."
[93] See, e.g., Arnold Sommerfeld, "Notes," in Einstein et al., *The Principle of Relativity*, p. 93.

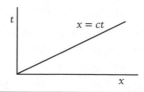

Figure 24. A ray of light in a time-space diagram

followed Einstein's work, as he employed the same symbol *x* to stand not only for *ct* but for *vt* as well.[94] The expression *vt* represented a distance, or a displacement in the positive direction along the X-axis, whereas the variable *x* was used to designate a position.

All the ambiguities that have been highlighted in the preceding paragraphs pertain to the mathematical description of motion within a single frame of reference. By contrast, most of the advances in kinematics that stemmed from special relativity pertained to the description of motion among multiple frames. In sum, basic symbols in algebraic kinematics played a variety of roles. The concepts of position, motion, trajectory, direction, and equality involved variations that were not explicitly conveyed in the algebraic notation. Owing to this flexibility in meaning, Einstein was able to derive a new theory of motion from mathematical procedures that were not conceived originally for such an application. Thus, we reach a paradox: Einstein's kinematic reformulation of physics involved algebraic ambiguities that yet enabled relativity theory to take shape. Such ambiguities continued to pervade the writings of physicists in the decades that followed.

Leaving It at That, or Not

In the early 1800s, the basic foundation of physics was considered to be mechanics, sometimes called "the science of motion." By the early 1900s, the basis of much physics was again considered to be "the science of motion," but by then that expression meant kinematics rather than mechanics. Several notions associated traditionally with mechanics had been denounced as metaphysical, to the extent that many physicists preferred not to give a central role to mechanics but to construe it as just another branch of physics.

[94] Einstein, "Zur Elektrodynamik bewegter Körper," pp. 899, 902; Einstein, *Relativity,* p. 116.

Many scientists had come to increasingly value knowledge based on perceptions. Kinematics had arisen as the study of motions as they appear
to observation. Meanwhile, the parallel tradition that misconstrued kinematics as the purely abstract study of motion continued to grow. Owing to
the rise of non-Euclidean geometries, physicists became less certain about
the ultimate validity of traditional geometry. Yet they employed traditional
algebra, and analytic geometry, virtually without hesitations. Accordingly,
the traditional principles of algebra, incorporated into kinematics, came to
receive a very high degree of trust, just as earlier trust had been placed on
metaphysical principles.

Soon after Ampère's formulation of kinematics, this would-be science
splintered into distinct studies delimited by the various specialists who
pursued them: physicists, mechanists, mathematicians, and theoretically
minded engineers. As the influential engineer Franz Reuleaux noted, "We
have found, on the one hand, a most unsatisfactory confusion of attempts
to find a form for one and the same circle of ideas. As many systems as
authors, no resting point reached, always new trying and seeking,—and as
a conclusion we have Ampère's well-defined science split into two, indeed
into four sciences, as if it were one of those Infusoria which are propagated
by division."[95] On the other hand, Reuleaux continued, observations had
become increasingly exact, and the systematic study of mechanisms had
progressed. Likewise, physical kinematics became more detailed, as new
concepts, formalisms, and theorems were devised. But various versions of
kinematics competed to occupy a prominent place in physics. Einstein's
kinematics, in particular, inherited the trend to emphasize the importance
of observations and measurement procedures. Thus, Newton's notions of
absolute space and time became widely rejected, because they lacked clear
connections to measurement procedures. Yet Einstein's kinematics also deviated from Ampère's original prescriptions, because it was based on highly
idealized measurement procedures and hardly on observations about the
motions of everyday objects. For example, the so-called kinematic effect of
length contraction was not something that had ever been observed. Thus,
Einstein's theory might seem to pertain not so much to kinematics proper
but to what was occasionally called phoronomy, the geometry of motion.

Ampère's fruitful notion of kinematics became neglected and forgot-

[95] Franz Reuleaux, *Theoretische Kinematik* (1875), reissued as *The Kinematics of Machinery*, trans. Alexander B. W. Kennedy (1876; repr., New York: Dover, 1963), p. 16.

ten. And likewise, whereas in the 1800s mathematicians had studied with interest the kinematics of mechanisms and of geometric figures, by the mid-1900s research in mathematical kinematics was virtually abandoned, as mathematicians turned their attention to more abstract domains.

Meanwhile, following Einstein's work, many contributions to physical kinematics shifted away from the question of the precise description of motion as it appears to observation. They shifted to mathematical deductions of invariant relations among coordinate systems, especially beyond the reach of observation. The idea that basic concepts should be defined in terms of feasible physical operations became widely repudiated. By the 1920s, Einstein himself slighted his procedural definition of simultaneity by saying that "a good joke should not be repeated too often." In many ways, the so-called kinematics of theoretical physics came to be less the geometry of motion and, rather, more the algebra of motion.

To a great extent, the algebraic methods that became employed in relativistic kinematics had not been designed originally for the analyses of motions. Accordingly, from early on, modern kinematics elicited a rise in debates about interpretation and meaning, just like algebra itself had engendered debates much earlier.

Most physicists admired Einstein's achievements in clarifying the concepts of time, though many continued to employ the elements of kinematics with constant carelessness. Too often, physicists did not employ properly even basic words such as *velocity*. Presumably, some may have felt that it would be irrelevant and even pedantic to discriminate systematically among kinematic concepts, at least beyond the confines of elementary textbooks and schoolrooms. Likewise, it is easy to imagine that some readers will skim over the preceding pages, with the same impatience that Hertz denounced in physicists' presentations of the elements of mechanics. A reflex reaction will be to assume that such ambiguities can have no serious significance, and that one should not nitpick what works well enough. One can only hope that such impressions are justifiable and that indeed no significant problems lurk in the layers of verbal inconsistencies, ambiguous notations, and special simplifying assumptions that are routinely employed in physical theories.

But that attitude is scarcely warranted by history. To the contrary, the negligence in rigor and clarity should be systematically inspected. We should strive to formulate physical relations with precision. And, to some extent, that labor involves even the formulation of new mathematical methods for physics. For example, to distinguish numerically between the physical concepts of distance and displacement, one might have to employ an artificial number system where signless numbers and positive numbers are not the same thing. One might distinguish systematically between what William Rowan Hamilton called tensors and scalars—for example, distinguish between 4 and +4. As another example, Hermann Grassmann touched upon the possibility of formulating varied concepts of equality: "In comparing two line segments one can assert equality of direction or length, of direction and length, of direction and position, and so forth; and in comparing other things, further relations of equality emerge."[96] Such notions might appear repulsive to anyone who greatly appreciates the structure of ordinary mathematics as it is handed down to us from tradition. Yet there were some mathematicians, like Richard Dedekind, who argued that even certain numbers are "free creations of the human mind." According to Dedekind, "It is only through the purely logical process of building up the science of numbers . . . that we are prepared to investigate our notions of space and time by bringing them into relation with this number-domain created in our mind."[97] If we devise new mathematical concepts, including numerical notions, in correspondence with empirical notions, such new concepts might serve as tools for representing and analyzing physical relations.

Einstein read Dedekind's "What Are and What Should Be Numbers?" together with his friends Moritz Solovine and Conrad Habicht in Bern. They also discussed parts of John Stuart Mill's *System of Logic*, where Mill cast doubt on the foundations of arithmetic by arguing that in the "Science of Numbers," just as in geometry, the "first principles are generalisations from experience . . . and that the peculiar certainty ascribed to it, on account of which its propositions are called necessary truths, is fictitious and hypothetical."[98] The extent to which such ideas influenced Ein-

---

[96] Hermann Grassmann, *Die lineale Ausdehnungslehre* (1844); translated by Lloyd C. Kannenberg in *A New Branch of Mathematics* (1995), p. 33.

[97] Richard Dedekind, *Was sind und was sollen Zahlen*, (1887), 2nd ed. (Braunschweig: Friedrich Vieweg und Sohn, 1893); *Essays on the Theory of Numbers*, I. *Continuity and Irrational Numbers*, II. *The Nature and Meaning of Numbers*, trans. Wooster Woodruff Beman (New York: Dover, 1963), pp. 31–32 (preface to the first edition).

[98] John Stuart Mill, *A System of Logic, Ratiocinative and Inductive: Being a Connected View of the Principles of Evidence and the Methods of Scientific Investigation*, 8th ed., vol. 2 (London: Longmans, Green, Reader, and Dyer, 1872), book II, chap. VI, p. 169.

stein remains unclear. Nonetheless, he did exhibit some skepticism toward even basic mathematical notions. For instance, he once expressed serious doubt that the total length of several matches, laid end to end, is really the arithmetical sum of their individual lengths—claiming: "I don't believe in mathematics."[99] Several other sources attest to Einstein's early reservations toward mathematics in general. His friend Solovine recounted that "Einstein was a skilled mathematician, but he often spoke out against the abuses of mathematics in the hands of physicists. 'Physics,' he would say, 'is basically a concrete, intuitive science. Mathematics is only a means to express the laws that govern phenomena.'"[100] Likewise, Hans Tanner, one of Einstein's first students at the University of Zurich from 1909 to 1911, recalled Einstein arguing that in physics "'the main thing is the content, not the mathematics. With mathematics one can prove anything.' That was his attitude towards mathematics."[101] Later in life, Einstein commented: "Inasmuch as the propositions of mathematics refer to reality, they are not certain, and inasmuch as they are certain, they do not refer to reality."[102]

In the end, it is not merely the precision of mathematical calculations that matters but also the precision of our physical concepts. All too often we operate with mathematical methods that do not capture significant physical distinctions. We disregard such shortcomings so long as we can recognize that the results of mathematical analyses yield the "correct" answers. But, in so doing, we succumb to an outlook where "the ends justify the means." There is a connection between this outlook and the physics of principles. Postulates are sometimes tailor-made and invoked to produce desired results. When a mathematical analysis by itself does not select a single physically correct answer, we appeal to such postulates in order to impose the

---

[99] Albert Einstein in discussion with the Geneva engineer Gustave Ferrière, circa 1910; quoted in Seelig in *Albert Einstein: A Documentary Biography*, p. 107.

[100] Solovine, *Letters to Solovine*, pp. 7–8.

[101] Seelig in *Albert Einstein: A Documentary Biography*, p. 107.

[102] Albert Einstein, "Geometrie und Erfahrung," an Address to the Prussian Academy of Sciences in Berlin, 27 January 1921; *Preussische Akademie der Wissenschaften, Mathematisch-Physikalische Klasse, Sitzungsberichte* (1921); reprinted in *Mein Weltbild*, new edition (Zurich: Europa Verlag, 1953), pp. 119–127, quotation on pp. 119–120; translation by A. Martínez; expanded version "Geometry and Experience," in Einstein, *Sidelights on Relativity*, trans. G. B. Jeffery and W. Perrett, (London: Methuen; New York: E. P. Dutton, 1922; repr., New York: Dover, 1983), p. 28.

expected results. This opportunism is useful as an elegant stopgap. But in the end, one hopes to formulate physical theories from the bottom up—constructive theories.

By the late 1800s, as we have seen, some influential physicists moved away from the efforts to causally explain phenomena in favor of just describing them. By the 1930s, physicists led by Niels Bohr renounced the efforts to describe even certain elementary physical structures and processes, as if it were just utterly impossible. Among some physicists there arose the jaded saying: "You never really understand a new theory. You just get used to it."[103]

Perhaps if physically meaningful concepts were employed systematically, the task of accounting for phenomena would not be quite as distant from that of describing phenomena. Perhaps then the task of describing phenomena would not be quite as distant from that of understanding them.

[103] Joseph Schwartz and Michael McGuiness, *Einstein for Beginners* (New York: Pantheon Books, 1979), p. 151. As another example, consider the following: "It is a question of habit. . . . In short, one does not understand a concept in physics, one gets used to it." Henri Arzeliès, *Relativistic Point Dynamics* (Oxford: Pergamon Press, 1972), p. xxxii.

Bibliography

Primary Sources

Abraham, Max, "Die Grundhypothesen der Elektronentheorie," *Physikalische Zeitschrift* **5** (September 1904), 576–579.

Ampère, André-Marie, "Suite d'une Classification naturelle pour les Corps simples," *Annales de Chimie et de Physique* **2** (1816), 5–32, 105–125.

———, *Essai sur la Philosophie des Sciences, ou exposition analytique d'une classification naturelle de toutes les connaissances humaines* (Paris: Bachelier, Imprimeur-Libraire pour les Sciences, 1834; reprint, Brussels: Culture et Civilisation, 1966).

———, *Essai sur la Philosophie des Sciences, Seconde Partie,* edited by Jean-Jacques Ampère (Paris: Bachelier, 1843).

Ampère, André-Marie, and Jean-Jacques Ampère, *Correspondance et Souvenirs de 1805 à 1864,* edited by H. Cheuvreux, vol. 1 (Paris: J. Hetzel, 1875).

Andrade, Jules, *Leçons de méchanique physique* (Paris, 1898).

Anonymous, "The Language of Physical Science," *The Engineer* **51** (25 February 1881), 143–144.

Antonmari, X., and C. A. Laisant, *Questions de Mécanique à l'Usage des Élèves de Mathématiques Spéciales* (Paris: Libraire Nony & Co., 1895).

Appell, Paul Émile, *Traité de Mécanique Rationelle; Cours de Mécanique de la Faculté de Sciences,* vol. 1, *Statique–Dynamique du Point* (Paris: Gauthier-Villars et Fils, 1893).

———, *Cours de Mécanique a l'usage des Candidats à l'École Centrale des Artes et Manufactures* (Paris: Gauthier-Villars, 1902).

Appell, Paul Émile, and James Chappuis, *Leçons de Mécanique Élémentaire à l'usage des Élèves des Classes de Première C et D, conformément aux Programmes du 31 mai 1902* (Paris: Gauthier-Villars, 1905).

Aquinas, Thomas, *Super Boetium de Trinitate, Expositio* (ca. 1256);*The Division and Methods of the Sciences: Questions V and VI of His Commentary on the De Trinitate of Boethius,* translated by Armand Maurer, 3rd rev. ed. (Toronto: Pontifical Institute of Mediaeval Studies, 1963).

Archimedes, *Geometrical Solutions Derived from Mechanics,* also known as "The Method," translated by J. L. Heiberg (Chicago: Open Court, 1909).

Aristotle, *Mathematics in Aristotle,* edited by Thomas L. Heath (1949; reprint, New York: Garland, 1980).

————, *Aristotle's Metaphysics, a new translation,* trans. Joe Sachs (Santa Fe: Green Lion Press, 1999).

Bacon, Francis, *The tvvoo bookes of Francis Bacon: of the Proficience and Aduancement of Learning, Divine and Humane* (London: Henrie Tomes, 1605).

[————], *Francisci Baronis de Verulamio, Vice-Comitis Sancti Albani, De Dignitate et Augmentis Scientiarum Libri IX,* edited by William Rawley (London, 1623; Paris: P. Mettayer, 1624).

————, *Of the Advancement and Proficience of Learning, or the Partitions of Sciences, IX Bookes,* translated by Gilbert Wats (Oxford: Lichfield, for Rob. Young and Ed. Forrest, 1640).

————, *Physical and Metaphysical Works of Bacon including his Dignity and Advancement of Learning, in nine books, and his Novum Organum, or, Precepts for the interpretation of nature,* edited by Joseph Devey (London: Henry G. Bohn, 1853).

Baker, Thomas, *Elements of Mechanism* (London, 1852); 2nd ed., (London, 1858).

Bélanger, Jean Baptiste, *Traité de cinématique* (Paris: Dunod, Gauthier Villars, Eugène Lacroix, 1864).

Beneke, Friedrich Eduard, *Neue Grundlegung zur Metaphysik; als Programm zu seinen Vorlesungen über Logik und Metaphysik* (Berlin: E. S. Mittler, 1822).

————, *Lehrbuch der Psychologie* (Berlin: E. S. Mittler, 1832).

————, *Die neue Psychologie; Erläuternde Aufsätze zur zweiten Auflage meines Lehrbuches der Psychologie als Naturwissenschaft* (Berlin: E. S. Mittler, 1845).

Bentham, George, *Essai sur la nomenclature et la classification des principales branches d'art-et-science. Ouvrage extrait du Chrestomathia de Jérémie Bentham* (Paris: Bossange Frères, Libraires, 1823).

Bernoulli, Jean, *Discours sur les loix de la communication du mouvement: qui a merité les Eloges de l'Academie Royale des Sciences aux années 1724. & 1726. & qui a concouru à l'occasion des Prix distribuez dans lesdites années* (Paris: Chez Claude Jombert, 1727).

Bernstein, Aaron, *Naturwissenschaftliche Volksbücher: Wohlfeile Gesammt-Ausgabe,* vol. 4 (Berlin: Franz Duncker, 1869).

Biot, Jean-Baptiste, *Essai de Géométrie Analytique, appliquée aux courbes et aux surfaces du second ordre* (1802; 8th ed., Paris: Bachelier, Imprimeur-Libraire, 1834).

Blumenthal, Otto, ed., *Fortschritte der Mathematischen Wissenschaften in Monographien,* no. 2: H. A. Lorentz, A. Einstein, H. Minkowski, *Das Relativitätsprinzip,* with notes by A. Sommerfeld, and a foreword by Otto Blumenthal (Stuttgart: B. G. Teubner, 1913).

Boltzmann, Ludwig, *Vorlesungen über die Prinzipe der Mechanik,* 2 vols. (Leipzig:

Barth, 1897, 1904); translated as "Lectures on the Principles of Mechanics," in Boltzmann, *Theoretical Physics and Philosophical Problems*, pp. 223–254.

———, "Über die Grundprinzipien und Grundgleichungen der Mechanik" (1899, Clark University 1889–1899, decennial celebration, Worcester, Mass.), in *Populäre Schriften*, Essay 16, pp. 253–307; "On the Fundamental Principles and Equations of Mechanics," in Boltzmann, *Theoretical Physics and Philosophical Problems*, pp. 101–128.

———, "Über die Entwickelung der Methoden der theoretischen Physik in neuerer Zeit," Address to the meeting of scientists at Munich (22 September 1899), in *Populäre Schriften*, Essay 14, pp. 198–277; "On the Development of the Methods of Theoretical Physics in Recent Times," in Boltzmann, *Theoretical Physics and Philosophical Problems*, pp. 77–100.

———, "Über die Prinzipien der Mechanik I," Inaugural lecture at Leipzig (November 1900), and Part II (Vienna, 1902), in *Populäre Schriften*, Essay 17, pp. 309–330; "On the Principles of Mechanics," in Boltzmann, *Theoretical Physics and Philosophical Problems*, pp. 129–158.

———, *Vorlesungen über die Principe der Mechanik*, part II (1904), §88; translated as "The Law of Inertia," in Boltzmann, *Theoretical Physics and Philosophical Problems*, pp. 261–265.

———, *Theoretical Physics and Philosophical Problems: Selected Writings*, translated by Paul Foulkes and edited by Brian McGuiness (Boston: D. Reidel, 1974).

Boscha, Johannes, ed., *Recueil des travaux offerts par les auteurs à H. A. Lorentz à l'occasion du 25ème anniversaire de son doctorat le 11 décembre 1900* (The Hague: Martinus Nijoff, 1900).

Bour, J. Edmond E., *Cours de mécanique et machines professé, à l'École polytechnique*, vol. 1, *cinétique;* vol. 2, *statique;* vol. 3, *dynamique et hydraulique* (Paris: Gauthier-Villars, 1865–1874).

Bresse, J. A. Charles, "Mémoire sur un théorème nouveau concernant les mouvements plans et l'application de la cinématique à la détermination des rayons de courbure," *Cahier du Journal de l'École Polytechnique*, vol. 20 (Paris: Imprimerie de la République, 1853), 89–115.

———, *Cours de mécanique et machines, professé à l'École polytechnique: I, cinématique, dynamique du point matériel, statique; II, dynamique des systèmes matériels en général, mécanique spéciale des fluides, étude des machines à l'état de mouvement* (Paris: Gauthier-Villars, 1885).

Browne, Walter R., *The Foundations of Mechanics*, reprinted from *The Engineer* (London: Charles Griffin and Co., 1882); originally published as a series of articles in *The Engineer*, in parts: I (30 September 1881), II (4 November 1881), III (2 December 1881), IV (16 December 1881), V (20 January 1882), VI (17 February 1882), VII (24 March 1882), VIII (12 May 1882).

Bucherer, Alfred, *Mathematische Einführung in die Elektronentheorie* (Leipzig: Teubner, 1904).

Burg, Adam, *Theoretische Principien der Mechanik nebst der Art und Weise ihrer Anwendung auf das Maschinen-Wesen nach den, auf höhere Mathematik basirten, ordentlichen Vorträgen*, edited by Karl Winternitz (Vienna: Doll, 1846).

Burmester, Ludwig E. H., *Lehrbuch der Kinematik: für Studirende der Maschinentechnik, Mathematik und Physik / Geometrisch Dargestellt* (Leipzig: Arthur Felix, 1888).

Byland, Hans, "Aus Einsteins Jugendtagen. Ein Gedenkblatt," *Neue Bündner Zeitung*, 7 February 1928.

Calinon, Auguste, "Étude critique sur la mécanique," *Bulletin de la Société des Sciences de Nancy*, ser. 2, **7** (1885), 87–180 (Paris: Berger-Levrault et Cie., Libraires-Éditeurs, 1886).

———, *Étude sur la Sphère, La Ligne Droite et le Plan* (Nancy: Imprimerie Berger-Levrault et Cie, 1888).

———, *Étude de Cinématique à deux et à Trois Dimensions* (Paris : Berger-Levrault et Cie., 1890).

———, *Introduction à la Géométrie des Espaces à Trois Dimensions* (Paris: Gauthier-Villars et Fils / Berger-Levrault et Cie., 1891).

———, "Étude sur l'indéterminacion géométrique de l'univers," *Revue Philosophique de la France et de l'Étranger* **36** (1893), 595–607.

———, *La Géométrie a deux Dimensions des Surfaces à Courbure Constante* (Paris: Gauthier-Villars et Fils / Berger-Levrault et Cie, 1895).

———, *Étude sur les diverses grandeurs en mathématiques,* (Paris: Gauthier-Villars / Beger-Levrault and Co., 1897).

Cardano, Girolamo, *Artis Magnae Sive de Regulis Algebraicis* (1545); *The Great Art, or the Rules of Algebra,* translated and edited by T. Richard Witmer (Cambridge, Mass.: MIT Press, 1968).

Carnot, Lazare N. M., *Essai sur les machines en général,* nouvelle édition (Dijon: Defay, 1786).

———, *Géométrie de Position* (Paris: J. Duprat, Libraire pour les Mathématiques, 1803).

———, *Principes Fondamentaux de l'Équilibre et du Mouvement* (Paris: Crapelet, 1803).

Carpenter, William B., *Mechanical Philosophy, Horology, and Astronomy* (London: Wm. S. Orr and Co., 1848).

Cayley, Arthur, "Coordinates Versus Quaternions," *Proceedings of the Royal Society of Edinburgh* **20** (1895), 271–275; read on 2 July 1894; reprinted in *The Collected Mathematical Papers of Arthur Cayley,* vol. 13 (Cambridge: University Press, 1897), doc. 962, pp. 541–544.

Chambers, Ephraim, *Cyclopaedia, or, An Universal Dictionary of Arts and Sciences,* 2nd ed., vol. 1 (London: D. Midwinter, A Bettesworth and C. Hitch, et al., 1738).

Clifford, William Kingdon, "Instruments Illustrating Kinematics, Statics, and Dynamics," in *Handbook to the Special Loan Collection of Scientific Apparatus* (1876); reprinted in Clifford, *Mathematical Papers*, edited by Robert Tucker, with an introduction by H. J. S. Smith (Bronx, N.Y.: Chelsea, 1968).

———, *Elements of Dynamic: An Introduction to the Study of Motion and Rest in Solid and Fluid Bodies, Part 1. Kinematic* (London: Macmillan, 1878).

———, *The Common Sense of the Exact Sciences*, edited and with a preface by Karl Pearson (1885), newly edited and with an introduction by James R. Newman, preface by Bertrand Russell (New York: Alfred A. Knopf, 1946).

Cohen, I. Bernard, "An Interview with Einstein," *Scientific American* **193**, no. 1 (July 1955), 68–73.

Cohn, Emil, "Ueber die Gleichungen des elektromagnetischen Feldes für bewegte Körper," *Annalen der Physik* **7** (1902), 29–56.

———, "Zur Elektrodynamik bewegter Systeme," Akademie der Wissenschaften zu Berlin, mathematische-physikalische Klasse, *Sitzungsberichte* (1904), 1294–1303, 1404–1416.

Collignon, Édouard, *Traité de Mécanique*, part 1: *Cinématique* (Paris: Libraire Hachette et Cie., 1873).

Comstock, J. L., *A System of Natural Philosophy in which the Principles of Mechanics, Hydrostatics, Hydraulics, Pneumatics, Acoustics, Optics, Astronomy, Electricity, and Magnetism are Familiarly Explained*, 28th ed. (New York: Robinson, Pratt, & Co., 1836).

Comte, Auguste, *Cours de Philosophie Positive*, 6 vols. (Paris: Bachelier, 1830–1842); as *The Philosophy of Mathematics*, translated by W. M. Gillespie (New York: Harper & Bothers, 1851).

Collignon, Edouard, *Cours élémentaire de Mécanique (Cinématique); Ouvrage répondant aux programmes officiels de 1866 pour l'enseignement secondaire spécial*, 3rd year, part 1 (Paris: L. Hachette et Cie., 1868).

Condillac, Étienne Bonnot abbé de, *Essai sur l'Origine des Connoissances Humaines* (Amsterdam: P. Mortier, 1746).

d'Alembert, Jean, *Traité de dynamique* (1743); 2nd ed. (Paris, 1758; reprint, New York: Johnson Reprint Corporation, 1968).

———, "Dimension," *Encyclopédie*, vol. 4 (Paris, 1754), p. 1010.

———, *Essai sur les élémens de Philosophie, ou sur les principes des connaissances humains, avec les éclaircissemens* (1759?); reprinted in *Oeuvres complètes de d'Alembert*, vol. 1, part 1 (Paris: A. Belin, et Bossange père, fils et frères, 1821).

Dedekind, Richard, *Was sind und was sollen Zahlen* (1887); 2nd ed. (Braunschweig: Friedrich Vieweg und Sohn, 1893); *Essays on the Theory of Numbers*, I. *Continuity and Irrational Numbers*, II. *The Nature and Meaning of Numbers*, translated by Wooster Woodruff Beman (New York: Dover, 1963).

Dee, John, "Mathematicall Præface specifying the chiefe Mathematicall Sciences," in *The Elements of Geometrie of the most Auncient Philosopher Euclide of Megara*, translated by H. Billingsley (London: John Daye, 1570).

De Morgan, Augustus, *On the Study and Difficulties of Mathematics,* Society for the Diffusion of Useful Knowledge (1831; reprint, Chicago: Open Court, 1902).

———, *A Budget of Paradoxes,* vol. 1 (1872); 2nd ed., edited by David Eugene Smith (Chicago: Open Court, 1915).

Descartes, René, *Discours de la Méthode pour Bien Conduire Sa Raison, et Chercher la Verité dans les Sciences; Plus La Dioptriqve. Les Meteores. Et La Géométrie. Qui sont des essais de cete methode* (Leyde: l'imprimerie de I. Maire, 1637); facsimile of the first edition, with translation and annotations by David Eugene Smith and Marcia L. Latham, *The Geometry of René Descartes* (New York: Dover, 1954). Also published in Descartes, *Discourse on Method, Optics, Geometry, and Meteorology,* translated by Paul J. Olscamp (Indianapolis: Bobbs-Merrill, 1965).

Develey, Isaac-Emmanuel-Louis, *Algèbre d'Emile,* 2 vols. (Lausanne: Imprimerie de Hignou, 1805).

Diderot, Denis, and Jean d'Alembert, eds., *Encyclopédie, ou Dictionnaire raisonné des sciences, des arts et des métiers, par une société de gens de lettres,* vol. 1 (Paris, 1751).

Dienger, J., *Studien zur analytischen Mechanik. Die allgemeinen Gesetze der Bewegung* (Stuttgart: J. B. Metzler'schen Buchhandlung, 1863).

Doherty, Joshua Joseph J., *An Elementary Text-book of Mechanics (Kinematics and Dynamics)* (Dublin, 1881).

Drude, Paul, *Physik des Aethers auf elektromagnetische Grundlage* (Stuttgart: Ferdinand Enke, 1894).

———, *Lehrbuch der Optik* (Leipzig: S. Hirzel, 1900).

Duhamel, M. [Jean Marie Constant], *Cours de Mécanique de l'École Polytechnique,* Part 1 (Paris: Bachelier, Imprimeur Libraire, 1845).

———, *Des méthodes dans les sciences de raisonnement,* 3rd ed. (Paris: Gauthier-Villars, 1896).

[Dühring, Eugen], *Nachrichten der Königliche Gesellschaft der Wissenschaften zu Göttingen* **13,** no. 8 (13 March 1872).

Dühring, Eugen Karl, *Cursus der Philosophie als streng wissenschaftlicher Weltanschauung und Lebensgestaltung* (Leipzig: E. Koschny, 1875).

———, *Kritische Geschichte der Nationalökonomie und des Socialismus* (Berlin: T. Grieben, 1871); 2nd ed. (Berlin: T. Grieben, 1875).

———, *Kritische Geschichte der allgemeinen Principien der Mechanik. Von der philosophischen Facultät der Universität Göttingen mit dem ersten Preise der Beneke-Stiftung gekrönte Schrift. Nebst einer Anleitung zum Studium mathematischer Wissenschaften* (Berlin: Theobald Grieben, 1873); 3rd ed. (Leipzig: Fues's Verlag/R. Reisland, 1887).

Einstein, Albert, "Über die Untersuchung des Aetherzustandes im magnetischen Feld" (1895), in *The Collected Papers of Albert Einstein,* vol. 1, pp. 6–9.

———, "Zur Elektrodynamik bewegter Körper," *Annalen der Physik* **17** (Leipzig, 1905), 891–921; reprinted in *The Collected Papers of Albert Einstein,* vol. 2, pp. 276–306. Several translations are available, including the following: H. A. Lorentz,

A. Einstein, H. Minkowski, and H. Weyl, *The Principle of Relativity*, translated by W. Perrett and G. B. Jeffery with notes by Arnold Sommerfeld (New York: Dover, 1952); Anna Beck (translator) and Peter Havas (consultant), *Collected Papers of Albert Einstein. English Translation* (Princeton: Princeton University Press, 1989), vol. 2, pp. 140–171; Arthur I. Miller, *Albert Einstein's Special Theory of Relativity: Emergence (1905) and Early Interpretation (1905–1911)* (Reading, Mass., Addison-Wesley, 1981; reprint, New York: Springer-Verlag, 1998); *Einstein's Miraculous Year: Five Papers That Changed the Face of Physics*, edited by John Stachel, with the assistance of Trevor Lipscombe, Alice Calaprice, and Sam Elworthy (Princeton: Princeton University Press, 1998). Note that the translations by Perrett and Jeffery and by Miller have some significant defects.

———, "Das Prinzip von der Erhaltung der Schwerpunktsbewegung und die Trägheit der Energie," *Annalen der Physik* **20** (1906), 627–633; reprinted in *The Collected Papers of Albert Einstein*, vol. 2, pp. 360–366.

———, "Bemerkungen zu der Notiz von Hrn. Paul Ehrenfest: 'Die Translation deformierbarer Elektronen und der Flächensatz,'" *Annalen der Physik* **23** (1907), 206–208; also in *The Collected Papers of Albert Einstein*, vol. 2, pp. 410–412.

———, "Über das Relativitätsprinzip und die aus demselben gezogenen Folgerungen," *Jahrbuch der Radioaktivität und Elektronik* **4** (1907), 411–462; reprinted in *The Collected Papers of Albert Einstein*, vol. 2, pp. 433–484.

———, "Le principe de relativité et ses conséquences dans la physique moderne," *Archives des sciences physiques et naturelles* **29** (1910), 5–28, especially sec. 2, pp. 7–10; reprinted in *The Collected Papers of Albert Einstein*, vol. 3, pp. 131–176.

———, "Zum Ehrenfestschen Paradoxon," *Physikalische Zeitschrift* **12** (1911), 509–510; reprinted in *The Collected Papers of Albert Einstein*, vol. 3, pp. 482–483.

———, "Die Relativitäts-Theorie," *Naturforschende Gesellschaft in Zürich. Vierteljahrsschrift* **56** (1911), 1–14; reprinted in *Collected Papers of Albert Einstein*, vol. 3, doc. 17, pp. 425–439.

———, "Relativität und Gravitation. Erwiderung auf eine Bemerkung von M. Abraham," *Annalen der Physik* **39** (1912), pp. 1059–1064; reprinted in *The Collected Papers of Albert Einstein*, vol. 4, pp. 181–186.

———, manuscript on the special theory of relativity (1912–1914), in *The Collected Papers of Albert Einstein*, vol. 4, doc. 1, pp. 9–108.

———, "Zum Relativitäts-Problem," *Scientia* **15** (1914), pp. 337–348; reprinted in *The Collected Papers of Albert Einstein*, vol. 4, pp. 608–622.

———, "Die Relativitätstheorie," *Die Kultur der Gegenwart. Ihre Entwickelung und ihre Ziele*, edited by Paul Hinneberg, part 3, sec. 3, vol. 1, *Physik*, edited by Emil Warburg (Leipzig: Teubner, 1915), pp. 703–713; reprinted in *The Collected Papers of Albert Einstein*, vol. 4, pp. 536–551.

———, *Über die spezielle und die allgemeine Relativitätstheorie* (Braunschweig: Vieweg & Son, 1917); Einstein, *La théorie de la rélativité restreinte et généralisé*, translated from the tenth German edition by J. Rouvière, with a preface by Émile

Borel (Paris: Gauthier-Villars, 1921); Einstein, *Relativity: The Special and the General Theory*, authorized translation by Robert W. Lawson (New York: P. Smith/ H. Holt, 1931; reprint, New York: Crown Publishers, 1961).

―――, "Grundgedanken und Methoden der Relativitätstheorie, in ihrer Entwicklung dargestellt," manuscript (January 1920), at the Pierpont Morgan Library, New York; a copy is in the Albert Einstein Archives, at the Einstein Papers Project, California Institute of Technology, Pasadena, California, item 2-070. The consequent edited article is shorter: "A Brief Outline of the Development of the Theory of Relativity," *Nature* **106** (17 February 1921), 782–784.

―――, "On the Theory of Relativity" (1921), *Mein Weltbild* (Amsterdam: Querido Verlag, 1934); reprinted in *Ideas and Opinions* (1954).

―――, "Geometrie und Erfahrung," an Address to the Prussian Academy of Sciences in Berlin, 27 January 1921; *Preussische Akademie der Wissenschaften, Mathematisch-Physikalische Klasse, Sitzungsberichte* (1921); reprinted in *Mein Weltbild*, new ed. (Zurich: Europa Verlag, 1953), pp. 119–127; expanded version "Geometry and Experience," in Einstein, *Sidelights on Relativity*, translated by G. B. Jeffery and W. Perrett (London: Methuen; New York: E. P. Dutton, 1922; reprint, New York: Dover, 1983).

―――, "Wie ich die Relativitätstheorie entdeckte," Lecture delivered at the University of Kyoto, Japan (14 December 1922), transcribed into Japanese by Jun Ishiwara, *Einstein Kyôzyu-Kôen-roku* [transcription of Professor Einstein's lecture], first published in the periodical *Kaizo* (1923); also (Tokyo: Kabushiki Kaisha, 1971), pp. 79–88. Translated by Y. A. Ono, as "How I Created the Theory of Relativity," *Physics Today* **35** (August 1982), 45–47. See also Seiya Abiko, "Einstein's Kyoto Address: 'How I Created the Theory of Relativity,'" *Historical Studies in the Physical Sciences* **31** (2000), 1–35. A partial translation is given by Tsuyoshi Ogawa, "Japanese Evidence for Einstein's Knowledge of the Michelson-Morley Experiment," *Japanese Studies in the History of Science*, no. 18 (1979), 73–81. Also available in German translation: Hans Joachim Haubold and Eiichi Yasui, "Jun Ishiwaras Text über Albert Einsteins Gastvortrag an der Universität zu Kyoto am 14. Dezember 1922," *Archive for History of Exact Sciences* **36**, no. 3 (1986), 271–279.

―――, *The Meaning of Relativity* (1922; reprint, Princeton: University Press, 1974).

―――, discographic recording of 6 February 1924, transcribed in Friedrich Herneck, "Zwei Tondokumente Einsteins zur Relativitätstheorie," *Forschungen und Fortschritte* **40** (1966), 133–135.

―――, untitled reflections, no translator noted, in Albert Einstein, John Dewey, James Jeans, et al., *Living Philosophies* (New York: Forum Publishing, 1930; reprint, New York: Simon and Schuster, 1931), pp. 3–7.

―――, Herbert Spencer Lecture, Oxford (1933), first published in *Mein Weltbild* (Amsterdam: Querido Verlag, 1934); translated by Alan Harris as "On the

Method of Theoretical Physics," in Einstein, *Essays in Science* (New York: Philosophical Library, 1934), pp. 12–21.

———, "Remarks on Bertrand Russell's Theory of Knowledge," in *The Philosophy of Bertrand Russell*, edited by Arthur Schilpp (1944); reprinted in Einstein, *Ideas and Opinions* (New York: Crown Publishers, 1954).

———, "Autobiographisches" (1946), translated by Paul Arthur Schilpp, in *Albert Einstein: Philosopher-Scientist*, edited by Schilpp, Library of Living Philosophers, vol. 7 (Evanston, Ill.: Library of Living Philosophers / George Banta, 1949).

———, *Ideas and Opinions* (New York: Bonanza Books, Crown Publishers, 1954).

———, "Erinnerungen-Souvenirs," *Schweizerische Hochschulzeitung* 28 *Sonderheft*, (1955), 145–153; reprinted as "Autobiographische Skizze," *Helle Zeit–Dunkle Zeit. In memoriam Albert Einstein*, edited by Carl Seelig (Zurich: Europa Verlag, 1956), pp. 9–17.

———, *Quatre conférences sur la théorie de la relativité* (Paris: Gauthier-Villars, 1971).

———, *Letters to Solovine*, with an introduction by Maurice Solovine (New York: Philosophical Library, 1986).

———, *The Collected Papers of Albert Einstein*, vol. 1, *The Early Years, 1879–1902*, edited by John Stachel (Princeton: Princeton University Press, 1987).

———, *Oeuvres choisies 4, Correspondances françaises*, edited by Michel Biezunski (Israel: Éditions du Seuil, 1989).

———, *The Collected Papers of Albert Einstein*, vol. 2, *The Swiss Years: Writings, 1900–1909*, edited by John Stachel (Princeton: Princeton University Press, 1989).

———, *The Collected Papers of Albert Einstein*, vol. 3, *The Swiss Years: Writings, 1909–1911*, edited by Martin J. Klein, A. J. Kox, Jürgen Renn, and Robert Schulman (Princeton: Princeton University Press, 1993).

———, *The Collected Papers of Albert Einstein*, vol. 4, *The Swiss Years: Writings, 1912–1914*, edited by Martin J. Klein, A. J. Kox, Jürgen Renn, and Robert Schulmann (Princeton: Princeton University Press, 1995).

———, *The Collected Papers of Albert Einstein*, vol. 5, *The Swiss Years: Correspondence, 1902–1914*, edited by Martin J. Klein, A. J. Kox, and Robert Schulmann (Princeton: Princeton University Press, 1993).

———, *The Collected Papers of Albert Einstein*, vol. 8, parts A, B, *The Berlin Years: Correspondence, 1918*, edited by Robert Schulmann, A. J. Kox, Michel Janssen, and József Illy (Princeton: Princeton University Press, 1998).

Einstein, Albert, and Michele Besso, *Correspondance, 1903–1955*, German transcriptions with French translations, notes, and introduction by Pierre Speziali (Paris: Hermann, 1972).

Einstein, Albert, H. A. Lorentz, H. Minkowski, and H. Weyl, *The Principle of Relativity: A Collection of Original Memoirs on the Special and General Theory of Relativity*, translated by W. Perrett and G. B. Jeffery, with notes by Arnold Sommerfeld (1923; reprint, New York: Dover, 1952).

Emerson, William, *The Principles of Mechanics; Explaining and Demonstrating The General Laws of Motion, The Laws of Gravity, Motion of Descending Bodies, Projectiles, Mechanic Powers, Pendulums, Centers of Gravity, &c. Strength and Stress of Timber, Hydrostatics, and, Construction of Machines* (London: W. Innys and J. Richardson, 1754).

Engels, Friedrich, *Herrn Eugen Dührings Umwälzung der Wissenschaft* (Leipzig, 1878); reissued in 1894 and translated in Karl Marx and Frederick Engels, *Collected Works*, vol. 25 (New York: International Publishers, 1987): *Anti-Dühring, Herr Eugen Dühring's Revolution in Science.*

Ettingshausen, Andreas ritter von, *Die Principien der heutigen Physik: bei der Feier der Ubernahme des ehemaligen Universitätsgebäudes von der Kaiserlichen Akademie der Wissenschaften am XXIX October MDCCCLVII* (Vienna: Aus der Kaiserl. königl. Hof- und Staatsdruckerei, [1857]).

Euler, Leonhard, *Introductio in analysin infinitorum* (1748); translated by John D. Blanton, *Introduction to Analysis of the Infinite* I (New York: Springer-Verlag, 1988).

———, "Solutio Problematis ad Geometriam Situs Pertinentis," *Commentarii Academiae Scientiarum Imperialis Petropolitanae* 8 (1741), 128–140; Euler, *Opera Omnia*, ser. 1, vol. 7 (1923), pp. 1–10; "Solution of a Problem Belonging to the 'Geometry of Position,'" reissued and translated by Peter Wolff, in *Breakthroughs in Mathematics* (New York: New American Library, 1970), pp. 197–206.

———, "Réflexions sur l'espace et le tems," presented on 1 February 1748, *Mémoires de l'Académie des Sciences de Berlin* 4 (1750), 324–333; also in: *Opera Omnia*, Series III, vol. 2, pp. 376–383. Translated by Link M. Lotter, "Reflections on Space and Time," reprinted in *The Changeless Order: The Physics of Space, Time and Motion*, edited by Arnold Koslow (New York: George Braziller, 1967).

———, *Theoria motus corporum solidorum seu rigidorum ex primis nostrae cognitionis principiis stabilita et ad omnes motus, qui in huiusmodi corpora cadere possunt, accommodata* (Rostochii et Gryphiswaldiae, 1765), reissued in *Opera Omnia*, ser. II, vols. 3–4 (1948, 1950).

———, "Formulae generales pro translatione quacunque corporum rigidorum" (1775), *Novi Commentarii Academiae Scientiarum Petropolitanae*, vol. 20 (1776), pp. 189–207; also in *Opera Omnia*, ser. II, vol. 9 (1968), pp. 84–98.

Fairbairn, William, *Treatise on mills and millwork*, part 1: *On the principles of mechanism and on prime movers*; part 2: *On machinery of transmission and the construction and arrangement of mills* (London: Longman, Green, Longman and Roberts, 1861–1863).

Fechner, Gustav Theodor, *Elemete der Psychophysik* (Leipzig: Breitkopf and Hartel, 1859).

FitzGerald, George Francis, "The Ether and the Earth's Atmosphere," *Science* 13 (1889), 390.

———, "Observation, Measurement, Experiment: Short Abstract of Methods of Induction: Measurement of Time, Mass, Length, Area, Volume" (1890s, unfinished); published in *The Scientific Writings of the Late George Francis FitzGerald,* edited by Joseph Larmor (Dublin and London: Hodges, Figgis, & Co., and Longmans, Green, & Co., 1902).

Föppl, August, *Einführung in die Maxwell'sche Theorie der Elektricität* (Leipzig: B.G. Teubner, 1894); 2nd ed.: Max Abraham and August Föppl (Leipzig: B. G. Teubner, 1904).

————, *Vorlesungen ueber Technische Mechanik,* vol. 1, *Einfuehrung in die Mechanick* (1899); 2nd ed. (Leipzig: B. G. Teubner, 1900).

————, *Die Mechanik im neunzehnten Jahrhundert. Ein akademischer Festvortrag gehalten in der Aula der k. techn. Hochschule in München am 4 Dezember 1901* (Munich: Ernst Reinhardt, 1902).

Gibbs, Josiah Willard, *Elements of Vector Analysis* (New Haven, Conn.: privately printed, 1881); reprinted in *The Scientific Papers of J. Willard Gibbs,* vol. 2 (New York: Dover, 1961), pp. 17–90.

————, "On Multiple Algebra," *Proceedings of the American Association for the Advancement of Science* **35** (1886), 37–66; reprinted in *The Scientific Papers of J. Willard Gibbs,* vol. 2, pp. 91–117.

————, "On the Determination of Elliptic Orbits from Three Complete Observations," in *Memoirs of the National Academy of Sciences* **4** (1889), part. II, 79–104; reprinted in *The Scientific Papers of J. Willard Gibbs,* vol. 2, pp. 118–147.

————, "Quaternions and the 'Ausdehnungslehre,'" *Nature* **44** (28 May 1891), 79–82; reprinted in *The Scientific Papers of J. Willard Gibbs,* vol. 2, pp. 161–168.

————, "Quaternions and the Algebra of Vectors," *Nature* **47** (16 March 1893), pp. 463–464; reprinted in *Scientific Papers of J. Willard Gibbs,* vol. 2, pp. 169–172.

————, *The Scientific Papers of J. Willard Gibbs,* 2 vols. (New York: Dover, 1961).

Girault, Charles François, *Éléments de géométrie appliquée à la transformation du mouvement dans les machines* (Caen, 1858).

Goodeve, Thomas Minchin, *The Elements of Mechanism* (London, 1860); 2nd ed., enlarged (London, 1865).

————, *Principles of Mechanics* (London, 1874); 2nd ed. (London: Longmans, Green, and Co., 1876).

Grassmann, Hermann, *Die lineale Ausdehnungslehre ein neuer Zweig der Mathematik dargestellt und durch Anwendungen auf die übrigen Zweige der Mathematik, wie auch auf die Statik, Mechanik, die Lehre vom Magnetismus und die Krystallonomie erläutert* (Leipzig: Otto Wigand, 1844); translated in Grassmann, *A New Branch of Mathematics,* pp. 9–297.

————, *Geometrische Analyse genüpft an die von Leibniz erfundene geometrische Charakteristik* (Leipzig: Weidmann'sche Buchhandlung, 1847); translated as *Geometric Analysis* in Grassmann, *A New Branch of Mathematics,* pp. 316–384.

————, "Der Ort der Hamilton'sche Quaternionen in der Ausdehnungslehre," *Mathematische Annalen* **12** (1877), 375–386; reissued in Grassmann, *A New Branch of Mathematics,* pp. 525–538.

————, *Die Ausdehnungslehre*, 2nd ed. (Berlin: Adolph Enslin, 1862); reissued in *Hermann Grassmanns Gesammelte Mathematische und Physikalische Werke*, edited by Friedrich Engel (Leipzig: B. G. Teubner, 1896).

————, *A New Branch of Mathematics, The Ausdehnungslehre of 1844 and Other Works*, translated by Lloyd C. Kannenberg (Chicago: Open Court, 1995).

Gross, Edward John, *An Elementary Treatise on Kinematics and Kinetics*, Rivington's Mathematical Series (London, 1876).

Habich, E. J., *Études Géométriques et Cinématiques* (Lima, Peru: Imprimerie de Carlos Paz Soldan, 1880).

Hachette, Jean-Nicolas, *Programme du Cours Élémentaire des Machines pour l'An 1808* (Paris: Imprimerie Impériale, 1808).

Hamilton, William, *Discussions on Philosophy and Literature, Education and University Reform, chiefly from the Edinburgh Review*, corrected, vindicated, enlarged, in notes and appendices (Edinburgh: Blackwood and Son, 1866).

Hamilton, William Rowan, Abstract to "Account of a System of Rays," presented to the Royal Irish Academy on 23 April 1827; reprinted in Robert Perceval Graves, *Life of William Rowan Hamilton*, vol. 1 (Dublin: Hodges, Figgis, & Co., 1882), pp. 228–229.

————, "Theory of Conjugate Functions, or Algebraic Couples; with a Preliminary and Elementary Essay on Algebra as the Science of Pure Time," *Transactions of the Royal Irish Academy* 17 (1837), 293–422; reprinted in *The Mathematical Papers of Sir William Rowan Hamilton*, vol. 3, edited by A. W. Conway and J. L. Synge (Cambridge: Cambridge University Press, 1967), pp. 3–96.

————, manuscript notebook 24.5 (entries for October 1843); published in *Mathematical Papers*, vol. 3, pp. 103–105.

————, "On a New Species of Imaginary Quantities Connected with the Theory of Quaternions," *Proceedings of the Royal Irish Academy* 2 (1844), pp. 424–434; reprinted in *Mathematical Papers*, vol. 3, pp. 111–116.

————, "On Quaternions; or on a New System of Imaginaries in Algebra," *Philosophical Magazine* 29 (1846), 26–31, 113–122, 326–328; reprinted in *Mathematical Papers*, vol. 3, pp. 227–297.

————, *Lectures on Quaternions: containing a systematic statement of a new mathematical method; of which the principles were communicated in 1843 to the Royal Irish Academy; and which has since formed the subject of successive courses of lectures, delivered in 1848 and subsequent years, in the halls of Trinity College Dublin: with numerous illustrative diagrams, and with some geometrical and physical applications* (Dublin: Hodges and Smith, 1853).

————, *Elements of Quaternions* (London: Longmans, Green, & Co., 1866) with a preface by William Edwin Hamilton; 2nd ed., edited by Charles Jasper Joly (London & New York: Longmans, Green, and Co., 1899–1901); 3rd ed., vol. 1, edited by Charles Jasper Joly (New York: Chelsea, 1969).

————, *The Mathematical Papers of Sir William Rowan Hamilton*, 3 vols., edited by

A. W. Conway and J. L. Synge (Cambridge: Cambridge University Press, 1931–1967).

Heaviside, Oliver, "On the Forces, Stresses, and Fluxes of Energy in the Electromagnetic Field," Royal Society: Abstract in *Proceedings* 50 (1891); *Transactions*, A. (1892); reprinted in Heaviside, *Electrical Papers*, vol. 2, pp. 521–574.

———, *Electrical Papers*, 2 vols. (London: Macmillan, 1892).

———, "Vectors versus Quaternions," *Nature* 47, no. 1223 (April 1893), 533–534.

———, *Electromagnetic Theory*, 3 vols. (London: Benn Brothers Ltd., The Electrician Printing and Publishing Company, 1893–1912; reprint, New York: Chelsea, 1971).

———, "The Elements of Vectorial Algebra and Analysis," *Electromagnetic Theory*, vol. 1 (1893), chap. III.

Heine, Eduard, *Handbuch der Kugelfunctionen*, vol. 1. *Theorie der Kugelfunctionen und der verwandten Functionen*, 2nd rev. ed. (Berlin: G. Reimer, 1878), vol. 2. *Anwendungen der Kugelfunctionen und der verwandten Functionen*, 2nd rev. ed. (Berlin: G. Reimer, 1881).

Helmholtz, Hermann von, "Ueber die tatsächlichen Grundlagen der Geometrie," *Verhandlungen des naturhistorisch-medicinischen Vereins zu Heidelberg* 4 (22 May 1866), 197–202.

———, "Über den Ursprung und die Bedeutung der geometrischen Axiome" (Lecture delivered in the Docenten Werein in Heidelberg in 1870); *Vorträge und Reden*, 5th ed., vol. 2 (Braunschweig: F. Vieweg, 1884), pp. 1–31; reissued in: Helmholtz, *Epistemological Writings*, pp. 226–248.

———, Speech held at the Commemoration-Day celebration of the Friedrich Wilhelm University of Berlin, 3 August 1878; revised, expanded and translated as "The Facts of Perception," in Helmholtz, *Science and Culture: Popular and Philosophical Essays*, pp. 342–380.

———, "Zählen und Messen," *Philosophische Aufsätze Eduard Zeller zu seinem fünfzijärigen Doktorjubiläum gewidmet* (Leipzig: Fues' Verlag, 1887), pp. 17–52; reprinted in Helmholtz, *Wissenschaftliche Abhandlungen*, vol. 3, pp. 356–391; translated by Malcolm F. Lowe, "Numbering and Measuring from an Epistemological Viewpoint," in Helmholtz, *Epistemological Writings*. Also published as *Counting and Measuring*, translated by Charlotte Lowe Bryan, with an introduction and notes by Harold T. Davis (New York: D. Van Nostrand, 1980).

———, *Epistemological Writings*, translated by Malcolm F. Lowe and edited by Robert S. Cohen and Yehuda Elkana, Boston Studies in the Philosophy of Science, vol. 37 (Boston: D. Reidel, 1977).

———, *Science and Culture: Popular and Philosophical Essays*, translated and edited by David Cahan (Chicago: University of Chicago Press, 1995).

Herder, Johann Gottfried von, *Auch eine Philosophie der Geschichte zur Bildung der Menschheit* (Riga, 1774); translated as *This Too; a Philosophy of History for the Formation of Humanity*, in *Johann Gottfried von Herder: Philosophical Writings*, translated and edited by Michael N. Forster (Cambridge: Cambridge University Press, 2002), pp. 272–360.

Hero of Alexandria, *Les mécaniques, ou, L'élévateur des corps lourds*, Arabic text of Qusta Ibn Luqa, established and translated by B. Carra de Vaux, introduction by D. R. Hill and commentaries by A. G. Drachman (Paris: Belles Lettres, 1998).

Herschel, John Frederick William, *A Preliminary Discourse on the Study of Natural Philosophy*, "a new edition" (Philadelphia: Lea Blanchard, 1840).

Hertz, Heinrich, "Über die Beziehungen zwischen den Maxwell'schen elektrody- namischen Grundgleichungen und den Grundgleichungen der gegnerischen Elektrodynamik," *Wiedemann's Annalen der Physik und Chemie* 23 (1884), 84–103; reprinted in Hertz, *Gesammelte Werke*, vol. 1. pp. 295–314.

———, Über die Grundgleichungen der Elektromagnetiks bewegter Körper," *Wie- demann's Annalen* 41 (1890); translated as "On the Fundamental equations of Electromagnetics for Bodies in Motion," in Hertz, *Electric Waves*, pp. 241–268.

———, "Über den Grundgleichungen Elektrodynamik ruhenden Körper," *Göttinger Nachrichten* (19 March 1890); also in *Wiedemann's Annalen* 40 (1890); translated as "On the Fundamental Equations of Electrodynamics for Bodies at Rest," in Hertz, *Electric Waves*, pp. 195–197.

———, *Die Prinzipien der Mechanik in neuem Zusammenhange dargestellt* (Leipzig: J. A. Barth, 1894), with a preface by Hermann von Helmholtz; authorized English translation by D. E. Jones and J. T. Walley as *The Principles of Mechanics, Presented in a New Form* (New York: Dover, 1956).

———, *Electric Waves: being Researches on the Propagation of Electric Action with Fi- nite Velocity Through Space*, authorized English translation by D. E. Jones with a preface by Lord Kelvin (London: Macmillan, 1894).

———, *Gesammelte Werke*, vol. 1, edited by Philip Lenard (Leipzig: Johann Ambro- sius Barth, 1895).

Hoüel, Jules, *Essai Critique sur les Principes Fondamentaux de la Géométrie Élémentaire* (Paris: Gauthier-Villars, 1867).

Hume, David, *An Enquiry Concerning Human Understanding*, edited by Tom L. Beau- champ (Oxford: Oxford University Press, 1999).

———, *A Treatise of Human Nature*, edited by David Fate Norton, and Mary J. Nor- ton (Oxford: Oxford University Press, 2000).

Huygens, Christiaan, *De Motu Corporum ex Percussione* (manuscript, mid-1650s), in Huygens, *Oeuvres complètes de Christiaan Huygens*, vol. 16 (La Haye: Société Hol- landaise des Sciences, Martinus Nijhoff, 1929), pp. 29–91.

———, *Horologium Oscillatorium* (1673): *Christiani Hvgenii Zvlichemii Const. f. Horologivm oscillatorivm, sive, De motv pendvlorvm ad horologia aptato demon- strationes geometricae* (Paris: Apud F. Muguet, 1673); reissued in *Oeuvres*, vol. 18 (1934), pp. 69–368.

Jevons, William Stanley, *The Principles of Science* (1874); 2nd ed. (London: Macmillan, 1877).

Jolly, Philipp Johannes Gustav von, *Die Principien der Mechanik* (Stuttgart: Franck, 1852).

Jussieu, Antonii Laurentii de, *Genera Plantarum secundum ordines naturales disposita juxta methodum in Horto regio Parisiensi exaratam anno MDCCLXXIV* (Paris: Herissant, 1789).

Kant, Immanuel Kant, *De Mundi Sensibilis atque Intelligibilis Forma et Principiis*, Dissertatio Pro Loco, 1770: *Kant's Inaugural Dissertation and Early Writings on Space*, translated by John Handyside (Chicago: Open Court, 1929).

————, *Kritik der reinen Vernuft* (Riga: Verlegts Johann Friedrich Hartknoch, 1781); *Critique of Pure Reason*, translated and edited by Paul Guyer and Allen W. Wood, Cambridge Edition of the Works of Immanuel Kant (Cambridge: Cambridge University Press, 1998).

————, *Prolegomena zu Einer Jeden Künftigen Metaphysik Prolegomena zu einer jeden künftigen Metaphysik die als Wissenschaft wird auftreten können* (Riga: Johann Friedrich Hartknoch, 1783), sec. 10; *Kant's Prolegomena to Any Future Metaphysics*, translated and edited by Paul Carus (Chicago: Open Court; London: Kegan Paul, Trench, Trübner & Co., 1902).

————, "Transcendental Aesthetics" (1781), translation by Norman Kemp Smith, *Critique of Pure Reason* (London: Macmillan, 1929), reprinted in J. J. C. Smart, *Problems of Space and Time* (New York: Macmillan, 1964).

————, *Neuer Lehrbegriff der Bewegung und Ruhe und der damit verknüpften Folgerungen in den ersten Gründen der Naturwissenschaft* (Königsberg: Johann Friedrich Driest, 1758); reissued in *Immanuel Kants Werke*, edited by Ernst Cassirer, vol. 2 (Berlin: Bruno Cassirer, 1922), pp. 15–28.

Kirchhoff, Gustav, *Lehrbuch der Mechanik* (1874); *Vorlesungen über Mathematische Physik*, vol. 1, *Mechanik* (Leipzig: B. G. Teubner, 1876); 3rd ed. (Leipzig: B.G. Teubner, 1883; reprint, Leipzig: B. G. Teubner, 1897).

Klein, Felix, and Conrad H. Müller, *Encyklopädie der mathematischen Wissenschaften*, vol. 4, *Mechanik*, 2 parts (Leipzig: Teubner, 1896–1935).

Klügel, Georg Simon, *Mathematisches Wörterbuch; oder, Erklärung der Begriffe, Lehrsätze, Aufgaben und Methoden der Mathematik mit den nöthigen beweisen und literarischen nachrichten Begleitet in alphabetischer ordnung*, 5 vols. (Leipzig: E. B. Schwickert, 1803).

Koenigs, Gabriel, *Leçons de cinématique: professées à la Sorbonne*, with notes by Gaston Darboux, Eugène Cosserat, and François Cosserat (Paris: A. Hermann, 1897).

Kohlrausch, Friedrich, *An Introduction to Physical Measurements, with Appendices on Absolute Electrical Measurements, Etc.*, translated from the seventh German edition by Thomas Hutchinson Waller and Henry Richardson Procter (London: J. & A. Churchill, 1894).

————, *Lehrbuch der Praktischen Physik*, 10th ed. (Leipzig: B. G. Teubner, 1905).

Krause, Albrecht, *Kant und Helmholtz: über den Ursprung und die Bedeutung der Raumanschauung und der geometrischen Axiome* (Lahr: Moritz Schauenburg, 1878).

Laboulaye, Charles, *Traité de cinématique ou théorie des mécanismes* (1849); rev. ed. (Paris: E. Lacroix / H. Plon, 1861).

Lacroix, S. F., *Élémens d'Algèbre, à l'Usage de l'École Centrale des Quatre-Nations;* 11th rev. ed. (Paris: Mme. Ve. Courcier, Imprimeur-Libraire pour les Mathématiques, 1815).

Lagrange, Joseph Louis, *Mécanique Analytique* (1788); 3rd ed., annotated by M. J. Bertrand, vol. 1 (Paris: Mallet-Bachelier, Gendre et Successeur de Bachelier [École Polytechnique], 1853).

———, *Théorie des Fonctions Analytiques, contenant les Principes du Calcul Différentiel, dégagés de toute considération d'Infiniment Petits ou d'Évanouissans, de Limites ou de Fluxions, et réduits a l'Analyse Algébrique des Quantités Finies* (Paris: Imprimerie de la République, 1799–1800).

Lamarle, Anatole Henri Ernest, and Ernest Lamarle, *Éxposé Géométrique du Calcul Différentiel et Intégral, précédé de la Cinématique du Point, de la Droit et du Plan, et Fondé tout entire sur les Notions les plus Élémentaires de la Géométrie Plane* (Paris: Mallet-Bachelier, Imprimeur-Libraire de l'École Impériale Polytechnique, du Bureau des Longitudes, 1861).

Lange, Ludwig, "Ueber die wissenschaftliche Fassung des Beharrungsgesetz," *Berichte der Königlichen Sachsischen Gesellschaft der Wissenschaften zu Leipzig, Mathematisch-Physische Klasse* 37 (1885), 331–351.

———, *Die Geschichtliche Entwickelung des Bewegungsbegriffes und Ihr Voraussichtliches Endergebniss. Ein Beitrag zur Historischen Kritik der Mechanischen Principien* (Leipzig: Wilhelm Engelmann, 1886).

Lanz, José María, and Agustín de Betancourt, *Essai sur la Composition des Machines* (Paris: Imprimerie Impériale, 1808).

———, *Essai sur la Composition des machines* (Paris: Bachelier, 1819).

———, *Ensayo sobre la Composición de las Máquinas,* translated by Manuel Díaz-Marta, with a prologue by José A. García Diego and commentaries by Aleksei N. Bogoliubov (Madrid: Colegio de Ingenieros de Caminos, Canales y Puertos, [1990]).

Larmor, Joseph, *Aether and Matter: A Development of the Dynamical Relations of the Aether to Material Systems on the Basis of the Atomic Constitution of Matter; including a Discussion of the Influence of the Earth's Motion on Optical Phenomena* (Cambridge: Cambridge University Press, 1900).

———, *Mathematical and Physical Papers,* vol. 1 (Cambridge: Cambridge University Press, 1929).

Lechalas, George, *Étude sur l'Espace et le Temps* (Paris: Alcan, 1898).

Lefebvre-Laboulaye, Charles Pierre, *Traité de Cinematique, ou théorie des mécanismes* (1849); 2nd ed. (Paris, 1861).

Legendre, Adrien Marie, *Eléments de Géométrie* (Paris, 1794); *Elements of Geometry,* translated by John Farrar (Cambridge: Hilliard and Metcalf, Cambridge University Press, 1819).

Leibniz, Gottfried Wilhelm, letter to Christiaan Huygens, 8 September 1679, in *Christiaani Hugenii aliorumque seculi XVII. virorum celebrium exercitationes mathematicae et philosophicae*, edited by Uylenbroek (1833), fasc. I, p. 10; translated in Hermann Grassmann, *A New Branch of Mathematics, The Ausdehnungslehre of 1844 and Other Works*, translated by Lloyd C. Kannenberg (Chicago: Open Court, 1995).

Lewes, G. H., *Comte's Philosophy of the Sciences: Being an Exposition of the Cours de Philosophie Positive of Auguste Comte* (London: Henry G. Bohn, 1853).

Lieber, Francis, ed., *Encyclopædia Americana. A Popular Dictionary of Arts, Sciences, Literature, History, Politics and Biography* (on the basis of the seventh edition of the German Conversations-Lexicon), with the editorial assistance of E. Wiggelsworth and T. G. Bradford, vol. 9 (Philadelphia: Carey and Lea, 1832).

Lodge, Oliver, *Elementary Mechanics; including Hydrostatics and Pneumatics* (London: W. & R. Chambers, Ltd., 1896).

Lorentz, Hendrik A., *Lehrbuch der Differential- und Integralrechnung nebst einer Einführung in andere Teile der Mathematik* (1882), revised by G. C. Schmidt (Leipzig: J. A. Barth, 1900).

———, "De l'influence du mouvement de la terre sur les phénomènes lumineux," *Verslagen van de gewone vergaderingen der wis-en natuurkundige afdeeling, Koninklijke Akademie van Wetenschappen te Amsterdam* 2 (1886); also in *Archives Néerlander* 21 (1887); reprinted in Lorentz, *Collected Papers*, vol. 4, pp. 153–214.

———, "The Relative Motion of the Earth and the Ether," *Verslagen van de gewone vergaderingen der wis- en natuurkundige afdeeling, Koninklijke Akademie van Wetenschappen te Amsterdam* 1 (1892); reprinted in Lorentz, *Collected Papers*, vol. 4., pp. 36–58.

———, *Versuch einer Theorie der elektrischen und optischen Erscheinungen in bewegten Körpern* (Leiden: E. J. Brill, 1895); reprinted in Lorentz, *Collected Papers*, vol. 5, pp. 1–137.

———, "Concerning the Problem of the Dragging along of the Ether by the Earth," *Verslagen van de gewone vergaderingen der wis- en natuurkundige afdeeling, Koninklijke Akademie van Wetenschappen te Amsterdam* 6 (1897); reprinted in *Collected Papers*, vol. 4., pp. 237–244.

———, "La Théorie de l'Aberration de Stokes dans l'Hypothèse d'un Éther n'ayant pas partout la Même Densité," *Verslagen van de gewone vergaderingen der wis- en natuurkundige afdeeling, Koninklijke Akademie van Wetenschappen te Amsterdam* 7 (1899); reprinted in *Collected Papers*, vol. 4., pp. 245–253.

———, "Vereenvoudigde theorie der electrische en optische verschijnselen in lichamen die zich bewegen," *Verslagen van de gewone vergaderingen der wis- en natuurkundige afdeeling, Koninklijke Akademie van Wetenschappen te Amsterdam* 7 (1899), 507–522; reissued in French translation as "Théorie Simplifiée des Phénomènes Électriques et Optiques dans des Corps en Mouvement," in Lorentz, *Collected Papers*, vol. 5., pp. 139–155.

————, *Zichtbare en onzichtbare bewegingen* (1901); German translation: *Sichtbare und unsichtbare Bewegungen; Vorträge auf Einladung des Vorstandes de Departements Leiden der Maatschappij tot nun van't algemeen im Februar und März 1901,* edited from the Dutch original by G. Siebert (Brauschweig, F. Vieweg und sohn, 1902).

————, "Some Considerations on the Principles of Dynamics, in Connection with Hertz's 'Prinzipien der Mechanik,'" *Proceedings of the Academy of Amsterdam* **3** (1902); reprinted in Lorentz, *Collected Papers,* vol. 4. pp. 36–58.

————, "Electromagnetische verschijnselen in een stelsel dat zich met willekeurige snelheid, kleiner dan die van het licht, beweegt," *Koninklijke Akademie van Wetenschappen te Amsterdam. Wis-en Natuurkundige Afdeeling. Verslagen van de Gewone Vergaderingen* **12** (1904), 986–1009; reissued as "Electromagnetic phenomena in a system moving with any velocity smaller than that of light," *Koninklijke Akademie van Wetenschappen te Amsterdam,* Section of Sciences, *Proceedings* **6** (1904), 809–831.

————, *The Theory of Electrons, and its Applications to the Phenomena of Light and Radiant Heat,* lectures delivered at Columbia University (1906); 2nd ed. (New York: G. E. Stechert & Co., 1923).

————, *Lectures on Theoretical Physics,* Delivered at the University of Leiden, authorized translation by L. Silberstein and A. P. H. Trivelli, vol. 3. *Maxwell's Theory and The Principle of Relativity for Uniform Translations,* edited by H. Bremekamp and A. D. Fokker (London: Macmillan, 1931).

————, *Collected Papers,* vol. 4, *Mechanics, Hydrodynamics and Wave Propagation* (The Hague: Martinus Nijhoff, 1937).

————, *Collected Papers,* vol. 5 (The Hague, Holland: Martinus Nijhoff, 1937).

————, *Selected Works of H. A. Lorentz,* vol. 5, ed. Nancy J. Nersessian and H. Floris Cohen (The Netherlands: Palm Publications, 1987).

Lorentz, H. A., A. Einstein, H. Minkowski, and H. Weyl, *The Principle of Relativity* (1923), with notes by Arnold Sommerfeld and translated by W. Perret and G. B. Jeffery (New York: Dover, 1952).

Lübsen, Heinrich Borchert, *Einleitung in die Infinitesimal-Rechnung (Differential- und Integral-Rechnung). Zum Selbstunterricht. Mit Rücksicht auf das Nothwendigste und Wichtigste* (Leipzig: Friedrich Brandstetter, 1862); 4th ed. (Leipzig: Friedrich Brandstetter, 1869).

Macfarlane, Alexander, "[Review of] *Utility of Quaternions in Physics.* By A. McAulay," in *Physical Review* **1** (1893), p. 389.

————, *Vector Analysis and Quaternions* (1896); 4th ed. (New York: John Wiley & Sons; London: Chapman & Hall, Ltd., 1906).

MacGregor, James Gordon, *An Elementary Treatise on Kinematics and Dynamics* (London, 1887).

Mach, Ernst, *Vorlesungen über Psychophysik* (Vienna: Sommer, 1863)

————, "Ueber die Definition der Masse," *Carl's Repertorium der Experimentalphysik*

4 (1868); reprinted in Mach's *Die Geschichte und die Wurzel des Satzes von der Erhaltung der Arbeit*, 2nd ed. (Leipzig: J. A. Barth, 1909).

———, address delivered on the anniversary meeting of the Imperial Academy of Sciences, at Vienna (25 May 1882); translated by T. J. McCormack, "The Economical Nature of Physical Inquiry," in Mach, *Popular Scientific Lectures* (1898; reprint, La Salle, Ill.: Open Court, 1986).

———, *Die Mechanik in ihrer Entwickelung historisch-kritisch dargestellt* (Leipzig: F. A. Brockhaus, 1883); 2nd ed. (1889), translated by T. J. McCormack, *The Science of Mechanics: A Critical and Historical Account of Its Development* (1893; revised in 1942 to include additional alterations up to the ninth German edition of 1933; reprint, La Salle, Ill.: Open Court, 1960).

———, *Beiträge zur Analyse der Empfindungen* (Jena: G. Fischer, 1886); *Contributions to the Analysis of the Sensations*, translated by C. M. Williams (Chicago: Open Court, 1897).

———, *Erkenntniss und Irrthum: Skizzen zur Psychologie der Forschung* (Leipzig: J. A. Barth, 1905); reissued in Mach, *Space and Geometry in the Light of Physiological, Psychological and Physical Inquiry*, translated by Thomas J. McCormack (La Salle, Ill., Open Court, 1906/1943).

Maclaurin, Colin *An Account of Sir Isaac Newton's Philosophical Discoveries* (1748); facsimile of the first edition, with a new introduction by L. L. Laudan, Sources of Science, no. 74 (New York: Johnson Reprint Corporation, 1968).

Mannheim, Amédée, *Cours de géométrie descriptive de l'École Polytechnique : comprenant les éléments de la géométrie cinématique* (Paris: Gauthier-Villars, 1880).

———, *Principes et développements de géométrie cinématique: ouvrage contenant de nombreuses applications a la théorie des surfaces* (Paris : Gauthier-Villars et fils, 1894).

Maxwell, James Clerk, "Address to the Mathematical and Physical Sections of the British Association," *British Association for the Advancement of Science Report* **40** (1870); reprinted in *The Scientific Papers of James Clerk Maxwell*, vol. 2, pp. 215–229.

———, "On the Mathematical Classification of Physical Quantities," *Proceedings of the London Mathematical Society* **3** (1871), 224–232; in *Scientific Papers*, vol. 2, 257–266.

———, *A Treatise on Electricity and Magnetism*, vol. 1 (1873); 2nd ed. (Oxford: Clarendon Press, 1881).

———, *Matter and Motion* (London: Society for promoting Christian knowledge, 1877; reprint, New York: Dover, 1991).

———, *The Scientific Papers of James Clerk Maxwell*, edited by W. D. Niven, vol. 2 (Cambridge: University Press, 1890; reprint, New York: Dover, 1965).

McAulay, Alexander, *Utility of Quaternions in Physics* (London: Macmillan, 1893).

Michelson, Albert A., and Edward W. Morley, "On the Relative Motion of the Earth and the Luminiferous Ether," *American Journal of Physics* **34** (1887), 333–345.

Mill, John Stuart, *A System of Logic, Ratiocinative and Inductive: Being a Connected View of the Principles of Evidence and the Methods of Scientific Investigation* (1846), 8th ed., vol. 2 (London: Longmans, Green, Reader, and Dyer, 1872).

Minkowski, Hermann, "Die Grundgleichungen für die elektromagnetischen Vorgaänge in bewegten Körpern," *Nachrichten der Gesellschaft der Wissenschaften zu Göttingen, Mathematisch-physikalische Klasse* (1908), 53–111; reprinted in Hermann Minkowski, *Gesammelte Abhandlungen*, edited by David Hilbert, Andreas Speiser, and Hermann Weyl, vol. 2 (New York: Chelsea, 1967), pp. 352–404.

———, "Raum und Zeit," *Physikalische Zeitschrift* **10** (1909), 104–111 [lecture delivered at the 80th Naturforscher-Versammlung in Cologne (21 September 1908)]; reprinted in *Gesammelte Abhandlungen*, vol. 2, pp. 432–434; translated by W. Perrett and G. B. Jeffery as "Space and Time," in H. A. Lorentz, A. Einstein, H. Minkowski, and H. Weyl, *The Principle of Relativity: A Collection of Original Memoirs on the Special and General Theory of Relativity*, with notes by Arnold Sommerfeld (New York: Dodd, Mead, 1923), pp. 75–91.

Monge, Gaspard, *Application de l'analyse à la géométrie* (1809); 5th ed., annotated by M. Liouville (Paris: Bachelier, Imprimeur-Libraire, 1850).

———, "Lessons," edited by B. Belhoste, R. Laurent, J. Sakarovitch, and R. Taton, in *L'École Normale de l'An III, Leçons de Mathématiques, Laplace, Monge*, edited by J. Dhombres (Paris: Dunod, 1992).

Morin, Arthur Jules, *Aide-mémoire de mécanique pratique à l'usage des officiers d'artillerie et des ingénieurs civils et militaires* (Metz: Mme. Theil, 1837).

———, *Hilfsbuch für praktische Mechanik*, translated into German by C. Holtzmann (Karlsruhe: Groos, 1838).

———, *Description des appareils chronométriques à style, propres à la présentation graphique et à la détermination des lois du mouvement, et des appareils dynamométriques propres à mesurer l'effort ou le travail développé par les moteurs animés ou inanimés et par les organes de transmission du mouvement dans les machines* (Metz: S. Lamort, 1838).

———, *Leçons de Mecanique Pratique. Notions Géométriques sur les Mouvements et leurs transformations, ou, Élements de Cinématique*, 2nd ed. (Paris: L. Hachette, 1857).

Nernst, W., and A. Schönflies, *Einführung in die mathematische Behandlung der Naturwissenschaften* (1895); 5th ed. (Munich: R. Oldenburg, 1907).

Neumann, Carl G., *Ueber die Principien der Galilei-Newton'schen Theorie. Akademische Antrittsvorlesung gehalten in der Aula der Universität Leipzig am 3. November 1869* (Leipzig: Teubner, 1870).

Neumann, Franz, *Einleitung in die Theoretische Physik, Vorlesungen gehalten an der Universität zu Königsberg*, edited by C. Pape (Leipzig: B. G. Teubner, 1883).

Newton, Isaac, *De Motu Corporum* (1684), in Newton, *The Preliminary Manuscripts for Isaac Newton's 1687 Principia, 1684–1686*, with an introduction by D.T. Whiteside (Cambridge: Cambridge University Press, 1989).

————, *Philosophia Naturalis Principia Mathematica* (1687), *Mathematical Principles of Natural Philosophy,* translated by Andrew Motte in 1729 and revised by Florian Cajori (Berkeley: University of California Press, 1946).

————, letter to Bentley, 28 February, 1692/3, *The Correspondence of Isaac Newton,* vol. 3, *1688–1694,* edited by H. W. Turnbull (Cambridge: Cambridge University Press, 1961).

————, *Geometry* (*ca.* 1693), published and translated in *The Mathematical Papers of Isaac Newton,* vol. 7, *1691–1695,* edited by Derek T. Whiteside (Cambridge: Cambridge University Press, 1976).

————, *Arithmetica Universalis* (1707); translated by Mr. Ralphson as *Universal Arithmetick: or, a Treatise of Arithmetical Composition and Resolution,* and revised by Mr. Cunn (London, 1728), reprinted in *The Mathematical Works of Isaac Newton,* vol. 2, assembled by Derek T. Whiteside, Sources of Science (New York: Johnson Reprint Corporation, 1967).

Nietzsche, Friedrich, *Unzeitgemässe Betrachtungen. Zweites Stück: Vom Nutzen und Nachtheil der Historie für das Leben* (1874), part 2; translated in *Untimely Mediations,* translated by R. J. Hollingdale, edited by Daniel Breazeale (Cambridge: Cambridge University Press, 1997), pp. 57–124.

Ostwald, Wilhelm, *Lehrbuch der allgemeinen Chemie,* 2nd rev. ed., vol. 2, part 1, *Chemische Energie,* (Leipzig: Wilhelm Engelmann, 1893).

Pappus of Alexandria, *Book 7 of the Collection,* edited with translation and commentary by Alexander Jones (New York: Springer-Verlag, 1986).

Pearson, Karl, *The Grammar of Science,* part I: *Physical,* Contemporary Science Series, (London: Walter Scott, 1892); 3rd ed. (London: Adam and Charles Black, 1911), p. 235, and also *Everyman's Library,* no. 939 (London: J. M. Dent & Sons, 1937).

Perigal, Henry, *Perigal's Contributions to Kinematics* (London, 1854).

————, *The Phases of Perigal's retrogressive kinematic parabola derived from the circle* (London, 1894).

Petersen, Julius, *Kinematik lehrbuch der statik fester kosper* (Copenhagen: Host, 1884).

————, *Kinematik,* German edition by R. von Fischer-Benzon (Copenhagen: Andr. Fred. Höst & Sohn, 1884).

Plato, *Republic* (ca. 380 BC), translated by G. M. A. Grube and revised by C. D. C. Reeve (Indianapolis: Hackett, 1992).

Playfair, John, *Outlines on Natural Philosophy,* vol. 1 (Edinburgh and London: A. Neil & Co. for Archibald Constable and Company, and Longman, Hurst, Rees, Orme and Brown, Cadell and Davies, and John Murray, 1812).

————, *Dissertation second, exhibiting a general view of the progress of mathematical and physical science, since the revival of letters in Europe* (Boston: Wells & Lilly, 1817). Issued as the second dissertation of the supplement to the fourth, fifth, and sixth editions of the *Encyclopaedia Britannica.* Reissued in vol. 1 (1835) of the seventh edition of the *Encyclopaedia Britannica*, as *Dissertations on the History of Metaphysical and Ethical and of Mathematical and Physical Science*, "Dissertation Third; Exhibiting a General View of the Progress of Mathematical and Physical Science, since the Revival of Letters in Europe."

Plutarch, "Life of Marcellus" (ca. 100 AD), in *Plutarch's Lives of Illustrious Men*, translated by John Dryden and others, and corrected from the Greek and revised by Arthur H. Clough (1876; reprint, Boston: Little, Brown, 1928).

Philoponus, *On Aristotle's Physics 2*, translated by A. R. Lacey (Ithaca: Cornell University Press, 1993).

Poincaré, Henri, *Théorie mathématique de la lumière*, Sorbonne lectures: 1887–1888, edited by J. Blondin (Paris, 1889).

———, "A Propos de la Théorie de M. Larmor," *Éclairge électrique* **3** (May 1895), 5–13, 289–295; reprinted in Poincaré, *Oeuvres* 9, pp. 383–394.

———, "Les idées de Hertz sur la mécanique," *Revue Générale des Sciences* **8** (30 September 1897), 734–743; reprinted in *Oeuvres de Henri Poincaré*, vol. 7 (Paris: Gauthier-Villars, Libraire du Bureau des Longitudes, de l'École Polytechnique, 1955), pp. 231–250.

———, "La mesure du temps," *Revue de Métaphysique et de Morale* **6** (January 1898), 1–13; reissued in Poincaré, *The Foundations of Science*, pp. 223–234.

———, *Cinématique et mécanismes: potentiel et mécanique des fluides / Cours professé à la Sorbonne*, edited by Amédée Guillet (Paris: G. Carré et C. Naud, 1899).

———, "La Théorie de Lorentz et le Principe de Réaction," in *Recueil de Travaux offerts par les Auteurs à H. A, Lorentz, professeur de Physique à l'Université de Leiden à l'occasion du 25e anniversaire de son Doctorat, le 11 décembre 1900;* also in *Archives néerlandaises des Sciences exactes et naturelles*, 2nd ser. **5** (1900), 252–278; reprinted in Poincaré, *Oeuvres* 9, pp. 466–488.

———, *Électricité et Optique: la Lumière et les Théories Électrodynamiques*, Leçons Professées a la Sorbonne en 1888, 1890 et 1899, Cours de la Faculté des Sciences de Paris: Cours de Physique Mathématique, Second Edition, second edition by Jules Blondin and Eugène Néculcèa (Paris: Gauthier-Villars, Imprimeur-Libraire, 1901).

———, *La Science et l'Hypothèse* (Paris: Flammarion, 1902).

———, "Létat actuel et l'avenir de la Physique mathématique" (1904); "The History of Mathematical Physics," *The Value of Science*, authorized translation by George Bruce Halsted, reprinted in Poincaré, *Foundations of Science*, pp. 297–302.

———, "L'état actuel et l'avenir de la Physique mathématique," lecture delivered at the International Congress of Arts and Science in Saint Louis, Missouri, in 1904, in *La valeur de la science* (1905); translated and reprinted as "The Present Crisis of Mathematical Physics," in *Foundations of Science*, pp. 303–313.

———, "Sur la dynamique de l'électron," *Comptes rendus des Seances de l'Académie des Sciences* [Paris] **40** (5 June 1905), 1504–1508; reprinted in Poincaré, *Oeuvres* 9, pp. 489–493.

————, "Sur la dynamique de l'électron," *Rendiconti del Circolo matematico di Palermo* **21** (1906), 129–176; reprinted in *Oeuvres* 9, pp. 494–550.

————, "La dynamique de l'électron," *Revue générale des Sciences pures et appliquées* **19** (1908), 386–402; reprinted in Poincaré, *Oeuvres* 9, pp. 551–586.

————, "L'Espace et le Temps," *Scientia* **12** (1912), pp. 159–170; also in Poincaré, *Dernières Pensées* (Paris: Flammarion, 1913).

————, "Analysis and Physics" (1905), in *The Value of Science*, authorized translation by George Bruce Halsted; reprinted in Poincaré, *Foundations of Science*, pp. 279–288.

————, *The Foundations of Science: Science and Hypothesis, The Value of Science, Science and Method*, authorized translation by George Bruce Halsted (New York: Science Press, 1913).

————, *Oeuvres de Henri Poincaré*, vol. 9, preface by Louis de Broglie (Paris: Gauthier-Villars, Éditeur-Imprimeur-Libraire, 1954).

Poisson, Siméon Denis, *Traité de Mécanique;* 2nd ed., vol. 1, Imprimeur-Libraire pour les mathématiques, no. 55 (Paris: Bachelier, 1833).

Poncelet, Jean Victor, *Mécanique Industrielle*, 2 parts (Brussels, 1839).

————, *Introduction à la mécanique industrielle, physique ou expérimentale*, 3rd ed. (Paris: François Xavier Kretz, 1870).

Popović, Milan, ed., *In Albert's Shadow: The Life and Letters of Mileva Marić, Einstein's First Wife*, translations partly by Bosko Milosavljević (Baltimore: Johns Hopkins University Press, 2003).

Proclus, *Eis proton Eukleidou stoicheion biblon* (ca. 450); *A Commentary on the First Book of Euclid's Elements*, translated and with an introduction and notes by Glenn R. Morrow (Princeton: Princeton University Press, 1970).

Prony, Gaspard Clair François Marie Riche de, Journal de l'École Polytechnique, Tome III, septième & 8me cahiers contenant les trois premières parties d'un ouvrage du C.en Prony, *Mécanique Philosophique, ou l'Analyse Raisonée des Diverses Parties de la Science de l'équilibre et du Mouvement* (Paris: Imprimerie de la République, 1800).

————, *Leçons de Mécanique Analytique, données à l'École Impériale Polytechnique*, 2 parts (Paris: Imprimerie de l'École Impériale des Ponts et Chaussées, 1810, 1815).

Rankine, William John Macquorn, *A Manual of Applied Mechanics; Encyclopedia Metropolitana*, 2nd ed., vol. 39 (London, 1858).

————, *A Manual of Machinery and Millwork* (London, 1869).

Redtenbacher, F., *Principien der Mechanik und des Maschinenbaues*, 2nd ed. (Mannheim: Bauermann, 1859).

Renn, Jürgen, and Robert Schulmann, eds., *Albert Einstein and Mileva Marić: The Love Letters*, translated by Shawn Smith (Princeton: Princeton University Press, 1992).

Resal, Henri, *Éléments de Mécanique* (Paris: MM. Didot frères and Mallet-Bachelier, libraires, 1851).

————, *Traité de Cinématique Pure* (Paris: Mallet-Bachelier, 1862).

Reuleaux, Franz, *Theoretische Kinematik: Grundzüge einer Theorie des Maschinenwesens* (Braunschweig: F. Vieweg und Sohn, 1875).

————, *The Kinematics of Machinery, Outlines of a Theory of Machines,* translated by Alexander Blackie W. Kennedy (London: Macmillan, 1876); reissued, with a new introduction by Eugene S. Ferguson (New York: Dover, 1963).

————, *Cinématique: principes fondamentaux d'une théorie générale des machines* (Paris: Savy, 1877).

Riemann, Bernhard, "Ueber die Hypothesen, welche der Geometrie zu Grunde liegen," lecture delivered in Göttingen, 1854, *Abhandlungen der Königlichen Gesellschaft der Wissenschaften zu Göttingen* 13 (1867), 132–152; reissued in Riemann, *Gesammelte mathematische Werke und wissentschaftlicher Nachlaß,* edited by Richard Dedekind and Heinrich Weber, 2nd ed. (Leipzig: B. G. Teubner, 1892); translated by William Kingdon Clifford, "On the Hypotheses which Lie at the Base of Geometry," *Nature* 8 (1873); parts I–II in 1 May: pp. 14–17, and part III in 8 May: pp. 36–37.

Ritter, August, *Lehrbuch der Technischen Mechanik,* 2nd ed. (Hannover: Carl Rümpler, 1870).

Ritz, Walther, *Oeuvres* (Paris: Gauthier-Villars, 1911).

Saint-Venant, Adhémar Jean Claude Barré de, *Principes de mécanique fondés sur la cinématique* (Paris: Bachelier, 1851).

Schopenhauer, Arthur, *Über die Vierfache Wurzel des Satzes von zureichenden Grunde* (Rudolstadt, 1813); translated by Karl Hillebrand as "On the Fourfold Root of the Principle of Sufficient Reason," in *Two Essays by Arthur Schopenhauer,* rev. ed. (London: George Bell and Sons, 1891).

————, *Die Welt als Wille und Vorstellung,* 2 vols. (1844); *The World as Will and Representation,* translated by E. F. J. Payne (New York: Dover, 1966).

————, *Parerga und Paralipomena* (1851); selections reissued in: Schopenhauer, *Essays and Aphorisms,* selected and translated with an introduction by R. J. Hollingdale (London: Penguin Books, 2004); and, Schopenhauer, *The Wisdom of Life and Counsels and Maxims,* translated by T. Bailey Saunders (Amherst, N.Y.: Prometheus Books, 1995).

Sigwart, Christoph, *Logik,* 2 vols. (Tübingen: Mohr, 1873 and 1878); *Logic,* translated by H. Dendy (London: Sonnenschein, 1895).

Solovine, Maurice, and Albert Einstein, *Lettres à Maurice Solovine* (Paris: Gauthier-Villars, 1956).

Spencer, Herbert, *The Classification of the Sciences, to which are added Reasons for Dissenting from the Philosophy of M. Comte,* 3rd ed. (London: Williams and Norgate, 1871).

Spieker, Theodor, *Lehrbuch der ebenen Geometrie mit Übungs-Aufgaben für höhere Lehranstalten,* 19th rev. ed. (Potsdam: Aug. Stein, 1890).

Spinoza, Baruch, *The Ethics* (1677); translated by Samuel Shirley, *The Ethics and Selected Letters*, edited by Seymour Feldman (Indianapolis: Hackett, 1982).

Stallo, Johann Bernhard, *The Concepts and Theories of Modern Physics* (New York: D. A. Appleton; London: Scientific Series/Kegan Paul, 1881); 3rd ed. (1888) reissued with an introduction by Percy W. Bridgman (Cambridge, Mass.: Belknap/ Harvard University Press, 1960).

Streintz, Heinrich, *Die Physikalischen Grundlagen der Mechanik* (Leipzig: Teubner, 1883).

Tait, Peter Guthrie, "On the Rotation of a Rigid Body about a Fixed Point," *Transactions of the Royal Society of Edinburgh* **25**; received on 13 October, read on 21 December 1868; reprinted in Tait, *Scientific Papers*, vol. 1, doc. XV.

———, "On Green's and Other Allied Theorems," *Transactions of the Royal Society of Edinburgh* **26**, received on 29 April, read on 16 May 1870; reprinted in Tait, *Scientific Papers*, vol. 1, doc. XIX.

———, "Address to Section A of the British Association," *British Association Report*, Edinburgh (3 August 1871); reprinted in Tait, *Scientific Papers*, vol. 1, doc. XXIII.

———, "Hamilton," *Encyclopaedia Britannica* (1880); reprinted in Tait, *Scientific Papers*, vol. 2, doc. CXXVIII.

———, "Quaternions," *Encyclopaedia Britannica* (1886); reprinted in Tait, *Scientific Papers*, vol. 2, doc. CXXIX.

———, *An Elementary Treatise on Quaternions*, 3rd ed. (Cambridge: Cambridge University Press, 1890).

———, "On the Importance of Quaternions in Physics," *Philosophical Magazine* (January 1890); reprinted in Tait, *Scientific Papers*, vol. 2, doc. XCVII.

———, "On the Intrinsic Nature of the Quaternion Method," *Proceedings of the Royal Society of Edinburgh* (2 July 1894); reprinted in Tait, *Scientific Papers*, vol. 2, doc. CXVI.

———, *Scientific Papers*, 2 vols. (Cambridge: University Press, 1898, 1900).

Tate, Thomas, *The Elements of Mechanism* (London: Longman, 1851).

Thomson, James, "On Dimensional Equations, and on some Verbal Expressions in Numerical Science," *Report of the British Association*, Section A, Dublin (1878), 451–455; reprinted in James Thomson, *Collected Papers in Physics and Engineering* (Cambridge: Cambridge University Press, 1912), doc. 56, pp. 375–379.

———, "On the Law of Inertia, the Principle of Chronometry and the Principle of Absolute Clinural Rest, and of Absolute Rotation," *Royal Society of Edinburgh Proceedings* **12** (1884), 568–578; reprinted in *Collected Papers*, doc. 57.

Thomson, William, and P. G. Tait, *Elements of Natural Philosophy*, vol. 1 (Oxford: Clarendon Press, 1873).

———, *Treatise on Natural Philosophy*, vol. 1, new ed. (Cambridge: Cambridge University Press, 1879).

Transon, Abel Étienne Louis, "Note sur les principes de la mécanique," *Journal de Mathématiques pures et appliquées* (*Journal de M. Liouville*), ser. 1, **10** (1845), 320–326.

Valson, C.-A., *La Vie et Travaux d'André-Marie Ampère*, 2nd ed. (Lyon: Imprimeur-Librarie de l'Archevêté et des Facultés Catholiques de Lyon, 1897).

Violle, Jules, *Lehrbuch der Physik*, part 1, *Mechanik*, vol. 1, *Allgemeine Mechanik und Mechanick der festen Körper;* German edition by E. Gumlich et al. (Berlin: Julius Springer, 1892).

Voigt, Woldemar, "Ueber das Doppler'sche Princip," *Königliche Gesellschaft der Wissenschaften und der Georg-Augusts-Universität zu Göttingen. Nachrichten* (1887), 41–51; reprinted in *Physikalische Zeitschrift* **16** (1915), 381–386.

Weierstrass, Karl, *Mathematische Werke*, vol. 1 (Berlin: Mayer & Müller, 1894).

Whewell, William, *An Introduction to Dynamics, Containing the Laws of Motion and The First Three Sections of the Principia* (Cambridge and London: Cambridge University, J. and J. J. Deighton; Whittaker, Treacher, & Arnot, 1832).

Wilkins, John, *Mathemagicall Magick or the Wonders that may be performed by Mechanicall Geometry* (London: M.F. for Sa:Gellibrand, 1648).

Willcock, Joseph, *Kinematics; or The Transformation of Motion* (London, 1860).

Willis, Robert, *Principles of Mechanism* (London: London &c., 1841); 2nd ed. (London: Longmans, Green, and Co., 1870).

Wilson, Edwin B., *Vector Analysis: A Text-Book for the Use of Students of Mathematics and Physics*, based on the lectures of J. Willard Gibbs (New Haven, Conn.: Yale University Press, 1901).

Wittenbauer, Ferdinand, *Kinematik des Strahles* (Graz, 1883).

Wolff, Christian, *Philosophia Rationalis, Sive Logica* (1728); 3rd ed. (Frankfurt, 1740).

Wood, James, *The Principles of Mechanics: Designed for the Use of Students in the University*, 3rd ed. (Cambridge: Cambridge University Press, 1803).

Secondary Sources

Some various secondary sources have been cited in the footnotes. The secondary literature on the history of Einstein's special theory of relativity is immense, so a comprehensive list of sources would require too many pages. Still, a selection of titles may help to guide researchers and students approaching the secondary literature on some of the main topics discussed in the present book.

Among the few histories of the classification of the sciences are the following two works: Robert Flint, *Philosophy as Scientia Scientiarum, and, A History of Classifications of the Sciences* (Edinburgh: W. Blackwood and Sons, 1904); and Henry Evelyn Bliss, *The Organization of Knowledge and the System of the Sciences*, with an introduction by John Dewey (New York: Holt, 1929). A brief overview, marred by a few inaccuracies, is Nicholas Fisher's "The Classification of the Sciences," in *Companion to the History of Modern Science*, edited by R. C. Olby, G. N. Cantor, J. R. Christie, and M. J. Hodge (London: Routledge, 1989).

A. Schönflies and M. Grubler authored an overview of kinematics in the 1800s, as the article "Kinematik," in Felix Klein's *Encyklopädie der mathematischen Wissenschaften*, vol. 4, part 3 (1902), article IV-3, pp. 190–278. A revised edition, in French, was authored by Gabriel Koenigs, "Cinématique," *Encyclopédie des sciences mathématiques pures et appliquées*, vol. 4, *Mécanique générale* (Paris: Gauthier-Villars, 1916), pp. 225–304. Another account of kinematics in the 1800s was issued by the Royal Society of London, *Catalogue of Scientific Papers, 1800–1900, Subject Index*, vol. 2, *Mechanics*, article on "Kinematics" (London: Royal Society, 1909). More recent histories of mechanics, which are relevant to the history of kinematics, are Ivor Grattan-Guiness, *Convolutions in French Mathematics, 1800–1844: From the Calculus and Mechanics to Mathematical Analysis and Mathematical Physics*, 3 vols. (Basel: Birkhäuser; Berlin: Deutscher Verlag der Wissenschaften, 1990); and also I. Grattan-Guiness, "The Varieties of Mechanics by 1800," *Historia Mathematica* **17** (1990), 313–338. Furthermore, Teun Koetsier wrote various studies on aspects of the history of kinematics: "A Contribution to the History of Kinematics," *Mechanism and Machine Theory* **18** (1983), 37–48; Koetsier, "From Kinematically Generated Curves to Instantaneous Invariants: Episodes in the History of Instantaneous Planar Kinematics," *Mechanism and Machine Theory* **21** (1986), 489–498; Koetsier, "The Centenary of Ludwig Burmester's 'Lehrbuch der Kinematik,'" *Mechanism and Machine Theory* **24** (1989), 37–38; and Koetsier, "Kinematics," in I. Grattan-Guiness, ed., *Companion Encyclopedia of the History and Philosophy of the Mathematical Sciences*, vol. 2 (London: Routledge, 1994), pp. 994–1001.

Regarding the history of the kinematics of machinery (mechanism) in particular, the following works are also available: M. Rühlmann, *Vorträge zur Geschichte der theoretischen Maschinenlehre und der damit in Zusammenhang stehenden mathematischen Wissenschaften*, 2 parts (Braunschweig: Schwetschke, 1881–1885); A. E. Richard De Jonge, "A Brief Account of Modern Kinematics," *American Society of Mechanical Engineers, Transactions* **65** (1943), 663–683; Eugene S. Ferguson, "Kinematics of Mechanisms from the Time of Watt," *Contributions from the Museum of History and Technology, United States National Museum, Bulletin 228*, paper 27 (Washington D.C.: Smithsonian Institution, 1962), pp. 185–230; Richard S. Hartenberg and Jacques Denavitt, "An Outline of Kinematics to 1900," in *Kinematic Synthesis of Linkages* (New York: McGraw-Hill, 1964), chap. 1; Arthur Guy Erdman, *Modern Kinematics: Developments in the Last Forty Years* (New York: Wiley, 1993); and Ferdinand Freudenstein, "Trends in the Kinematics of Mechanisms," *Applied Mechanics Reviews* **12** (1959), 587–590.

On the kinematics of the special theory of relativity, the most comprehensive account is still that by Henri Arzèlies, *Relativistic Kinematics* (Oxford: Pergamon, 1966), which, though being essentially a treatise on physics, includes annotations on many articles from the 1900s that now have historical value. For beginning readers, these two books are recommended accounts of special relativity and its origins:

Max Born, *Die relativitätstheorie Einsteins und ihre physikalischen Grundlagen geme-inverständlich dargestellt* (Berlin: J. Springer, 1920); reissued as *Einstein's Theory of Relativity*, translated by Henry L. Bose (1924); revised edition, prepared with the collaboration of Günther Leibfried and Walter Biem (New York: Dover, 1962); and Stanley Goldberg, *Understanding Relativity: Origin and Impact of a Scientific Revolution* (Boston: Birkhäuser, 1984).

An authoritative and comprehensive account of the history of electrodynamics, leading to Einstein's theory, and a work that fairly places special relativity as just one among several plausible theories (without ascribing to it the supposedly evident superiority that characterizes most accounts ever since Einstein became famous) is found in Olivier Darrigol's *Electrodynamics from Ampère to Einstein* (New York: Oxford University Press, 2000). More recently, Robert Rynasiewicz provides a short, focused account in "The Optics and Electrodynamics of 'On the Electrodynamics of Moving Bodies,'" *Annalen der Physik* (Leipzig) **14**, suppl. (2005), 38–57. Also worthy of attention is Rynasiewicz's earlier article, "The Construction of the Special Theory: Some Queries and Considerations," in *Einstein: The Formative Years, 1879–1909*, edited by Don Howard and John Stachel, Einstein Studies, vol. 8 (Boston: Birkhäuser, 2000).

As for Einstein's original paper of 1905, Arthur I. Miller's older but extensive study is still a useful resource, despite some defects: *Albert Einstein's Special Theory of Relativity: Emergence (1905) and Early Interpretation (1905–1911)* (Reading, Mass.: Addison-Wesley, 1981; reprinted, New York: Springer Verlag, 1998). On the history of the concept of length contraction, see Harvey R. Brown, "The Origins of Length Contraction: I. The FitzGerald-Lorentz Deformation Hypothesis," *American Journal of Physics* **69**, no. 10 (October 2001), 1044–1054; and Brown, *Physical Relativity: Space-Time Structure from a Dynamical Perspective* (Oxford: Oxford University Press, 2005).

On the history of the concept of simultaneity, see Max Jammer's *Concepts of Simultaneity: From Antiquity to Einstein and Beyond* (Baltimore: Johns Hopkins University Press, 2006), which includes extensive and useful references to the secondary literature over the past several decades. Another accessible but shorter introduction to the topic of the conventionality of simultaneity and the problem of measuring the one-way speed of light is Wesley C. Salmon, "The Philosophical Significance of the One-Way Velocity of Light," *Noûs* **11** (1977), 253–292. Also useful for its more technical discussions and its review of the literature is the article by Ronald Anderson, I. Vetharaniam, and G. E. Stedman, "Conventionality of Synchronisation, Gauge Dependence, and Test Theories of Relativity," *Physics Reports* **295** (1998), 93–180. A typical example of the difficulties involved in attempts to devise methods to measure the one-way speed of light, from an immense literature, are the articles by Charles Nissim-Sabat: "Can One Measure the One-Way Velocity of Light?" *American Journal of Physics* **50** (1982), 533–536; "A Gedanken Experiment to Measure the One-Way Velocity of Light," *British Journal of the Philosophy of Science* **35** (1984), 62–64;

and "A Retraction of 'A Gedanken Experiment to Measure the One-Way Velocity of Light,'" *British Journal of the Philosophy of Science* 38 (1987), 75. Regarding Einstein's actual outlook on conventions in special relativity, see Alberto A. Martínez, "Conventions and Inertial Reference Frames," *American Journal of Physics* 73, no. 5 (May 2005), 452–454.

Against the abundant popular biographies of Albert Einstein, the richer and more reliable sources on Einstein's youth are documentary biographies that have been published in Switzerland. First, Carl Seelig gathered informative recollections by Einstein and his peers, which he edited and issued in his works: *Albert Einstein und die Schweiz* (Zurich: Europa Verlag, 1952); *Helle Zeit–Dunkle Zeit. In memoriam Albert Einstein* (Zurich: Europa Verlag, 1956); and *Albert Einstein: Eine dokumentarische Biographie* (Zurich: Europa Verlag, 1960), reissued as Seelig, *Albert Einstein, a Documentary Biography*, translated by Mervyn Savill (London: Staples Press, 1956). Likewise, Max Flückiger collected additional original documents that he published in *Albert Einstein in Bern: Das Ringen um ein neues Weltbild, Eine dokumentarische Darstellung über den Aufstieg eines Genies* (Bern: Paul Haupt, 1974). Most recently, Ann Hentschel and Gerd Graßhoff produced a valuable commemorative synthesis and collection of materials, in *Albert Einstein: "Those Happy Bernese Years"* (Bern: Stämpfli, 2005). Alongside such documentary biographies, several works convey well-balanced accounts of Einstein's youth, including John Stachel's "The Young Einstein: Poetry and Truth," in *Einstein from 'B' to 'Z'* (Boston: Birkhäuser, 2002), pp. 21–38; and Lewis Pyenson, *The Young Einstein: The Advent of Relativity* (Boston: Adam Hilger, 1985).

Einstein's own writings and correspondence, listed in the preceding bibliography of primary sources, are supplemented by various recollections from his peers. Among them are the following. His sister, Maja Winteler-Einstein, drafted a biographical manuscript: "Albert Einstein—Beitrag für sein Lebensbild" (1924), which was published in part in *The Collected Papers of Albert Einstein*, vol. 1, pp. xlviii–lxvi. Max Talmey knew the Einstein family well and later wrote about "Einstein as a Boy Recalled by a Friend," *New York Times*, 10 February 1929, p. 145; which was reissued in an expanded form in Talmey, *The Relativity Theory Simplified; And the Formative Period of Its Inventor* (New York: Falcon Press, Darwin Press, 1932). Julia Niggli knew the young Albert when he finished high school in Switzerland, and she later published "Nochmals Albert Einstein. Begegnungen und Briefe," *Aargauer Tagblatt* (Aarau, Switzerland), year 106, no. 142 (Friday, 20 June 1952), Beilage "150 Jahre Kantonsschule," pp. 2–4. Margarete J. T. Niewenhuis-von Uexküll knew Albert when he studied at Zurich and she wrote a draft (unpublished), "Albert Einstein nach meiner Erinnerung," manuscript (1953/54?), 14 pages, held at the Bibliothek of the Eidgenössische Technische Hochschule, Zurich, item no. Hs 304:99. Likewise, Louis Kollros, one of Einstein's fellow students at Zurich, wrote some recollections: "Erinnerungen-Souvenirs," *Schweizerische Hochschulzeitung* 28 *Sonderheft* (1955), 169–173 (reissued as "Einnerungen eines Kommilitionen," in Seelig, *Helle*

Zeit, pp. 17–31). One of Einstein's co-workers at the Swiss Patent Office, Josef Sauter, shared some recollections at the Conference "50 Jahre Relativitätstheorie" in Bern (1955); reprinted in Flückiger (1974).

Alexander Moszkowski interviewed Einstein in the mid-1910s and published an unauthorized account of their conversations in *Einstein: Einblicke in seine Gedankenwelt. Gemeinverständliche Betrachtungen über die Relativitätstheorie und ein neues Weltsystem / Entwickelt in Gesprächen mit Einstein* (Hamburg: Hoffmann & Campe; Berlin: F. Fontane, 1921); reissued as *Einstein, the Searcher: His Work Explained from Dialogues with Einstein*, translated by Henry L. Brose (London: Methuen, 1921), reprinted as *Conversations with Einstein* (New York: Horizon Press, 1970). Despite Einstein's discomfort with Moszkowski's decision to publish that overly flattering book, Einstein noted that the account was roughly accurate. Also, in 1916, the psychologist Max Wertheimer interviewed Einstein at length about the creative process that led him to his theory of 1905, and decades later he published his account in *Productive Thinking* (New York: Harper & Brothers, 1945). Einstein read drafts of the document and judged it to be mostly accurate. On this topic of the psychology of creativity, a relevant letter by Einstein appears in Jacques Hadamard, *An Essay on the Psychology of Invention in the Mathematical Field* (Princeton: Princeton University Press, 1945; reprinted, New York: Dover, 1954).

Einstein's stepson-in-law, Rudolf Kayser, wrote an authorized biography, which Einstein proofread, and published it under the pseudonym Anton Reiser, *Albert Einstein: A Biographical Portrait*, with a preface by Albert Einstein (New York: A. & C. Boni, 1930). Also, David Reichinstein was a physical chemist at Zurich who knew Einstein pretty well and wrote an early biography: *Albert Einstein, sein Lebensbild und seine Weltanschaunng* (Prague: Enst Ganz, 1935). Also, Einstein was dismayed that his other stepson-in-law also chose to write a biography, unauthorized, full of flattery and hyperbole: Dimitri Marianoff, with Palma Wayne, *Einstein: An Intimate Study of a Great Man* (Garden City, N.Y.: Doubleday, Doran, 1944). Einstein opposed and complained against the publication of this biography for including "falsehoods" and "sentimental balderdash" (see Einstein to Carl Seelig, 25 February 1952, Einstein Archives, item 39-011). Also, Philipp Frank was a physics colleague and friend who published an authorized biography (but not proofread by Einstein): *Einstein, Sein Leben und seine Zeit* (Munich: P. List, 1949); reprinted, with a foreword by Einstein from 1942 (Braunschweig: Friedr. Vieweg & Sohn, 1979); and note that an English translation appeared before the original German text: Frank, *Einstein: His Life and Times* (New York: Knopf, 1947; London: Jonathan Cape, 1948).

Another biographer who knew Einstein well was Antonina Vallentin, a writer who was a close friend of Einstein's second wife, Elsa. She published *Le Drame d'Albert Einstein* (Paris: Libraire Plon, 1954), reissued as *Einstein, a Biography*, translated from the French by Moura Budberg (London: Weidenfeld and Nicolson, 1954). Most of Vallentin's biography pertains to Einstein's later years and his character, again in a celebratory vein. Furthermore, very valuable are the interviews by R. S.

Shankland, which dealt especially with Einstein's formulation of the special theory and his prior relevant experimental knowledge: "Conversations with Albert Einstein," *American Journal of Physics* **31** (1963), 47–57, and "Conversations with Albert Einstein. II," *American Journal of Physics* **41** (1973), 895–901. Finally, the classic scientific biography is still that by Einstein's later colleague, Abraham Pais, *"Subtle is the Lord . . .": The Science and the Life of Albert Einstein* (New York: Oxford University Press, 1982).

Regarding Einstein's relationship with his first wife, Mileva Marić, see the collections of letters, with commentaries, by Renn and Schulmann, and by Popović, in the primary sources. Also, John Stachel has authored a valuable account of their relationship: "Albert Einstein and Mileva Marić: A Collaboration That Failed to Develop," in *Creative Couples in the Sciences,* edited by Helena M. Pycior, Nancy G. Slack, and Pnina G. Abir-Am (Rutgers University Press, 1996); reissued in Stachel, *Einstein from 'B' to 'Z'* (2002), 39–55. For critical discussions of the recent and imaginative claims about Marić's supposed role as secret coauthor of Einstein's works, see Alberto A. Martínez, "Handling Evidence in History: The Case of Einstein's Wife," *School Science Review* **86**, no. 316 (March 2005), 49–56; and, the "New Introduction" to the new edition of John Stachel, *Einstein's Miraculous Year: Five Papers That Changed the Face of Physics,* with a foreword by Roger Penrose (Princeton: Princeton University Press, 2005).

More indirect accounts of Einstein's life and insights are available in recollections from his son, Hans Albert. The writer Bela Kornitzer briefly met with Einstein and then interviewed Hans Albert at length: Kornitzer, "Einstein Is My Father," *Ladies' Home Journal* **68** (April 1951), 47, 134, 136, 139, 141, 255–256. Transcripts of an interview with Hans Albert appear in G. J. Whitrow, *Einstein, the Man and His Achievement* (1967; reprinted, New York: Dover, 1973). Less reliable are the two following books. Peter Michelmore interviewed Hans Albert for two days in 1962 and wrote a biography of Einstein, which Hans Albert did not proofread (and which includes various errors): Michelmore, *Einstein: Profile of the Man* (New York: Dodd, Mead, 1962). Likewise, Aylesa Forsee interviewed Hans Albert and wrote *Albert Einstein: Theoretical Physicist* (New York: Macmillan; London: Collier-Macmillan, 1963), which includes plausible but dubious quotations.

Index